Principles and Applications of RF/Microwave in Healthcare and Biosensing

Principles and Applications of RF/Microwave in Healthcare and Biosensing

Edited by

Changzhi Li

Mohammad-Reza Tofighi

Dominique Schreurs

Tzyy-Sheng Jason Horng

AMSTERDAM • BOSTON • HEIDELBERG • LONDON
NEW YORK • OXFORD • PARIS • SAN DIEGO
SAN FRANCISCO • SINGAPORE • SYDNEY • TOKYO

Academic Press is an imprint of Elsevier

Academic Press is an imprint of Elsevier
125 London Wall, London EC2Y 5AS, United Kingdom
525 B Street, Suite 1800, San Diego, CA 92101-4495, United States
50 Hampshire Street, 5th Floor, Cambridge, MA 02139, United States
The Boulevard, Langford Lane, Kidlington, Oxford OX5 1GB, United Kingdom

Notices
Knowledge and best practice in this field are constantly changing. As new research and experience broaden
our understanding, changes in research methods, professional practices, or medical treatment may become
necessary.

Practitioners and researchers must always rely on their own experience and knowledge in evaluating
and using any information, methods, compounds, or experiments described herein. In using such information
or methods they should be mindful of their own safety and the safety of others, including parties for
whom they have a professional responsibility.

To the fullest extent of the law, neither the Publisher nor the authors, contributors, or editors, assume
any liability for any injury and/or damage to persons or property as a matter of products liability, negligence
or otherwise, or from any use or operation of any methods, products, instructions, or ideas contained
in the material herein.

British Library Cataloguing-in-Publication Data
A catalogue record for this book is available from the British Library

Library of Congress Cataloging-in-Publication Data
A catalog record for this book is available from the Library of Congress

ISBN: 978-0-12-802903-9

For Information on all Academic Press publications
visit our website at https://www.elsevier.com

Working together
to grow libraries in
developing countries

www.elsevier.com • www.bookaid.org

Publisher: Joe Hayton
Acquisition Editor: Tim Pitts
Editorial Project Manager: Charlotte Kent
Production Project Manager: Melissa Read
Designer: Greg Harris

Typeset by MPS Limited, Chennai, India

Contents

List of Contributors

M. Baboli
Columbia University, NY, United States

O. Borić-Lubecke
University of Hawaii, Manoa, HI, United States

J.-C. Chiao
University of Texas at Arlington, Arlington, TX, United States

R. Gómez-García
University of Alcalá, Madrid, Spain

T.-S. J. Horng
National Sun Yat-Sen University, Kaohsiung, Taiwan

C. Li
Texas Tech University, Lubbock, TX, United States

V. Lubecke
University of Hawaii, Manoa, HI, United States

M. Mercuri
Holst Centre, IMEC, Eindhoven, Netherlands

J.-M. Muñoz-Ferreras
University of Alcalá, Madrid, Spain

J. Oberhammer
KTH Royal Institute of Technology, Stockholm, Sweden

A. Rahman
University of Hawaii, Manoa, HI, United States

D.M.M.-P. Schreurs
KU Leuven, Leuven, Belgium

M.-R. Tofighi
Pennsylvania State University, Harrisburg, PA, United States

F. Töpfer
KTH Royal Institute of Technology, Stockholm, Sweden

F.-K. Wang
National Sun Yat-Sen University, Kaohsiung, Taiwan

Introduction

BACKGROUND

Healthcare is a top global challenge due to the ever-increasing demand for higher quality of life, the aging population, and the various social, cultural, political, and economic impacts. While the productivity and stability of society are dependent on the outcomes of healthcare, the rising costs of medical care are impacting almost everyone in the world. Due to the aging of populations and the prevalence of chronic diseases, there is a strong demand to use advanced technologies to shoulder the burdens of individuals and healthcare systems in both developing and developed countries.

In recent years, a significant growth of research that uses engineering innovations to improve the efficacy while reducing the cost of health provision has been taking place. Among these efforts, radio frequency (RF) and microwave technologies play a critical role in disease diagnosis, care delivery, and telemedicine. Researchers have devoted a lot of efforts to the monitoring and imaging applications of RF/microwave technologies, some of which have entered or are being tested for clinical use. They also apply wireless sensing and communication over or through body tissues to transmit physiological or biochemical information. In addition, based on the biological effects caused by electromagnetic waves interacting with tissues, microwave diagnosis and treatment of diseases in living systems are under rapid development, generating significant impacts in fields such as biosensing and hyperthermia.

Although RF/microwave engineering is a well-established discipline with many applications such as wireless communication and wireless power transfer, using RF/microwave technology for healthcare and biosensing involves special grand challenges. Most conventional RF/microwave systems rely on electromagnetic waves that propagate in homogeneous and isotropic media. However, when applying the same discipline to biomedical applications, tremendous puzzles are presented due to the anatomical, physiological, and biochemical variations in human bodies. Moreover, the problems are compounded by realistic issues such as the lack of simulation models and experimental data, difficulty in conducting tests because studies strongly depend on human subjects, unmeasurable effects in biological systems, difficulty unifying protocols and conducting experiments, safety issues in human studies, as well as practical constraints in clinical implementation. In addition, another barrier that is difficult to break is the large gap between the languages and mindsets of engineers and healthcare practitioners.

Therefore, there is a strong demand for both engineering and clinical researchers to understand the achievements and future directions in the development of microwave technologies for the fast-growing biomedical fields. It is believed that more clinical, pharmaceutical, and biochemical aspects can be addressed if researchers and practitioners from different fields are made aware of the multitude of possibilities, potential new applications, and challenges in future innovative healthcare solutions.

RECENT PROGRESS ON RF/MICROWAVE BIOMEDICAL RESEARCH

Important recent progress on RF/microwave biomedical research can be roughly divided into two categories. One is the biological interaction and effects; the other is RF systems and instrumentation for healthcare applications.

Researchers have been intensively studying the mechanisms, effects, and applications of the interaction of electromagnetic waves with biological materials at molecular, cellular, and tissue levels. On one hand, a large number of devices and components on circuit-board and integrated-chip levels have been developed for microwave characterization of biological materials and living systems. Based on these devices, therapeutic and diagnostic applications have been proposed and some of them are being clinically tested. Examples of typical applications include the diagnosis of malignant tumors and lab on chip for real-time complex body fluid analysis. On the other hand, RF/microwave energy is used to treat diseases in therapeutic and surgical applications. Examples include hyperthermia and minimally invasive natural orifice transluminal endoscopic surgery, which in general can improve treatment efficiency, reduce pain, and shorten the recovery time for patients.

In the meantime, there are abundant systems and instrumentations for healthcare applications developed based on RF/microwave technologies. Prime examples are magnetic resonance imaging (MRI) and microwave imaging. MRI, whose invention has been honored by several Nobel Prizes, has been widely adopted in clinical practice to benefit the well-being of human society. Microwave and millimeter-wave imaging have shown their advantages and daily uses not only in health applications, but also in security systems. Another area that has caught the attention of many researchers and practitioners is that of wireless sensors and systems for health monitoring and telemedicine. For example, wireless signals are utilized to noninvasively measure physiological signals and biological signals at doctors' offices, in health-care facilities, or even at home; they can also transmit signals between sensors and stimulators implanted in the human body and external wearable controllers for further transmission, processing, and control, forming a closed-loop system for continuous and autonomous management of disease symptoms.

ABOUT THIS BOOK

This book aims to help authors learn the multiple directions in which RF/micro-wave technologies are heading toward healthcare and biosensing applications, the achievements that have been made so far, and the challenges for researchers to solve in the near future. Chapter 1 by Li and Schreurs reviews the fundamentals of microwave engineering, which will be used in the other chapters of this book. Since microwave engineering is a discipline that has gone through many years of development, Chapter 1 only illustrates basic knowledge and is aimed at helping readers from other disciplines to understand the basic principles relevant to this book. Therefore, readers with good microwave engineering background can skip that chapter. In Chapter 2, Interaction between electromagnetic waves and biolog-ical materials, by Tofighi, the modeling and measurement procedure for electro-magnetic properties of biological materials, namely complex permittivity, is described, which serves as the foundation for the interaction between RF/micro-wave and biological materials. The chapter also covers issues related to interfac-ing with tissue through sensing probes, such as coaxial, microstrip, and coplanar waveguide transmission lines, providing a useful perspective that is beneficial to some other medical and biological applications discussed in this book. Chapter 3, Microwave cancer diagnosis, by Töpfer and Oberhammer, focuses on microwave diagnosis of malignant tumors, based on the fact that a special microwave signa-ture has been observed for many malignant tumors. A multitude of techniques, including free-space quasi-optical techniques, near-field probes, microwave tomography, ultra-wideband radar, and passive microwave imaging, some of which have already entered clinical trials, are presented for the diagnosis of breast cancer, skin cancer, and brain tumors. Chapter 4, Wireless closed-loop stimulation systems for symptom management, by Chiao, presents closed-loop systems for autonomous management of disease symptoms, using examples of neural and gas-tric electrical stimulation applications that target the management of neurological and gastric disorders. Wireless signal transduction and wireless power transfer mechanisms across tissues make it possible to eliminate batteries in the implants for long-term use and to reduce the implant size for endoscopic implementation. As a result, these systems can provide better care for patients who have chronic illness and improve the healthcare system with personalized medicine and lower costs. Starting from Chapter 5, Human-aware localization using linear-frequency-modulated continuous-wave radars, the rest of the book discusses biomedical radars for various healthcare applications. In Chapter 5, Human-aware localiza-tion using linear-frequency-modulated continuous-wave radars, Muñoz Ferreras, Gómez García and Li present coherent linear-frequency-modulated continuous-wave radars. With a relatively simple hardware front-end, the solution is capable of both relative displacement and absolute range measurements, thus enabling versatile health care applications based on physiological motion sensing and human-aware localization. Chapter 6, Biomedical radars for monitoring health, by

Wang, Mercuri, Horng, and Schreurs. discusses in more depth specific applications, such as vital signs monitoring, exercise assistance, and fall detection, showing the wide applicability of injection-locked radar and step-frequency continuous-wave radar. Finally, Chapter 7, RF/wireless indoor activity classification, by Rahman, Borić-Lubecke, Lubecke and Baboli, applies biomedical radar to indoor activity classification.

We hope that this book will inspire the readers to develop their own ideas to leverage advanced RF/microwave technologies and devise innovative solutions in the highly important field to benefit the well-being of all human beings.

We dedicate our respects to pioneers who started to apply RF/microwave technologies to healthcare and biosensing. We would also like to sincerely acknowledge the important supports provided by funding agencies, including the Belgian National Fund for Scientific Research (FWO), the European Institute of Innovation & Technology (EIT)—Health, the Ministry of Science and Technology (MOST) of Taiwan, and the US National Science Foundation (NSF).

Fundamentals of microwave engineering

1

C. Li[1] and D.M.M.-P. Schreurs[2]

[1]Texas Tech University, Lubbock, TX, United States [2]KU Leuven, Leuven, Belgium

CHAPTER OUTLINE

1.1 INTRODUCTION

The principles and applications of radio frequency (RF)/microwave in healthcare and biosensing are strongly based on the analysis, design, and use of microwave components, circuits, and systems. "Microwave" refers to electromagnetic waves from 1 to 300 GHz, while RF is normally defined as the electromagnetic wave frequencies from about 3 kHz to 300 GHz [1,2]. Therefore, RF and microwave combined cover a very large range of the electromagnetic spectrum and have many potential applications that interact with our daily life from the cell level to the human body level. To help readers better understand the analysis and the engineering techniques used in other chapters of this book, fundamentals of microwave engineering are presented in this chapter. The discussion starts from transmission lines, which are used in almost any RF/microwave circuit to carry high-frequency signals. It will then review S-parameters, which are the basic "language" for microwave measurements. Based on that, the Smith Chart and impedance matching will be introduced as important microwave circuits and systems design tools. With the help of these tools, microwave passive components (e.g., power dividers, combiners, and hybrids) and active building blocks (e.g., amplifiers, mixers, and oscillators) will be discussed. Finally, some fundamental RF/microwave system architectures will be presented.

If a reader has been trained with microwave knowledge, she/he can skip this chapter. On the other hand, if the reader is from another discipline such as a healthcare profession, reading this chapter would help her/him to understand the technologies involved in various applications covered by the other chapters. However, it should be noted that microwave engineering and systems is based on a large engineering discipline that has gone through many years of development. Therefore, the contents of this chapter only illustrate the basic knowledge sets and principles. As readers go further into specific healthcare and biosensing topics in other chapters, they are encouraged to use the references or contact the chapter authors to get relevant information and suggestions on self-learning methods for the corresponding microwave theory and techniques topics.

1.2 TRANSMISSION LINES THEORY

1.2.1 A SIMPLE MODEL OF TRANSMISSION LINES

From coaxial TV cables to signal traces on on-chip biosensors, transmission lines widely exist in various RF/microwave systems. At microwave frequencies, most of the traces on ordinary circuit boards can no longer be treated as an ideal line that passes exactly the same voltage and current from one end to the other. Instead, the signal changes when it propagates along the traces. The transmission line model serves as a useful tool to describe the voltage and current along a microwave signal path. Let us start from a simple model by analyzing the voltage and current relationship along the transmission line of Fig. 1.1.

As a two-port network, a transmission line receives a signal from the source at the input port, and delivers power to the load at the output port. To apply fundamental circuit analysis, we can divide the transmission line into many identical small sections with lengths of Δx. When the signal passes each section, it experiences a resistive loss modeled as a resistance R per unit length (Ω/m). At high frequencies, a conductive wire also introduces an inductance to the signal that passes through it. Therefore, each section has an inductance L per unit length (H/m), which is in series with the resistance. In addition, there is a capacitance C per unit length (F/M) and a conductance G per unit length (S/m), which model the coupling between the upper and lower path and the leakage among the two paths.

Due to the loss, leakage, capacitive coupling, and inductive effect, the voltages and currents along the transmission line are functions of position and time.

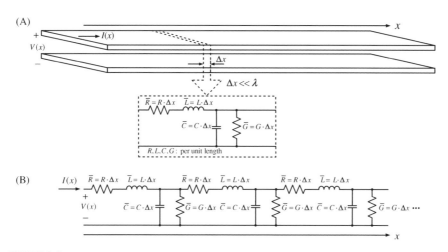

FIGURE 1.1

Transmission line model. (A) A realistic transmission line, where a small section with length of Δx, much smaller than the wavelength, is modeled using discrete components. (B) Modeling of transmission line.

Applying Kirchhoff's voltage law and Kirchhoff's current law to a small section of the above model results in the following relationship:

$$
\begin{cases}
\dfrac{V(x) - V(x + \Delta x)}{\Delta x} = R \cdot I(x) + L \cdot \dfrac{\partial I(x)}{\partial t} \\[3mm]
\dfrac{I(x) - I(x + \Delta x)}{\Delta x} = G \cdot V(x + \Delta x) + C \cdot \dfrac{\partial V(x + \Delta x)}{\partial t}
\end{cases}
\tag{1.1}
$$

As Δx approaches zero, the left hand side is recognized as the derivative of $V(x)$ and $I(x)$. When the circuit is excited with sinusoidal signals and is in a steady state, the differentiation operator $\partial/\partial t$ in the time domain can be replaced with $j\omega$ in the frequency domain. Therefore, when a transmission line is divided into very small sections with Δx approaching zero, the voltage and current along the transmission line can be represented in the frequency domain as:

$$
\begin{cases}
-\dfrac{dV(x)}{dx} = R \cdot I(x) + j\omega L \cdot I(x) & (1.2a) \\[3mm]
-\dfrac{dI(x)}{dx} = G \cdot V(x) + j\omega C \cdot V(x) & (1.2b)
\end{cases}
$$

If we apply the differentiation operation d/dx to Eq. (1.2a), we can obtain the expression of $dI(x)/dx$ in terms of $V(x)$. Then, substituting the result into Eq. (1.2b) leads to the classical wave equation, of which the solutions to the voltage and current are [1]:

$$
\begin{cases}
V(x) = V_0^+ e^{-\gamma x} + V_0^- e^{\gamma x} \\
I(x) = I_0^+ e^{-\gamma x} + I_0^- e^{\gamma x}
\end{cases},
\tag{1.3}
$$

where $\gamma = \partial + j\beta = \sqrt{(R + L\omega j)(G + C\omega j)}$ is the propagation constant. The $V_0^+ e^{-\gamma x}$ term represents the "forward-propagating voltage," whereas the $V_0^- e^{\gamma x}$ term represents the "backward-propagating voltage." Likewise, the $I_0^+ e^{-\gamma x}$ term represents the "forward-propagating current" and the $I_0^- e^{\gamma x}$ term represents the "backward-propagating current." From here, we see that there are two waves traveling in opposite directions in a transmission line. It should be noted that V_0^+, V_0^-, I_0^+, and I_0^- are complex constants that are independent of time and position. Their values can be evaluated using the boundary conditions at the input and output ports of the transmission line. If we insert the equation for $V(x)$ into Eq. (1.2a), another form for the current along the transmission line can be obtained as:

$$
I(x) = \frac{1}{\sqrt{\frac{R + L\omega j}{G + C\omega j}}} \left[V_0^+ e^{-\gamma x} - V_0^- e^{\gamma x} \right] = I_0^+ e^{-\gamma x} + I_0^- e^{\gamma x},
\tag{1.4}
$$

where $\sqrt{(R + L\omega j)/(G + C\omega j)}$ is defined as the characteristic impedance of the transmission line, and will be denoted as Z_0 in the future. If we specify Z_0 and γ, we can completely specify a transmission line, with V_0^+ and V_0^- dependent on the two terminals of the transmission line.

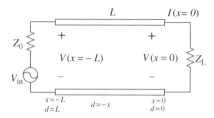

FIGURE 1.2

A transmission line with a source and a load.

1.2.2 IMPEDANCE AND REFLECTION COEFFICIENT

Fig. 1.2 shows the fundamental configuration of a transmission line in a micro-wave system. The left side of the transmission line is connected to a microwave signal source with source impedance Z_0. The right side of the transmission line is connected to a load impedance Z_L. The end that is connected to the source will be referred to as the "source-end," while the end that is terminated with load Z_L will be referred to as the "load-end" in this book. Both x and d are popular variables used to denote the location along a transmission line [1,3,4]. At the load-end, $x = d = 0$. As we move from the load-end back to the source-end, x becomes increasingly negative while d becomes increasingly positive, and $x = -d$.

With this setup, the impedance of a transmission line at location x can be calculated as the voltage-to-current ratio at that location:

$$Z(x) = \frac{V(x)}{I(x)} = Z_0 \cdot \frac{V_0^+ e^{-\gamma x} + V_0^- e^{\gamma x}}{V_0^+ e^{-\gamma x} - V_0^- e^{\gamma x}}. \tag{1.5}$$

Likewise, we can define the reflection coefficient of a transmission line at location x as the ratio between the backward-propagating voltage wave and the forward-propagating voltage wave, as follows:

$$\Gamma(x) = \frac{V_0^- e^{\gamma x}}{V_0^+ e^{-\gamma x}} = \frac{V_0^-}{V_0^+} e^{2\gamma x} = \Gamma_0 \cdot e^{2\gamma x}, \tag{1.6}$$

where $\Gamma_0 = \Gamma(x = 0) = V_0^-/V_0^+ = (Z_L - Z_0)/(Z_L + Z_0)$ represents the reflection coefficient at the load-end of the transmission line. On the other hand, if Γ_0 is given at the load-end of the transmission line, then the load impedance can be found as $Z_L = Z_0(1 + \Gamma_0)/(1 - \Gamma_0)$. Finally, the impedance and reflection coefficient along a transmission line can be correlated as:

$$Z(x) = Z_0 \frac{e^{-\gamma x} + \Gamma_0 e^{\gamma x}}{e^{-\gamma x} - \Gamma_0 e^{\gamma x}} = Z_0 \frac{1 + \Gamma(x)}{1 - \Gamma(x)}. \tag{1.7}$$

1.2.3 SPECIAL CASES OF TRANSMISSION LINES

Several special cases of transmission lines are worth studying because they are frequently used in modern biomedical microwave systems. When the load-end of a transmission line is matched, i.e., when the load impedance Z_L equals the characteristic impedance Z_0 in Fig. 1.2, then there will be no reflection and $\Gamma_0 = 0$. Under these circumstances, the voltage and current along the transmission line only have forward-propagating components, and the impedance $Z(x)$ along the transmission line is a constant that is always equal to Z_0. On the other hand, in order to achieve $|\Gamma_0| > 1$, which leads to $|(Z_L - Z_0)/(Z_L + Z_0)| > 1$, we need Re $(Z_L) < 0$. This means having a reflection coefficient amplitude that is greater than one requires active components such as power sources to be in the circuit.

If a transmission line is lossless, then both the series resistance and parallel conductance will be zero, i.e., $R = G = 0$. In that case, the propagation constant will be $\gamma = j\beta$, where $\beta = \omega\sqrt{LC}$. This is based on the general definition of the wave equation, $\beta = 2\pi/\lambda$, where λ is the wavelength of the signal in the given transmission line. The characteristic impedance of the transmission line is thus simplified as $Z_0 = \sqrt{L/C}$, which is a real number. In that case, the impedance along the transmission line is simplified to:

$$Z(d) = Z_0 \frac{e^{j\beta d} + \Gamma_0 e^{-j\beta d}}{e^{j\beta d} - \Gamma_0 e^{-j\beta d}} = Z_0 \frac{Z_L + jZ_0 \cdot \tan(\beta d)}{Z_0 + jZ_L \cdot \tan(\beta d)}. \tag{1.8}$$

As in most microwave applications, transmission lines are implemented to have minimal attenuation (or loss) to the signal carried. It is useful to analyze the transmission line by assuming it is lossless. Here we will discuss several special cases of lossless transmission lines that are frequently used in practical microwave systems. The following special cases of lossless transmission lines are frequently used in engineering practice.

When a short-circuit terminates a transmission line, the load impedance $Z_L = 0$. As a result, when moving along the transmission line from the load-end toward the source-end, the impedance changes in the following manner: $Z(d) = jZ_0\tan(\beta d)$. From this equation, it is straightforward that the impedance looking into the transmission line from the signal source will be purely reactive. It could be either inductive or capacitive, depending on the sign of the $\tan(\beta d)$ term. Fig. 1.3 shows the plot of impedance change along the transmission line. When $d = 0$, the impedance is zero, because a short circuit is present at the load-end. As we move from the load-end toward the source-end, the impedance looking toward the load-end will become increasingly inductive. When we move to the position where $\beta d = \pi/2$, the impedance will reach its maximum and form an effective open circuit. After that, the impedance transitions to become capacitive. The capacitive impedance reduces to 0 (an equivalent short) at $\beta d = \pi$, and becomes inductive again after that. In this manner, the impedance periodically changes between inductive and capacitive as we move along the transmission line, with

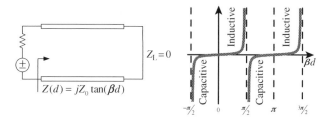

FIGURE 1.3

The change of impedance along a transmission line with short-ended load.

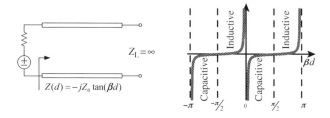

FIGURE 1.4

The change of impedance along a transmission line with an open-ended load.

either a zero-impedance (short) or an infinite-impedance (open) at the transition point between inductive and capacitive regions. Therefore, with a short-terminated transmission line, we can design the length of the transmission line to effectively synthesize either an inductor or a capacitor with an arbitrary value. We can also realize a short circuit or an open circuit when βd is chosen to be even or odd multiples of $\pi/2$.

Similar to the case of a short-ended transmission line, an open-ended transmission line ($Z_L = \infty$) presents either an inductive or a capacitive impedance at the source-end. Of course, designers can also realize an effective open or short circuit when βd is a multiple of $\pi/2$ where the impedance transitions between inductive and capacitive regions. When moving along the transmission line from the load-end toward the source-end, the impedance changes in the following manner: $Z(d) = -jZ_0\cot(\beta d)$. The impedance varies with βd as shown in Fig. 1.4.

A quarter-wavelength transmission line refers to the special case in which the length of the transmission line is chosen to be $\lambda/4$. In such a case, $\beta d = (2\pi/\lambda) \cdot (\lambda/4) = \pi/2$, which results in the input-impedance at the source-end of the transmission line being $Z(d) = Z_0^2/Z_L$. This is often called a quarter-wavelength transformer, because designers can change a load resistance of any value R_L to a desired input resistance R_{in} by selecting the proper transmission line-characteristic impedance of

$Z_0 = \sqrt{R_L \cdot R_{in}}$. A special example is a quarter-wavelength transmission line that can convert a short circuit at the load-end to an open circuit at the input side. Vice versa, a quarter-wavelength transmission line can convert an open circuit at the load-end to a short circuit at the input side.

1.2.4 VOLTAGE STANDING WAVE RATIO

As we have seen, there are two waves (i.e., the forward-propagating wave and the backward-propagating wave) traveling in opposite directions on a transmission line. The sum of the two waves produces a so called standing-wave pattern, which means the sum of the waves is a sinusoidal function of time and its amplitude is a function of position. Let us start from the special case of a lossless transmission line terminated with a short circuit at the load-end, which has $\Gamma_0 = -1$. Going back from the phasor representation to the real time expression, the voltage wave along the transmission line with short-circuit termination can be expressed as [3]:

$$V(d,t) = \mathrm{Re}\left[V(d) \cdot e^{j\omega t}\right] = \mathrm{Re}\left[V_0^+ (e^{+j\beta d} + \Gamma_0 e^{-j\beta d}) \cdot e^{j\omega t}\right] = -2\sin(\beta d) \cdot \sin(\omega t + \varphi) \cdot \left|V_0^+\right|.$$
(1.9)

where $V_0^+ = |V_0^+| \cdot e^{-j\varphi}$ is expressed using its amplitude $|V_0^+|$ and phase φ. It should be noted that in the above result, $\sin(\beta d)$ changes the envelope information of the complete waveform, while $\sin(\omega t + \varphi)$ determines the time-varying behavior of the waveform. The corresponding waveform is plotted in Fig. 1.5. As we can see, the term with the location information d is decoupled with the term with the time information t. This means the envelope of the waveform is a function of location only. When $\omega t + \varphi$ is $\pi/2$, the waveform reaches its maximum amplitude (the envelope shown in Fig. 1.5). As time changes, the waveform just shrinks its amplitude, but is not moving along the transmission line at all. This is a real standing wave.

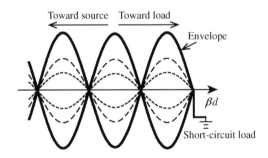

FIGURE 1.5

Standing wave formed on a transmission line with a short-circuit termination.

In the general case with an arbitrary load termination, $\sin(-\beta d + \omega t)$ appears in the expression of $V(d,t)$. The voltage wave can be written as:

$$V(d) = |V_0^+| e^{j\varphi} \cdot e^{j\beta d} [1 + |\Gamma_0| \cdot e^{j\phi_{\Gamma_0}} \cdot e^{-2j\beta d}], \qquad (1.10)$$

where ϕ_{Γ_0} is the phase of Γ_0. It indicates that at different locations along the transmission line, the phase will change the value of $|V(d)|$. Thus the voltage standing wave ratio (VSWR) is defined as the ratio between the maximum and the minimum values of $|V(d)|$ [1]:

$$\text{VSWR} = \frac{|V(d)|_{max}}{|V(d)|_{min}} = \frac{|V_0^+| \cdot [1 + |\Gamma_0|]}{|V_0^+| \cdot [1 - |\Gamma_0|]} = \frac{1 + |\Gamma_0|}{1 - |\Gamma_0|}. \qquad (1.11)$$

On the other hand, if the VWSR of a transmission line is given, we can use the above equation to find the corresponding reflection coefficient magnitude $|\Gamma_0|$. Although datasheets for today's microwave components use S-parameters more often, the VSWR still has wide applications in microwave systems. Specifications on the input/output matching of microwave components can be given using the VSWR. For example, an amplifier may need a VSWR smaller than 1.4. According to the relationship of the VSWR and the reflection coefficient derived above, this corresponds to the magnitude of the reflection coefficient being lower than 0.17, or -15.39 dB.

1.3 **RF MATRICES**

1.3.1 **MATRICES FOR CIRCUIT ANALYSIS**

In circuit analysis, there are various matrices that can be used to model the voltage and current relationship of a given network. For example, the "impedance matrix" (or "Z-parameters") of a two-port network describes the voltage at the two ports in Fig. 1.6 as a function of the current flowing into the two ports [5]. Once the Z-parameters of a two-port network are given, the voltage at the input and output ports (v_1 and v_2) can be calculated based on the currents (i_1 and i_2) that flow into the ports:

$$\begin{bmatrix} v_1 \\ v_2 \end{bmatrix} = \begin{bmatrix} z_{11} & z_{12} \\ z_{21} & z_{22} \end{bmatrix} \begin{bmatrix} i_1 \\ i_2 \end{bmatrix}. \qquad (1.12)$$

Likewise, an "admittance matrix" (or "Y-parameters") describes the admittances that convert voltage at the two ports (v_1 and v_2) into the current (i_1 and i_2) [5]. A "hybrid matrix" (or "h-parameters matrix") correlates the currents and voltages [5]. Finally, a "transmission matrix," sometimes called an "ABCD-matrix" (or "ABCD-parameters"), can be used to calculate the voltage and current at one port using the voltage and current at the other port [5].

FIGURE 1.6

A two-port network.

All the Z, Y, h, and $ABCD$-parameters are used to define the voltage (V) and current (I) relationships. They are very useful at low frequencies because the parameters are readily measured using short- and open-circuit tests at the terminals of a two-port network. For example, we can easily find parameter y_{11} by measuring with a short circuit at port 2, because $y_{11} = i_1/v_1$ when $v_2 = 0$. Moreover, the above matrices can be conveniently used to analyze circuits with several building blocks. For example, when multiple two-port networks are cascaded, the overall Z-parameters are the sum of individual Z-parameters. When multiple two-port networks are connected in parallel, the overall Y-parameters are the sum of individual Y-parameters. When multiple two-port networks are cascaded, the overall $ABCD$-matrix can be obtained by multiplying individual $ABCD$-matrices together.

Unfortunately, the above parameters are very difficult to measure at microwave frequencies. This is mainly because ideal open and short circuits are difficult to implement at microwave frequencies, as any additional signal trace introduces a transmission line effect from the open/short circuit point, as discussed above. Also, a two-port network may oscillate at microwave frequencies when terminated with an open/short condition, causing stability issues. Based on the previous section, it is found that V^+, V^-, I^+, and I^- are very convenient to use for the analysis of microwave structures. This necessitates a new representation of a two-port network, using the so-called scattering matrix (or S-parameters).

1.3.2 S-PARAMETERS

To start, let us first normalize the forward- and backward-propagating voltage waves to the square root of "power," by defining normalized incident and scattered voltage waves as:

$$\begin{cases} a(x) = \dfrac{V_0^+ e^{-\gamma x}}{\sqrt{Z_0}} \\[2mm] b(x) = \dfrac{V_0^- e^{+\gamma x}}{\sqrt{Z_0}} \end{cases} \qquad (1.13)$$

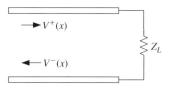

FIGURE 1.7

A one-port network based on a transmission line terminated by Z_L.

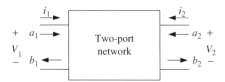

FIGURE 1.8

A two-port network.

It should be noted that $|a(x)|^2$ and $|b(x)|^2$ correspond to the powers of the forward- and backward-propagating waves, respectively. S-parameters are defined based on normalized waves as follows:

A typical transmission line-based one-port network is shown in Fig. 1.7, where the load-end of the transmission line is terminated with impedance Z_L. Because the powers of the forward- and backward-propagating waves are $|a(x)|^2$ and $|b(x)|^2$, respectively, the power that flows into the one-port network, which is named as the net forward power flow, can be found as $|a(x)|^2 - |b(x)|^2$. The S-parameter in this one-port case is defined as $S_1(x) = b(x)/a(x) = \Gamma(x)$. As we can see, the S-parameter here represents how much signal is reflected from the one-port network, which is also referred to as the reflection coefficient of the network.

A typical two-port network for S-parameter analysis is shown in Fig. 1.8. Let us look at the interface immediately at the input/output of the two-port network first. Similar to the case of a one-port network, the normalized voltage waves can be defined as:

$$a_1 = \frac{V_1^+}{\sqrt{Z_0}}, \; a_2 = \frac{V_2^+}{\sqrt{Z_0}}, \; b_1 = \frac{V_1^-}{\sqrt{Z_0}}, \text{ and } b_2 = \frac{V_2^-}{\sqrt{Z_0}} \qquad (1.14)$$

Now, we can define the 2×2 S-parameters (scattering) matrix as:

$$\begin{bmatrix} b_1 \\ b_2 \end{bmatrix} = \begin{bmatrix} S_{11} & S_{12} \\ S_{21} & S_{22} \end{bmatrix} \cdot \begin{bmatrix} a_1 \\ a_2 \end{bmatrix}, \qquad (1.15)$$

where $S_{11} = b_1/a_1$ when $a_2 = 0$. To make $a_2 = 0$, we need to terminate port 2 in a certain way so that there is no power reflected from the load of port 2 back to the two-port network. Furthermore, if we only look at port 1 and treat port 2 as part of the network, then the equation obtained in the one-port analysis is still valid. Therefore, S_{11} is equivalent to:

$$S_{11} = \frac{\dfrac{V_1 - Z_0 I_1}{2\sqrt{Z_0}}}{\dfrac{V_1 + Z_0 I_1}{2\sqrt{Z_0}}} = \Gamma_1, \tag{1.16}$$

when port 2 is terminated so that $a_2 = 0$. This tells us that although the S-parameter is defined using forward- and backward-propagating waves (e.g., V_1^+ and V_1^-), it can be calculated using the sum of voltage waves (e.g., V_1). This makes it easy to convert between circuit analysis and wave-based analysis.

With this background, let us look at the real case of a two-port network driven by a signal source at one port and terminated by load impedance at the other port, as shown in Fig. 1.9. Typically, transmission lines ("signal trace" in the view of analog circuit analysis) are used to connect between the signal source and the actual two-port network, as well as the actual two-port network and the load impedance. In our discussion, we model the transmission lines to be lossless, to have characteristic impedance of Z_{o1} and Z_{o2}, and lengths of l_1 and l_2, respectively. Now, the 2×2 S-parameter for the "entire two-port" network between the $x_1 = 0$ and $x_2 = 0$ planes can be defined as

$$\begin{bmatrix} b_1' \\ b_2' \end{bmatrix} = \begin{bmatrix} S_{11}' & S_{12}' \\ S_{21}' & S_{22}' \end{bmatrix} \cdot \begin{bmatrix} a_1' \\ a_2' \end{bmatrix}. \tag{1.17}$$

At the two ends (i.e., $x_1 = 0$ and $x_2 = l_1$) of the input transmission line, the normalized voltage waves are different by a phase shift along the transmission line, which leads to $a_1 = a_1' \cdot e^{-j\beta_1 l_1}$. Similarly, it can be derived that $a_2 = a_2' \cdot e^{-j\beta_2 l_2}$, $b_1 = b_1' \cdot e^{j\beta_1 l_1}$, and $b_2 = b_2' \cdot e^{j\beta_2 l_2}$. Plug the above equations into the $[S]$ parameter of the "original two-port component" between $x_1 = l_1$ and $x_2 = l_2$, and we can obtain:

$$\begin{cases} b_1' \cdot e^{j\beta_1 l_1} = S_{11} \cdot a_1' \cdot e^{-j\beta_1 l_1} + S_{12} \cdot a_2' \cdot e^{-j\beta_2 l_2} \\ b_2' \cdot e^{j\beta_2 l_2} = S_{21} \cdot a_1' \cdot e^{-j\beta_1 l_1} + S_{22} \cdot a_2' \cdot e^{-j\beta_2 l_2} \end{cases}. \tag{1.18}$$

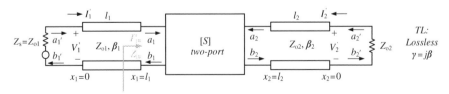

FIGURE 1.9

A realistic setup of a two-port network in a microwave system.

Comparing this with the 2×2 S-parameter for the "entire two-port" network between the $x_1 = 0$ and $x_2 = 0$ planes derived earlier, we can express S'-parameters as a function of S-parameter and phase delay:

$$[S] = \begin{bmatrix} S_{11} & S_{12} \\ S_{21} & S_{22} \end{bmatrix} = \begin{bmatrix} S'_{11} \cdot e^{2j\beta_1 l_1} & S'_{12} \cdot e^{j(\beta_1 l_1 + \beta_2 l_2)} \\ S'_{21} \cdot e^{j(\beta_1 l_1 + \beta_2 l_2)} & S'_{22} \cdot e^{2j\beta_2 l_2} \end{bmatrix}. \tag{1.19}$$

This is because of the phase delay associated with the signal going from one place to another. An example that highlights the importance of shifting reference planes for two-port S-parameters is the calibration of vector network analyzer measurement: any cable that connects the device under test (DUT) with the network analyzer can be modeled as a transmission line. Due to the phase shift of the transmission lines, the S-parameter directly "seen" by the network analyzer will be different from that of the DUT unless calibration is performed to correlate the two sets of S-parameters. It should be noted that:

a. Since $\Gamma_{in} = (Z_{in} - Z_{o1})/(Z_{in} + Z_{o1})$, the input reflection coefficient is dependent on Z_{o1}. Furthermore, since Z_{in} will change if Z_{o2} on the load side changes, both the input reflection coefficient and the input impedance depend on Z_{o1} and Z_{o2}.

b. Because both Γ_{in} and Z_{in} are dependent on Z_{o1} and Z_{o2}, the S-parameters are also dependent on Z_{o1} and Z_{o2}. In many applications and analyses, systems are designed so that Z_{o1} and Z_{o2} are 50Ω.

1.3.3 CONVERSION AMONG THE MATRICES

A two-port network can be described in terms of several parameters at a given frequency. Therefore, it is desirable to have equations to convert from one set of parameters to another. For example, given the Z-parameter of a two-port network, it can be shown that the scattering matrix in terms of Z-parameters is given by $[S] = [b]/[a] = ([z] + [Z_0])^{-1}([z] - [Z_0])$. Solving for $[z]$, we can also find $[z] = [Z_0]([U] + [S])([U] - [S])^{-1}$, where $[U]$ represents the unit diagonal matrix. The conversions among the Z, Y, h, ABCD, and S-parameters can be found as a compact table in Section 1.9 of [3].

1.4 SMITH CHART

The Smith chart is a useful graphical tool for RF and microwave engineers. From calculating the reflection coefficient and impedance at various points on a transmission line to designing the matching network of a microwave system, the Smith chart is a handy tool that is even included in lots of modern computer-aided design software and test equipment. The Smith chart is based on a polar plot of the complex reflection coefficient $\Gamma(x)$ overlaid with the corresponding impedance $Z(x)$.

1.4.1 REFLECTION COEFFICIENT ON THE SMITH CHART

To understand Smith chart, let us revisit the network in Fig. 1.2. According to Eq. (1.6), the reflection coefficient at location x along a lossless transmission line ($\gamma = j\beta$) can be represented as $\Gamma(x) = \Gamma_0 e^{j2\beta x} = \Gamma_r + j\Gamma_i$, where Γ_r and Γ_i represent the real and imaginary parts of the reflection coefficient, respectively. It should be noted that except for Γ_0, which always refers to the reflection coefficient at $x = 0$, other parameters such as Γ, Γ_r, and Γ_i are specified at the location x, but x may not show explicitly in this section for simplicity. Also, Γ_0 could be a complex value if the load has reactive components, e.g., inductor or capacitor. If we map Γ to a complex plane, as shown in Fig. 1.10, then every point on the "reflection coefficient plane" represents a given value of Γ. The distance from that point to the origin represents the magnitude of Γ, and the angle enclosed represents the phase of Γ. This is the fundamental principle of the Smith chart.

What happens if we move along the transmission line? This can be understood by checking the reflection coefficient: $\Gamma(x) = \Gamma_0 e^{j2\beta x}$. When moving from the load to the generator (signal source), x will change from 0 to $-L$, which produces a negative phase shift in the reflection coefficient because of the "$e^{j2\beta x}$" term. This means with a given x, we just need to rotate CW from Γ_0 by an angle of $2\beta x$ on the $|\Gamma_0|$ circle, as shown in Fig. 1.11. In general, when moving from one point to another point on a transmission line, clockwise rotation corresponds to moving toward the generator (signal source), while counterclockwise rotation corresponds to moving toward the load. If the physical distance between two points is x on a transmission line, then we need to rotate through an angel of $2\beta x$ degrees on the Smith chart. Because of this, each revolution on the Smith chart represents a movement of one-half wavelength along the transmission line.

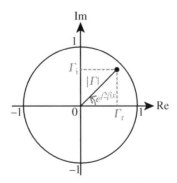

FIGURE 1.10

The reflection coefficient on a Smith chart. The center of the Smith chart represents $\Gamma = 0$, and the outer circle of the chart represents $|\Gamma| = 1$.

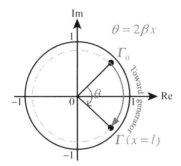

FIGURE 1.11

Smith chart representation of moving along a transmission line.

1.4.2 RELATIONSHIP BETWEEN REFLECTION COEFFICIENT AND IMPEDANCE

The reflection coefficient for a lossless transmission line of characteristic impedance Z_0, terminated with an impedance Z_L, is given by:

$$\Gamma = \frac{Z - Z_0}{Z + Z_0} = \frac{(Z/Z_0) - 1}{(Z/Z_0) + 1} = \frac{z - 1}{z + 1}, \tag{1.20}$$

where Z is the impedance at location x on the transmission line when looking toward the load. $z = Z/Z_0$ is the normalized impedance, which is a complex number that can be represented as $z = r + jx$. We use "r" to represent normalized resistance, and "x" to represent normalized reactance. It should be noted that first, Z, z, r, and x are all functions of the location x along the transmission line; second, the variable x is used twice with different meanings: one is reactance, the other is position. If we plug the complex representation of $\Gamma = \Gamma_r + j\Gamma_i$ and $Z = r + jx$ back into the above equation and equalize the real part and imaginary part on both sides of the equation, we have the following equations to correlate the reflection coefficient and impedance on the Smith chart:

$$\left(\Gamma_r - \frac{r}{r+1}\right)^2 + \Gamma_i^2 = \left(\frac{1}{r+1}\right)^2 \tag{1.21a}$$

$$(\Gamma_r - 1)^2 + \left(\Gamma_i - \frac{1}{x}\right)^2 = \left(\frac{1}{x}\right)^2 \tag{1.21b}$$

On a Smith chart, Eq. (1.21a) corresponds to a circle centered at $[r/(r+1), 0]$ with a radius of $1/(r+1)$. This is called constant resistance circle because for a fixed resistance r, all possible points of Γ are located on such a circle no matter what the value of x is. Similarly, Eq. (1.21b) corresponds to a circle centered at $[1, 1/x]$ with a radius of $1/x$. This is called constant reactance circle because for a fixed reactance x, all possible points of Γ are located on the circle no matter what the value of r is. The constant resistance circles and constant reactance circles are shown in Fig. 1.12.

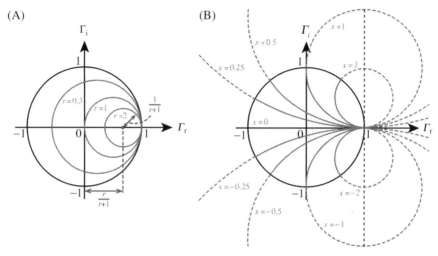

FIGURE 1.12

(A) Constant resistance circles with different values of r. (B) Constant reactance circles with different values of x.

1.4.3 ADMITTANCE ON THE SMITH CHART

In the previous sub-section, the reflection coefficient Γ and the normalized impedance of z were correlated on the Smith chart based on their mathematical relationship. Likewise, a new parameter Γ_{adm} can be defined based on the normalized admittance:

$$\Gamma_{adm} = \frac{y-1}{y+1},\qquad(1.22)$$

where the normalized admittance y consists of normalized conductance g and normalized susceptance b: $y = Y/Y0 = (G + jB)/Y_0 = g + jb$. Note the similarity between Eq. (1.22) and the definition of Γ in last section. Therefore, the same Smith chart can also be used to describe the relationship between Γ_{adm} and normalized admittance y. In this case, the Smith chart is called an admittance chart, or "Y Smith chart." Actually, it can be easily shown that $\Gamma_{adm} = -\Gamma = \Gamma e^{j\pi}$. In practice, people do not use Γ_{adm} quite often. Instead, we just need to rotate the Smith chart by 180 degrees so that both y and z can be found using Γ on a single Smith chart.

1.4.4 USEFUL RULES OF SMITH CHART AND IMPEDANCE/ ADMITTANCE CALCULATION

Several useful rules of the Smith chart can be summarized as follows. The corresponding graphic summary is shown in Fig. 1.13.

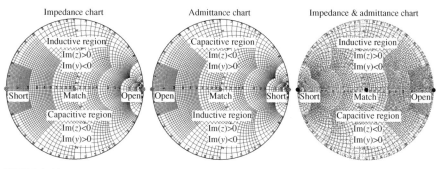

FIGURE 1.13

Comparison of impedance chart, admittance chart, and impedance and admittance chart.

1. There are five types of constant-valued circles on the Smith chart:
   a. A constant-$|\Gamma|$ circle, also called a constant standing wave ratio (SWR) circle, consists of points that have the same magnitude in reflection coefficient. Moving along such a circle corresponds to moving along a lossless transmission line.
   b. A constant resistance r circle consists of points that have the same normalized resistance. Moving along such a circle corresponds to adding a series inductor or capacitor.
   c. A constant reactance x circle consists of points that have the same normalized reactance. Moving along such a circle corresponds to changing the real part of impedance.
   d. A constant conductance g circle consists of points that have the same normalized conductance. Moving along such a circle corresponds to adding a parallel inductor or capacitor.
   e. A constant susceptance b circle consists of points that have the same normalized susceptance. Moving along such a circle corresponds to changing the real part of admittance.
2. A Smith chart is divided into an inductive region and an capacitive region:
   a. For an "impedance chart" or an "impedance & admittance chart": the upper half plane is the inductive region, corresponding to $\text{Im}(z) > 0$ and $\text{Im}(y) < 0$; the lower half plane is the capacitive region, corresponding to $\text{Im}(z) < 0$ and $\text{Im}(y) > 0$.
   b. When the Smith chart is used as an admittance chart: the upper half plane is the capacitive region, corresponding to $\text{Im}(z) < 0$ and $\text{Im}(y) > 0$; the lower half plane is the inductive region, corresponding to $\text{Im}(z) > 0$ and $\text{Im}(y) < 0$.
3. Moving along a lossless transmission line toward the generator corresponds to moving clockwise along a constant SWR circle.
4. Moving along a lossless transmission line toward the load corresponds to moving counterclockwise along a constant SWR circle.

5. Moving a physical distance of L along a transmission line corresponds to rotating $2\beta L$ along a constant SWR circle.
6. Each revolution on a Smith chart corresponds to moving $\lambda/2$ along a transmission line.
7. The real axis on a Smith chart has $x = 0$, which means purely resistive.
8. The $|\Gamma| = 1$ circle is the boundary of a Smith chart, which means $r = 0$ and purely reactive.
9. It is useful to remember some special points on an impedance chart or an impedance and admittance chart:
   a. The left end on the real axis is the short circuit point.
   b. The right end on the real axis is the open circuit point.
   c. The center of the Smith chart is the impedance matching point.
10. Likewise, it is useful to remember some special points on the Smith chart when used as an admittance chart:
   a. The left end on the real axis is the open circuit point.
   b. The right end on the real axis is the short circuit point.
   c. The center of the Smith chart is the impedance matching point.

1.5 IMPEDANCE MATCHING

Let us consider the power P_{in} that is fed into the system shown in Fig. 1.14, which can be expressed as:

$$P_{in} = \mathrm{Re}\left[V_{in} \cdot I_{in}^*\right] = |I_{in}|^2 \cdot \mathrm{Re}[Z_{in}] = \frac{|E_1|^2}{|Z_S + Z_{in}|^2} \cdot \mathrm{Re}[Z_{in}] = \frac{|E_1|^2 R_{in}}{(R_{in} + R_S)^2 + (X_{in} + X_S)^2} \quad (1.23)$$

The condition to maximize the input power P_{in} can be mathematically derived from the condition of $\partial P_{in}/\partial X_{in} = 0$ and $\partial P_{in}/\partial R_{in} = 0$, which leads to [1,3] $X_{in} = -X_S$ and $R_{in} = R_S$. This indicates that in order to maximize the power flow, complex conjugate matching, i.e., $Z_{in} = Z_S^*$, is needed. Under such a condition, the maximum power flow is $P_{in,max} = |E_1|^2/[4\mathrm{Re}(Z_S)]$. It should be noted that this is the available power from the source and that it depends only on the source, not

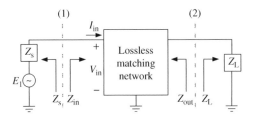

FIGURE 1.14

A lossless matching network between a signal source and a load.

on the load or on the transmission line. In general, if the matching network is lossless, the complex conjugate matching condition to maximize power flow at any place in the system is $Z_{\text{toward load}} = Z^*_{\text{toward source}}$. The matching network can be either a transmission line, a combination of multiple transmission lines, or a lumped element network. Moreover, it can be proven that for a lossless matching network, either matching at (1) or matching at (2) leads to the same conjugate matching result, achieving maximum power flow. This means either $Z_{\text{in}} = Z^*_S$ at (1) or at $Z_{\text{out}} = Z^*_L$ at (2) will result in maximum power flow. However, this does not hold true for a lossy matching network.

1.5.1 MATCHING WITH LUMPED ELEMENTS

Typical lumped element matching networks include L-matching, T-matching, and π-matching networks. Different types of matching networks have different bandwidths and design complexities. An L-matching network typically has two discrete components, and is one of the simplest matching networks. As an example, Fig. 1.15 shows L-matching network designs on the Smith chart to match a low impedance and a high impedance to 50 Ω, respectively. The basic design concept is that in order to move a load to the center of the Smith chart, which represents the impedance matching point, a designer can move the load twice along constant resistance and constant reactance circles, respectively. Adding a series

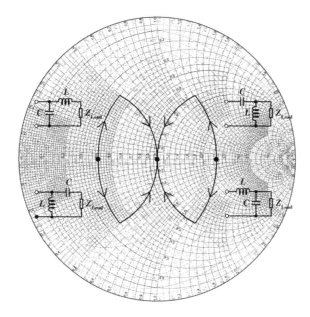

FIGURE 1.15

Impedance matching using L-matching networks, illustrated on an impedance and admittance chart.

inductor or capacitor moves the load along a constant resistance circle clockwise (i.e., further into the inductive region) or counterclockwise (i.e., further into the capacitive region), respectively. On the other hand, adding a shunt inductor or capacitor moves the load along a constant conductance circle counterclockwise (i.e., further into the inductive region) or clockwise (i.e., further into the capacitive region), respectively.

1.5.2 MATCHING BY TRANSMISSION LINES

Different from lumped element components that introduce relatively large parasitic capacitive and inductive effects, transmission lines are fully modeled and thus are more appropriate for impedance matching at high microwave frequencies. Also, transmission line matching networks do not require discrete commercial components such as inductors and capacitors to be mounted in a microwave system, and thus may be easier for board level implementation and tuning after fabrication. However, the tradeoff of using a transmission line matching network is that transmission lines need to be a specific length to achieve impedance matching, and thus this may lead to a larger physical size of the microwave system. Therefore, for RF/microwave integrated circuit designs where layout size is an important consideration, transmission line matching networks are only used at high millimeter wave frequencies.

Fig. 1.16 shows an example of a single-stub transmission line matching network. The design procedure is shown in Fig. 1.17. The normalized load impedance z_L shown in an impedance chart is first rotated by 180 degrees, to be transformed into the admittance y_L on an admittance chart. After that, the design is completed, with the chart treated as an admittance chart. The shunt stub is positioned a distance l_1 from the load. This distance is chosen so that the series transmission line with a length of l_1 just moves the load to a point with normalized reactance being 1. Then, the length of the shunt stub l_2 is chosen so that it generates a normalized input admittance y_3 that exactly cancels the susceptance part of the admittance (y_2 in Fig. 1.18) presented at the input of the series transmission

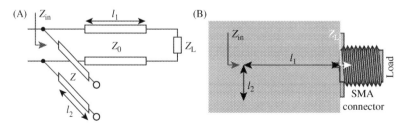

FIGURE 1.16

Impedance matching using transmission lines. (A) Circuit diagram. (B) Actual microstrip line realization.

(A)   (B)

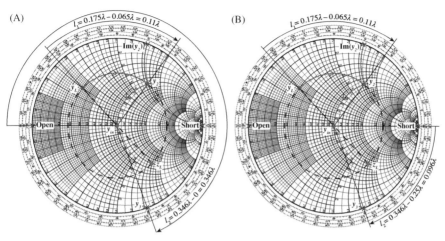

FIGURE 1.17

Design of a single-stub transmission line matching network on an admittance chart.
(A) Matching with open-ended stub. (B) Matching with short-ended stub.

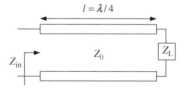

FIGURE 1.18

Impedance matching using a quarter-wave transformer.

line l_1. In order to achieve this, the shunt stub is made capacitive or inductive according to whether the main transmission line is presenting an inductive or capacitive admittance, respectively. The shunt stub could be either open-ended or short-ended, which will result in different lengths of l_2. A single stub will only achieve a perfect match at a specific frequency, resulting in a narrowband matching network. For wideband matching, several stubs may be used, spaced along the main transmission line. Filter design techniques can be applied to result in wideband shunt stub matching. Interested authors should refer to [6] for details.

1.5.3 MATCHING BY A QUARTER-WAVE TRANSFORMER

A quarter-wave transformer impedance matching network is shown in Fig. 1.18. Since the input impedance of a quarter-wavelength transmission line is $Z_{in} = Z_0^2/Z_L$, in order to match a load impedance of Z_L to impedance Z_{in}, we just need to design a transmission line with characteristic impedance of $Z_0 = \sqrt{Z_{in} Z_L}$. Since a transmission line is only a quarter-wave length long for a certain operating frequency, matching by a basic quarter-wave transformer is normally narrowband.

1.6 MICROWAVE PASSIVE COMPONENTS

Let us start from practical components, instead of mathematics. Consider one of the most useful and fundamental three-port devices that we can build: a power divider, shown in Fig. 1.19. To efficiently divide power, the input port (port 1) must first be matched so that $S_{11} = 0$. Likewise, the divided power must be efficiently delivered to the output ports of the divider (i.e., without loss, $P_1 = P_2 + P_3$): $|S_{21}|^2 + |S_{31}|^2 = 1$. In addition, it is desirable that ports 2 and 3 be matched such that $S_{22} = S_{33} = 0$, and that ports 2 and 3 be isolated such that $S_{23} = S_{32} = 0$. The latter condition ensures that no signal incident on port 2 will "leak" into port 3—and vice versa. The above conditions completely describe an ideal power divider—conditions that can be met—and thus represent the specifications of an ideal power divider. As an example, an ideal power divider could take the input signal into port 1 and divide into two equal parts, with half of the input power exiting port 2, and the other half of the power exiting port 3, which means $|S_{21}|^2 = |S_{31}|^2 = 0.5$. But it is not necessary that a lossless power divider has to divide the input power into two equal parts.

1.6.1 THREE-PORT POWER COUPLERS

Three-port couplers are also known as T-junction couplers or T-junction dividers. A rule that can be proven for any three-port device is that it is impossible to simultaneously achieve: (1) matching at all three ports, (2) lossless, and (3) reciprocal. A reciprocal network is a network by which the power losses are the same between any two ports regardless of direction of propagation (scattering parameter $S_{21} = S_{12}$, $S_{13} = S_{31}$, etc.). A network is known to be reciprocal if it is passive

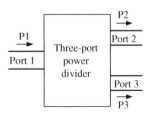

FIGURE 1.19

A three-port power divider.

and contains only isotropic materials. Examples of reciprocal networks include cables, attenuators, and all passive power splitters and couplers. Several typical types of T-junction power dividers will be discussed as follows.

1.6.1.1 The resistive power divider: matched, reciprocal, but lossy

The basic configuration for a resistive divider is shown in Fig. 1.20. When the source impedance and the load impedance are both Z_0, this symmetrical power divider will be matched at port 1 if R is selected so that $Z_0 = R + (R + Z_0)\|$ $(R + Z_0) = 1.5R + Z_0/2$. Solving this equation, it can be found that the condition for port 1 to be matched is $R = Z_0/3$. From the symmetry of the circuit, we find that all the other ports will be matched as well (i.e., $S_{11} = S_{22} = S_{33} = 0$). Moreover, it can be shown that: $S_{12} = S_{21} = S_{31} = S_{13} = S_{23} = S_{32} = \frac{1}{2}$ in an ideal case. It should be noted that $|S_{21}|^2 + |S_{31}|^2 = \frac{1}{2} < 1$. The corresponding physical meaning is that a resistive power divider is lossy. In fact, the power that comes out from each port is just one-quarter of the input power. As an example, for any signal that enters port 1, $P_{2\_out} = P_{3\_out} = P_{1\_in}/4$, which can be expressed in dB as $P_{2\_out}(dB) = P_{3\_out}(dB) = P_{1\_in}(dB) - 6dB$. In other words, half of the input power is absorbed by the divider because of the resistive components used. As a result, a resistive power divider that is reciprocal and matched at all the three ports can be easily designed and implemented, but this is achieved at the cost of extra power loss for each output port. Assuming equal power division ratio, the extra power loss will be at least 3 dB for each of the output ports.

1.6.1.2 The lossless divider: lossless, can be reciprocal, but not matched

In contrast to the resistive divider that depends on three resistors for impedance matching, a lossless power divider can be realized by three transmission lines intersecting at a T-junction as shown in Fig. 1.21. To be ideal, we want $S_{11} = 0$. Thus,

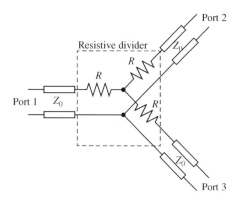

FIGURE 1.20

A resistive divider that is matched at three ports and is reciprocal but lossy ($R = Z_0/3$).

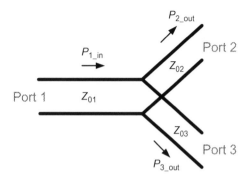

FIGURE 1.21

A lossless divider based on a transmission line T-junction that is reciprocal but not simultaneously matched at all the three ports.

when ports 2 and 3 are terminated with matched loads, the input impedance at port 1 must be equal to Z_{01}. This will only be true if the values of Z_{02} and Z_{03} are selected such that $Z_{01} = Z_{02} \| Z_{03} = Z_{02} \cdot Z_{03}/(Z_{02} + Z_{03})$. Note, however, that this circuit is not symmetrical; thus, we find that $S_{22} \neq 2$ and $S_{33} \neq 3$. It is evident that this divider is lossless (no resistive components), so that $P_{1\_in} = P_{2\_out} + P_{3\_out}$, where P_{1_in} is the power incident at port 1, and P_{2_out} and P_{3_out} are the powers absorbed by matched loads of ports 2 and 3, respectively. Unless $Z_{02} = Z_{03}$, the power will not be divided equally between P_{2_out} and P_{3_out}. With a little mathematical analysis, it can be shown that the division ratio is $\alpha = P_{2\_out}/P_{3\_out} = Z_{03}/Z_{02}$. Thus, if we desire an ideal ($S_{11} = 0$) divider with a specific division ratio of α, we will need $Z_{02} = Z_{01}(1 + 1/\alpha)$ and $Z_{03} = Z_{01}(1 + \alpha)$.

In order to eventually let all the transmission lines have the same value of characteristic impedance, we can use a quarter-wave transformer on ports 2 and 3, as shown in Fig. 1.22. As an example, in order to design a lossless 3 dB divider that has equal output at ports 2 and 3, the value of α will be 1, which results in $Z_{02} = Z_{03} = 2Z_{01}$. Based on that, the characteristic impedance for the quarter-wave transformer will be $\sqrt{2}Z_0$ at both port 2 and port 3.

1.6.1.3 Circulators: matched, lossless, but nonreciprocal

A circulator is a device that is nonreciprocal and can be ideally lossless and matched simultaneously at all the three ports. "Nonreciprocal" means that the signal flows in a certain way that cannot be reversed. One example is shown in Fig. 1.23, where $P_{2out} = P_{1in}$, $P_{3out} = P_{2in}$, and $P_{1out} = P_{3in}$. One way to realize a circulator is to use anisotropic ferrite materials, which are often "biased" by a permanent magnet. The result is a nonreciprocal device. Circulators are frequently used in microwave/wireless solutions such as transceivers and radar sensors. Fig. 1.24 shows a transceiver with circulator to split the transmitter and receiver signal paths.

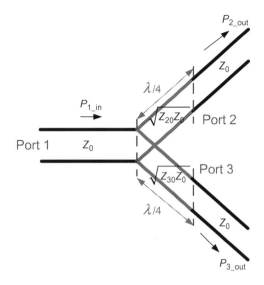

FIGURE 1.22

Design of a lossless divider based on a transmission line T-junction and quarter-wave transformers.

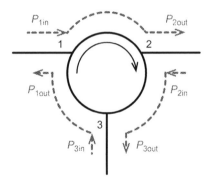

FIGURE 1.23

A nonreciprocal circulator that can be matched at all three ports and is ideally lossless.

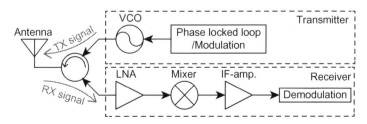

FIGURE 1.24

Use of a circulator to split the transmitter and receiver signal paths.

1.6.1.4 The Wilkinson power divider/combiner

A Wilkinson power splitter is an example of a reciprocal, matched three-port network that is only lossy between ports 2 and 3, which does not affect its efficiency as a combiner or splitter if the summation of power appears at port 1 [7,8]. Actually, when combining power from ports 2 and 3, or splitting power into ports 2 and 3, we do not want ports 2 and 3 to cross-talk. We even want the attenuation of the signal between ports 2 and 3 to be as great as possible. The Wilkinson power divider achieves all these requirements based on two quarter-wave transmission lines and a resistor that isolates ports 2 and 3. It is a specific class of power divider circuits, which can achieve isolation between the output ports while maintaining a matched condition on all ports. The Wilkinson power divider/combiner relies on quarter-wave transformers to match the split ports to the common port. The resistor not only helps all the three ports get matched, but also isolates port 2 from port 3 at the operating frequency. In the meantime, the resistor adds no resistive loss to the power split.

A basic equal-amplitude, two-way split, single-stage Wilkinson divider is shown in Fig. 1.25. The arms are quarter-wave transformers of impedance $\sqrt{2} \times Z_0$. It should be noted that the "Wilkinson" concept can also be applied to an N-way divider. When a signal enters port 1, it splits into equal-amplitude, equal-phase output signals at ports 2 and 3. Since each end of the isolation resistor between ports 2 and 3 is at the same potential, no current flows through it and therefore the resistor is decoupled from the input. The two output port terminations will add in parallel at the input, so they must be transformed to $2 \times Z_0$ each at the input port to combine to Z_0. The quarter-wave transformers in each leg accomplish this. Without the quarter-wave transformers, the combined impedance of the two outputs at port 1 would be $Z_0/2$. The characteristic impedance of the quarter-wave lines must be equal to $\sqrt{2} \times Z_0$ so that the input is matched to Z_0 when ports 2 and 3 are terminated with Z_0.

Consider a signal input at port 2. In this case, it splits equally between port 1 and the resistor R, with none appearing at port 3. The resistor thus serves the important function of decoupling ports 2 and 3. Note that for a signal input at either port 2 or 3, half the power is dissipated in the resistor and half is delivered to port 1. But why is port 2 isolated from port 3 and vice versa? Consider that the

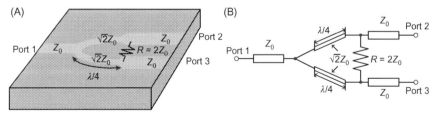

FIGURE 1.25

Wilkinson power divider/combiner (A) and its transmission line model (B).

signal splits when it enters port 2. Part of it goes clockwise through the resistor and part goes counterclockwise through the upper arm, then splits at the input port, then continues counterclockwise through the lower arm toward port 3. The recombining signals at port 3 end up being equal in amplitude. Moreover, they are 180 degrees out of phase due to the half-wavelength that the counterclockwise signal travels, while the clockwise signal doesn't. The two signal voltages subtract to zero at port 3 and the signal disappears, at least under ideal circumstances. In real couplers, there is a finite phase through the resistor that will limit the isolation of the output ports.

1.6.2 FOUR-PORT POWER COUPLERS

Besides the power dividers/combiners described in the last section, there are some power couplers with four ports. In some applications, one of the four ports needs to be terminated with the system impedance (e.g., 50 Ω), rendering the other three ports as input/output ports. As examples, this section will discuss two types of four-port power couplers: the 90 degree hybrid and the 180 degree hybrid.

The branch-line is the simplest type of four-port quadrature coupler that provides 90 degrees of phase difference between the two output ports when serving as a divider. An ideal branch-line coupler is shown in Fig. 1.26A. Each transmission line in the center square is a quarter-wavelength long. In order to achieve impedance matching, an even-odd mode analysis [1] indicates that the two horizontal transmission lines should have a characteristic impedance of $Z_0/\sqrt{2}$, whereas the two vertical transmission lines should have an characteristic impedance of Z_0. A signal entering the top left port (port 1 in the figure) is split into two quadrature signals (90 degrees out-of-phase) on the right (ports 2 and 3), with the remaining port 4 fully isolated from the input port at the center frequency. In this case, port 4 has to be properly terminated with impedance Z_0 to achieve impedance matching at all the four ports. Remember that the lower output port (port 3) has the most negative transmission phase since it has the farthest path to travel.

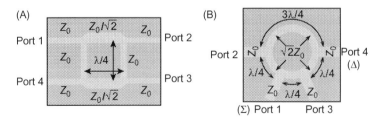

FIGURE 1.26

Microstrip branch-line coupler (A) and rat-race 180 degree hybrid (B). The unused port should be terminated with a system impedance of Z_0.

The rat-race shown in Fig. 1.26B provides either 0 degrees or 180 degrees of phase difference at its two outputs when serving as a divider. It is named because of its shape. The circumference is 1.5 wavelengths. For an equal-split rat-race coupler, the impedance of the entire ring is fixed at $\sqrt{2} \times Z_0$, or 70.7 Ω for a 50 Ω system. If there are two input signals, one at port 2 and one at port 3, then the difference between these two signals is at port 4 (the Δ case), and the sum of the two input signals is at port 1 (the $\sum$ case). Either used in the Δ or the $\sum$ case, the isolated port that is "unused" for signal input/output should be terminated with a resistive load of $R = Z_0$.

1.7 MICROWAVE ACTIVE CIRCUITS

Typical microwave active circuits include amplifiers, mixers, oscillators, etc. They use active devices such as bipolar junction transistors, field-effect transistors, and high-electron-mobility transistor to either convert direct current energy into microwave signals, or provide signal conditioning (e.g., amplification and filtering) to input microwave signals. The theory and popular types of transistors are covered in many fundamental electrical engineering books [9,10], and are used not only in RF/microwave applications, but also other areas such as digital circuits and sensors. Therefore, interested readers are referred to [11] and [12] for fundamentals on microwave active devices. This section will introduce three types of active circuits that are frequently used in the healthcare and biosensing solutions presented in other chapters of this book, namely amplifiers, mixers, and oscillators.

1.7.1 AMPLIFIERS: GAIN, NOISE FIGURE, LINEARITY, STABILITY, AND EFFICIENCY

An amplifier is one of the most fundamental building blocks of microwave systems. It takes energy from a signal source and generates output to match the input signal shape but with a larger amplitude. The types of frequently used microwave amplifiers include low-noise amplifiers, power amplifiers, and gain blocks. To evaluate the performance of an amplifier, its gain, noise figure, linearity, stability, and efficiency are frequently examined.

1.7.1.1 Amplifier gain and constant gain circles

Several power gain equations appear in the literature and are used in the design of microwave amplifiers [1,3]. To understand the physical meanings of the power gain definitions, let us start from the generalized model of a microwave amplifier in Fig. 1.27, where a transistor is the main active device that amplifies the signal, and two passive networks are added at the input/output for impedance and, if applicable, noise matching purposes. First, let us use P_{IN}, P_L, P_{AVS}, and P_{AVN} to

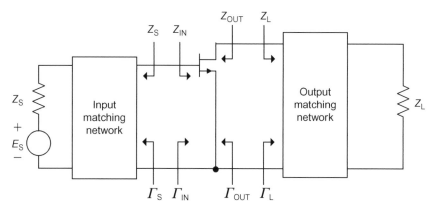

FIGURE 1.27

A generalized model of microwave amplifier.

denote the power input to the network, power delivered to the load, power available from the source, and power available from the network, respectively. The transducer power gain G_T, power gain G_P (also called operating power gain), and available power gain G_A can be defined and derived as [3]:

$$G_T = \frac{P_L}{P_{AVS}} = \frac{1 - |\Gamma_S|^2}{|1 - \Gamma_{IN}\Gamma_S|^2}|S_{21}|^2\frac{1 - |\Gamma_L|^2}{|1 - S_{22}\Gamma_L|^2} = \frac{1 - |\Gamma_S|^2}{|1 - S_{11}\Gamma_S|^2}|S_{21}|^2\frac{1 - |\Gamma_L|^2}{|1 - \Gamma_{OUT}\Gamma_L|^2}$$

$$G_P = \frac{P_L}{P_{IN}} = \frac{1}{1 - |\Gamma_{IN}|^2}|S_{21}|^2\frac{1 - |\Gamma_L|^2}{|1 - S_{22}\Gamma_L|^2}$$

$$G_A = \frac{P_{AVN}}{P_{AVS}} = \frac{1 - |\Gamma_S|^2}{|1 - S_{11}\Gamma_S|^2}|S_{21}|^2\frac{1}{1 - |\Gamma_{OUT}|^2}$$

$$(1.24)$$

where Γ_{IN} and Γ_{OUT} can be calculated using the transistor S-parameters and the reflection coefficients looking from the transistor toward the source (Γ_S) and the load (Γ_L) [3]:

$$\Gamma_{IN} = S_{11} + \frac{S_{12}S_{21}\Gamma_L}{1 - S_{22}\Gamma_L}$$

$$\Gamma_{OUT} = S_{22} + \frac{S_{12}S_{21}\Gamma_S}{1 - S_{11}\Gamma_S}$$

$$(1.25)$$

It should be noted that:

1. G_T is a function of Γ_S and Γ_L (both input and output matching) and the S-parameters [**S**] of the transistor,
2. G_P is a function of Γ_L and [**S**] (not related to input matching), and
3. G_A is a function of Γ_S and [**S**] (not related to output matching).

In many cases the gain, noise, and bandwidth cannot be optimized simultaneously. There are trade-offs among them. It is sometimes preferable to design for less than the maximum obtainable gain to improve bandwidth, or to obtain a specific value of amplifier noise performance. This can be done by designing the input and output matching sections to have less than maximum gains; in other words, mismatches are purposely introduced to reduce the overall gain while improving other aspects of the performance. The design procedure is facilitated by plotting constant gain circles on the Smith chart, to represent loci of Γ_S and Γ_L that give fixed values of gain.

We will discuss the case of a unilateral device ($S_{12} = 0$), because in many applications $|S_{12}|$ of a transistor is small. The case of a bilateral device must sometimes be considered in practice and is discussed in detail in Sections 3.5 and 3.6 of [3]. A two-port network is unilateral when $S_{12} = 0$. Based on Eq. (1.25), $\Gamma_{in} = S_{11}$, $\Gamma_{out} = S_{22}$, and the unilateral transducer power gain (denoted as G_{TU}) is simplified as $G_{TU} = G_S G_O G_L$ in linear scale or $G_{TU}(\text{dB}) = G_S(\text{dB}) + G_O(\text{dB}) + G_L(\text{dB})$ in dB scale, where:

$$G_S = \frac{1 - |\Gamma_S|^2}{|1 - S_{11}\Gamma_S|^2}, \quad G_O = |S_{21}|^2, \quad G_L = \frac{1 - |\Gamma_L|^2}{|1 - S_{22}\Gamma_L|^2}. \tag{1.26}$$

As a result, the microwave amplifier can be represented by three gain blocks G_S, G_O, and G_L cascaded to each other. The input matching network determines Γ_S and therefore the value of G_S according to the above equations; the output matching network determines Γ_L and therefore the value of G_L. The maximum gain is obtained under conjugate matching when $\Gamma_s = S_{11}^*$ and $\Gamma_L = S_{22}^*$, resulting in the maximum values given by:

$$G_{S,\text{max}} = \frac{1}{1 - |S_{11}|^2} \text{ and } G_{L,\text{max}} = \frac{1}{1 - |S_{22}|^2}. \tag{1.27}$$

Now, define normalized gain factor as $g_i = G_i/G_{i,\text{max}}$, with "$I$" being either "$S$" or "$L$," representing "source" or "load," and $0 \le g_i \le 1\,0$ dc It can be shown that [1,3] the values of Γ_i that produce a constant value of g_i lie in a circle determined by $|\Gamma_i - C_{gi}| = r_{gi}$, where the center and radius of the circle are given by:

$$C_{gi} = \frac{g_i S_{ii}^*}{1 - |S_{ii}|^2(1 - g_i)} \text{ and } r_{gi} = \frac{\sqrt{1 - g_i}(1 - |S_{ii}|^2)}{1 - |S_{ii}|^2(1 - g_i)}, \tag{1.28}$$

where $ii = 11$ when $i = S$, and $ii = 22$ when $i = L$. Each given value of g_i corresponds to a constant-G_i circle. It should be noted that constant G_S circles describe the matching at the input port, which corresponds to Γ_S, and constant G_L circles describe the matching at the output port, which corresponds to Γ_L. Since we have seen that the available power gain G_A only depends on the input matching network, another way of analyzing the input matching condition is to plot the constant G_A circles instead of the constant G_S circles [3]. Some modern EDA design tools such as the AWR Microwave Office simulate and plot constant G_A circles to represent the input matching condition. For example, Fig. 1.28 shows

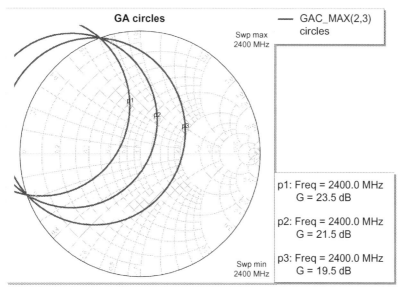

FIGURE 1.28

Simulated constant available power gain circles for an Infineon BFP740 Silicon Germanium RF Transistor operating at 3 V, 10 mA, and 2.4 GHz.

the constant available power gain G_A contours in the input reflection plane for an Infineon BFP740 Silicon Germanium RF Transistor biased at 3 V and 10 mA operating at 2.4 GHz. The results are obtained using the NI AWR Design Environment (MWO). The three circles p1, p2, and p3 represent available power gains of 23.5 dB, 21.5 dB, and 19.5 dB, respectively, where the 23.459 dB gain of p1 is the maximum available power gain of the transistor (i.e., $|S_{21}/S_{12}|$ when the transistor is not unconditionally stable).

1.7.1.2 Amplifier stability

Passive matching networks produce values of Γ_S and Γ_L such that $|\Gamma_S| < 1$ and $|\Gamma_L| < 1$. In other words, the resistive part associated with Z_S and Z_L is positive. However, from the above equations, it is possible that certain values of S-parameters will result in $|\Gamma_{IN}| > 1$ and/or $|\Gamma_{OUT}| > 1$ (even if $|\Gamma_S| < 1$ and $|\Gamma_L| < 1$). When $|\Gamma_{IN}| > 1$ or $|\Gamma_{OUT}| > 1$, the input or output ports of the transistor present a negative resistance, and oscillations can occur. Obviously, i.e., a situation that we must avoid in amplifier design, which leads to the stability considerations. The stability of an amplifier, or its resistance to oscillations, can be determined from the S-parameters, the matching networks, and how the input and output are terminated (i.e., terminations). In a two-port network, oscillations are possible when either the input or output port presents a negative resistance. This occurs when $|\Gamma_{IN}| > 1$ or $|\Gamma_{OUT}| > 1$, which for a unilateral device occurs when $|S_{11}| > 1$ or $|S_{22}| > 1$.

For example, a unilateral transistor is a transistor where $S_{12} = 0$ (or its effect is so small that it can be set equal to zero). If $S_{12} = 0$, it follows from Eq. (1.25) that $|\Gamma_{IN}| = |S_{11}|$ and $|\Gamma_{OUT}| = |S_{22}|$. Therefore, if $|S_{11}| > 1$, the transistor presents a negative resistance at the input; if $|S_{22}| > 1$, the transistor presents a negative resistance at the output.

A two-port network as shown in Fig. 1.27 is said to be unconditionally stable at a given frequency if the real parts of Z_{IN} and Z_{OUT} are greater than zero for all passive load and source impedances. If the two-port network is not unconditionally stable, it is potentially unstable. That is, some passive load and source terminations can produce input and/or output impedances having a negative real part. In terms of reflection coefficients, the requirement for unconditional stability at a given frequency is as follows: when $|\Gamma_S| < 1$ and $|\Gamma_L| < 1$, then $|\Gamma_{IN}| < 1$ and $|\Gamma_{OUT}| < 1$, by which all reflection coefficients are normalized to the same characteristic impedance Z_0. A graphical analysis of the above equations is especially useful in the analysis of potentially unstable transistors. First, the regions where values of Γ_L and Γ_S produce $|\Gamma_{IN}| = 1$ and $|\Gamma_{OUT}| = 1$ can be determined. Setting the magnitude of $|\Gamma_{IN}|$ and $|\Gamma_{OUT}|$ equal to 1 and solving for the values of Γ_L and Γ_S shows that the solutions for Γ_L and Γ_S lie on circles (called stability circles). Assuming $\Delta = S_{11}S_{22} - S_{12}S_{21}$, then $|\Gamma_{IN}| = 1$ corresponds to an output stability circle on the Γ_L chart with radius r_L and center C_L [3]:

$$
r_L = \left| \frac{S_{12}S_{21}}{|S_{22}|^2 - |\Delta|^2} \right|
$$
$$
C_L = \frac{(S_{22} - \Delta S_{11}^*)^*}{|S_{22}|^2 - |\Delta|^2}
\tag{1.29}
$$

On the other hand, $|\Gamma_{OUT}| = 1$ corresponds to an input stability circle on the Γ_S chart with radius r_L and center C_L [3]:

$$
r_S = \left| \frac{S_{12}S_{21}}{|S_{11}|^2 - |\Delta|^2} \right|
$$
$$
C_S = \frac{(S_{11} - \Delta S_{22}^*)^*}{|S_{11}|^2 - |\Delta|^2}
\tag{1.30}
$$

With the S-parameters of a two-port device at one frequency, the expressions can be calculated and plotted on a Smith chart, and the set of values of Γ_L and Γ_S that produce $|\Gamma_{IN}| = 1$ and $|\Gamma_{OUT}| = 1$ can be easily observed as shown in Fig. 1.29. This figure illustrates the graphical construction of the stability circles. On one side of the stability circle boundary, in the Γ_L plane, we will have $|\Gamma_{IN}| < 1$ and on the other side $|\Gamma_{IN}| > 1$. Similarly, in the Γ_S plane on one side of the stability circle boundary, we will have $|\Gamma_{OUT}| < 1$ and on the other side $|\Gamma_{OUT}| > 1$.

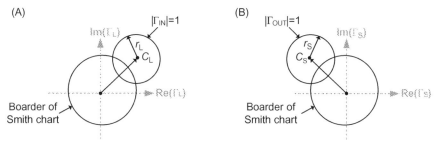

FIGURE 1.29

(A) Γ_L circle (left) and (B) Γ_S circle (right), which lead to $|\Gamma_{IN}| = 1$ and $|\Gamma_{OUT}| = 1$, respectively.

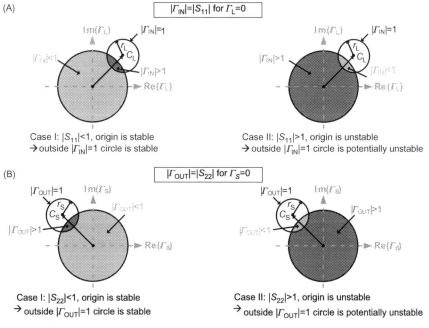

FIGURE 1.30

Based on the values of S_{11} and S_{22}, the stable region can be decided on the Γ_L chart and the Γ_S chart, respectively. (A) Γ_L chart for input stability analysis; (B) Γ_S chart for output stability analysis.

Next, we need to determine which area in the Smith chart represents the stable region—in other words, the region where values of $|\Gamma_L| < 1$ produce $|\Gamma_{IN}| < 1$, and where values of $|\Gamma_S| < 1$ produce $|\Gamma_{OUT}| < 1$. It turns out that stable and unstable regions are decided by the range of S-parameters. On the Γ_L chart shown in Fig. 1.30A, the center of the Smith chart represents $\Gamma_L = 0$, which results in

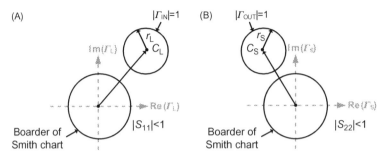

FIGURE 1.31

Graphical explanation of unconditional stability. (A) Input; (B) Output.

$\Gamma_{IN} = S_{11}$. Therefore, if $|S_{11}| < 1$, then the origin of the Smith chart is stable. Based on continuity, this means the entire region outside the $|\Gamma_{IN}| = 1$ circle is stable. On the other hand, if $|S_{11}| > 1$, then the origin of the Smith chart is unstable. In order to satisfy continuity, the region inside the $|\Gamma_{IN}| = 1$ circle should be stable. On the Γ_S chart shown in Fig. 1.30B, the chart origin represents $\Gamma_S = 0$, which results in $\Gamma_{OUT} = S_{22}$. Therefore, if $|S_{22}| < 1$, then the origin of the Smith chart is stable, which further means the entire region outside the $|\Gamma_{OUT}| = 1$ circle is stable. On the other hand, if $|S_{22}| > 1$, then the origin of the Smith chart is unstable. In order to satisfy continuity, the region inside the $|\Gamma_{OUT}| = 1$ circle should be stable. For unconditional stability, any passive load or source matching network must produce a stable condition. From a graphical point of view, for $|S_{11}| < 1$ and $|S_{22}| < 1$, we want the stability circles to fall completely outside the Smith chart, as shown in Fig. 1.31.

There is also a mathematical description of the necessary and sufficient conditions for a two-port to be unconditionally stable. It has been shown that a straightforward but lengthy manipulation of existing equations results in the following necessary and sufficient conditions for unconditional stability [1]:

$$K = \frac{1 - |S_{11}|^2 - |S_{11}|^2 + |\Delta|^2}{2|S_{12}S_{21}|} > 1 \tag{1.31}$$

$$|\Delta| = |S_{11}S_{22} - S_{12}S_{21}| < 1$$

K is called the Rollet stability factor, which can be simulated with many CAD tools such as the National Instruments (NI) AWR Design Environment and the Keysight Advanced Design System.

A potentially unstable transistor can be made unconditionally stable by adding resistive load to stabilize the transistor and negative feedback to provide the proper gain and input/output matching [3]. These techniques are not recommended in narrowband amplifiers because of the resulting degradation in power gain, noise figure, and VSWRs. Narrowband amplifier design with potentially unstable transistors is best done by proper selection of Γ_S and Γ_L to ensure stability. On the other hand, the techniques are popular in the design of some broadband amplifiers by which the transistor is potentially unstable.

1.7.1.3 Amplifier noise figure

The total noise output power of an amplifier is composed of the amplified noise input power plus the noise power produced by the amplifier itself. The noise input power can be modeled by a source resistor that produces thermal or Johnson noise caused by random fluctuations of the electrons due to thermal agitation. The rms value of the noise voltage produced by a noisy resistor R over a frequency range is given by $v_{n,\mathrm{rms}} = \sqrt{4kTBR}$, where k is Boltzmann's constant (i.e., $k = 1.374 \times 10^{-23}$ J/K), T is the resistor noise temperature in Kelvin, and B is the noise bandwidth [3]. The thermal noise power depends on the bandwidth and not on a given center frequency. Such a distribution of noise is called white noise. The available noise power from resistor R is $P_N = kTB$. The noise figure of an amplifier is defined by comparing the input/output signal-to-noise ratio (SNR):

$$F = \frac{\mathrm{SNR}_{\mathrm{input}}}{\mathrm{SNR}_{\mathrm{output}}} = \frac{P_{No}}{P_{Ni}G_A} = \frac{P_{Ni}G_A + P_n}{P_{Ni}G_A}, \tag{1.32}$$

where P_{No} is the total available noise power at the output of the amplifier, $P_{Ni} = kTB$ is the available noise power due to R in a bandwidth of B, G_A is the available power gain, and P_n represents the noise power appearing at the output due to internal amplifier noise. It is desirable to have a low noise figure, so that the SNR is not deteriorated too much after the amplifier. For cascade stage, it can be shown that the total noise figure is determined by [13,14]:

$$F = F_1 + \frac{F_2 - 1}{G_{A1}} + \frac{F_3 - 1}{G_{A1}G_{A2}} + \frac{F_4 - 1}{G_{A1}G_{A2}G_{A3}} + \dots, \tag{1.33}$$

where F_n (n = 1, 2, 3 ...) is the noise figure of the nth stage and G_{An} is the available power gain of the nth stage. From this equation, it is shown that the noise figure and gain of the first stage are the most important for the overall noise of the system. That is why, from a system design point of view (e.g., the biomedical radar that will be discussed in later chapters), we usually place a "low-noise amplifier" at the first stage of a receiver.

Now, let us focus on the noise figure of a single amplifier. It has been shown that the noise figure of a two-port amplifier can be expressed as [15]:

$$F = F_{\min} + \frac{R_N}{G_S}|Y_S - Y_{\mathrm{OPT}}|^2 = F_{\min} + \frac{4R_N}{Z_0} \frac{|\Gamma_S - \Gamma_{\mathrm{OPT}}|^2}{(1 - |\Gamma_S|^2)|1 + \Gamma_{\mathrm{OPT}}|^2}, \tag{1.34}$$

where $Y_S = G_S + jB_S$ is the source admittance presented to the transistor, Y_{OPT} is the optimum source admittance that results in the minimum noise figure, $F_{\min}$ is the minimum noise figure of the transistor attained when $Y_S = Y_{\mathrm{OPT}}$, R_N is the equivalent noise resistance of the transistor, and G_S is the real part of source admittance. Instead of the admittances Y_S and Y_{OPT}, we can use the reflection coefficients Γ_S and Γ_{OPT}, because there is a one-to-one correspondence between the admittance and the reflection coefficients. The quantities $F_{\min}$, Γ_{OPT}, and R_N are characteristics of the particular transistor being used, and are called the noise parameters of the device. The noise parameters may be given by the manufacturer or measured. The bottom line is, once the transistor of the amplifier is selected, the noise parameters will be settled. For a fixed noise figure F, it can be shown that this corresponds to a circle in the Γ_S plane [3]. First define the noise figure parameter N as:

$$N = \frac{|\Gamma_S - \Gamma_{\mathrm{OPT}}|^2}{1 - |\Gamma_S|^2} = \frac{F - F_{\min}}{4R_N/Z_0}|1 + \Gamma_{\mathrm{OPT}}|^2, \tag{1.35}$$

which is a constant for a given noise figure F and a given transistor with a set of noise parameters. After some mathematical manipulation, the noise figure of a two-port amplifier can be rewritten as [3]:

$$\left|\Gamma_S - \frac{\Gamma_{\mathrm{OPT}}}{N+1}\right| = \sqrt{N(N+1-|\Gamma_{\mathrm{OPT}}|^2)/(N+1)} \tag{1.36}$$

This result defines circles of constant noise figure with centers and radii according to:

$$C_F = \frac{\Gamma_{\mathrm{OPT}}}{N+1}$$

$$R_F = \frac{\sqrt{N(N+1-|\Gamma_{\mathrm{OPT}}|^2)}}{(N+1)}. \tag{1.37}$$

For example, Fig. 1.32 shows the simulated constant power gain G_A circles and the constant noise figure circles for an Infineon BFP740 Silicon Germanium RF Transistor. As we can see, the location of Γ_S to achieve the maximum available power gain is different from the location of Γ_S to achieve the best noise performance. Therefore, on the designer side, there will always be a tradeoff to select Γ_S (and thus the input matching network) to ccommodate the design goals on the amplifier gain and noise performance. As an example, Fig. 1.32 shows the constant noise figure contours together with the constant available power gain contours in the input reflection plane for an Infineon BFP740 Silicon Germanium RF Transistor biased at 3 V and 10 mA operating at 2.4 GHz. The results are obtained using the NI AWR Design Environment (MWO). As explained before, circles p1, p2, and p3 represent available power gains of 23.5 dB, 21.5 dB, and 19.5 dB, respectively. In the meantime, circles p5 and p6 represent contours in the source plane which provide constant noise figures with a 0.2 dB step from $F_{\min}$, which is represented by marker p4.

1.7.1.4 Dynamic range and intermodulation distortion

In the ideal case, we have a linear amplifier with a fixed gain. For example, an ideal 20 dB gain amplifier amplifies the signal by 20 dB, no matter what the input signal level is. However, in practice, all real components become nonlinear at high power levels. These effects set a minimum and maximum realistic power range, or dynamic range, over which a given component or circuit will operate as

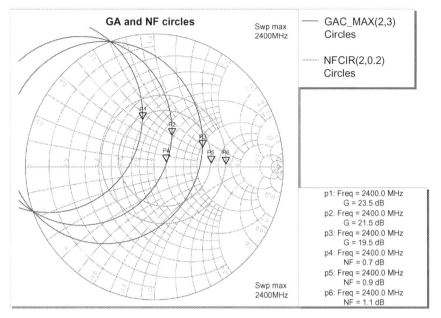

FIGURE 1.32

Simulated constant available power gain circles and constant noise figure circles for an Infineon BFP740 Silicon Germanium RF Transistor.

desired. Let us use a nonlinear amplifier as example. In the most general sense, the output response v_o of a nonlinear circuit can be modeled as a Taylor series in terms of the input signal voltage v_i: $v_o = a_0 + a_1 v_i + a_2 v_i^2 + a_3 v_i^3 + \cdots$. Consider the case where a single frequency sinusoid $v_i = V_0 \cos\omega_0 t$ is applied to the input of a general nonlinear network, such as the amplifier considered above. The output can be calculated as:

$$v_o = a_0 + a_1 V_0 \cos\omega_0 t + a_2 V^2{}_0 \cos^2\omega_0 t + a_3 V_0^3 \cos^3\omega_0 t + \cdots$$

$$= a_0 + \frac{1}{2} a_2 V_0^2 + \left(a_1 V_0 + \frac{3}{4} a_3 V_0^3 \right) \cos\omega_0 t + \frac{1}{2} a_2 V_0^2 \cos 2\omega_0 t + \frac{1}{4} a_3 V_0^3 \cos 3\omega_0 t + \cdots$$

(1.38)

This result leads to the voltage gain of the signal component at frequency ω_0:

$$G_v = \frac{v_o^{(\omega_0)}}{v_i^{(\omega_0)}} = \frac{a_1 V_0 + \frac{3}{4} a_3 V_0^3}{V_0} = a_1 + \frac{3}{4} a_3 V_0^2 \qquad (1.39)$$

The result shows that the voltage gain is equal to the a_1 coefficient, as expected, but with an additional term proportional to the square of the input voltage amplitude. In most practical cases, a_3 is typically negative, so the gain of the amplifier tends to decrease for large values of V_0. This effect is called gain compression, or

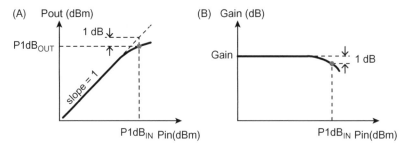

FIGURE 1.33

Definition of P1dB using (A) Plot of P_{OUT} as a function of P_{IN}. (B) Plot of gain as a function of P_{IN}.

saturation. Physically, this is usually due to the fact that the instantaneous output voltage of an amplifier is limited by the power supply voltage used to bias the active device. Smaller magnitudes of a_3 will lead to higher output voltages. A typical amplifier response is shown in Fig. 1.33. We define the 1 dB compression point as the power level at which the output power has decreased by 1 dB from the ideal, linear characteristic. This power level is usually denoted by P1dB, and can be stated in terms of either input power or output power, i.e., input-referred P1dB and output-referred P1dB. At 1 dB compression point:

$$20 \log \left(1 + \frac{3 a_3}{4 a_1} V_0^2 \right) = -1 \tag{1.40}$$

Based on the above equation, the input-referred P1dB in a 50-Ω system can be found as:

$$P_{1dB,in}(\text{dBm}) = 10 \log \left(\frac{V_0^2}{2 \times 50} \times 1000 \right) = 0.364 + 10 \log \left(\frac{4}{3} \left| \frac{a_1}{a_3} \right| \right). \tag{1.41}$$

Besides the 1 dB compression point, another important figure of merit related to the dynamic range is the third-order intercept point. Consider a two-tone input voltage consisting of two closely spaced frequencies: $v_i = V\cos\omega_1 t + V\cos\omega_2 t$. The output spectrum consists of harmonics at frequencies of $m\omega_1 + n\omega_2$, where m, $n = 0, \pm 1, \pm 2, \pm 3, \ldots$ We are interested in the original frequency tones at ω_1 and ω_2, as well as the third-order intermodulation tones of $2\omega_1 - \omega_2$ and $2\omega_2 - \omega_1$. This is because the third-order intermodulation tones have frequencies very close to the original frequencies. It can be shown that the output is:

$$
\begin{aligned}
v_o &= a_0 + a_1 v_i + a_2 v_i^2 + a_3 v_i^3 + \cdots \\
&= a_0 + a_1 V\cos\omega_1 t + a_1 V\cos\omega_2 t + \frac{3a_3}{4} V^3 \cos(2\omega_1 - \omega_2)t + \frac{3a_3}{4} V^3 \cos(2\omega_2 - \omega_1)t + \cdots,
\end{aligned}
$$

$$\tag{1.42}$$

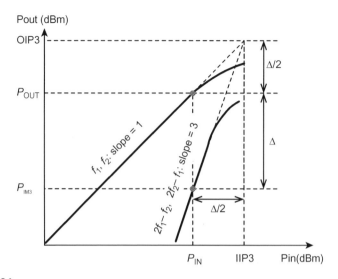

FIGURE 1.34

Definition of input- and output-referred IP3.

where the standard trigonometric identities have been used to expand the initial expression. The output power of the first-order, or linear, product is proportional to the input power. Therefore the line describing this response has a slope of unity (before the onset of compression). The line describing the response of the third-order products has a slope of three. Both linear and third-order responses will exhibit compression at high input powers. If we extend their idealized responses with dotted lines, the two lines will intersect, as shown in Fig. 1.34. This hypothetical intersection point, where the first-order and third-order powers are equal, is called the third-order intercept point, denoted as IP3, and specified as either an input or an output power (OIP3 or IIP3).

1.7.1.5 Efficiency

There are other important metrics that evaluate the performance of amplifiers. For example, RF power amplifiers convert a low-power radio-frequency signal into a higher-power signal and typically drive the antenna of a transmitter. One important design goal for power amplifiers is their efficiency. One metric to rate the efficiency of a power amplifier is the power-added efficiency, which takes into account the effect of the gain of the amplifier and is calculated as a percentage as follows: $(P_{RF\text{-}OUT} - P_{RF\text{-}IN})/P_{DC}$, where $P_{RF\text{-}OUT}$, $P_{RF\text{-}IN}$, and P_{DC} stand for the RF output power, the RF input power, and the DC power consumption of the amplifier, respectively. Interested authors are encouraged to refer to [16] for more information about amplifier efficiency.

1.7.2 MIXERS: ACTIVE VERSUS PASSIVE, NOISE FIGURE, LINEARITY

A mixer is a three-port device that uses nonlinear elements to achieve frequency conversion. The symbol and functional diagram of a mixer are shown in Fig. 1.35. For the up-converter (Fig. 1.35A), a local oscillator (LO) signal $v_{LO}(t) = \cos(2\pi f_{LO}t)$ at a relatively high frequency f_{LO} is connected to one of the input ports. A lower frequency baseband or intermediate frequency (IF) signal $v_{IF}(t) = \cos(2\pi f_{IF}t)$ is applied to the other mixer input. The IF signal typically contains the information or data to be transmitted.

The output of the idealized mixer is given by the product of the two input signals:

$$v_{RF}(t) = K \cdot v_{LO}(t) \cdot v_{IF}(t) = K \cdot \cos(2\pi f_{LO}t) \cdot \cos(2\pi f_{IF}t)$$
$$= \frac{K}{2}\left\{\cos[2\pi(f_{LO} - f_{IF})t] + \cos[2\pi(f_{LO} + f_{IF})t]\right\}, \tag{1.43}$$

where K is a constant accounting for the voltage conversion gain (or loss) of the mixer. The RF signal is seen to consist of the sum and difference of the input signal frequencies $f_{RF} = f_{LO} \pm f_{IF}$. From the frequency spectra, we see that the mixer has the effect of modulating the LO signal with the IF signal. The sum and difference frequencies at $f_{LO} \pm f_{IF}$ are called the sidebands of the carrier frequency f_{LO}, with $f_{LO} + f_{IF}$ being the upper sideband (USB), and $f_{LO} - f_{IF}$ being the lower

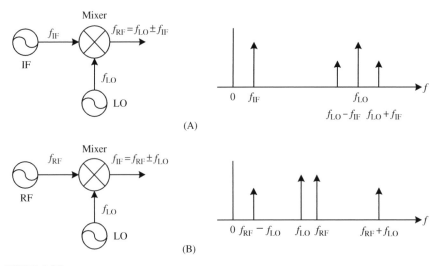

FIGURE 1.35

A mixer functioning as (A) An up-converter; (B) A down-converter.

Based on Pozar DM. Microwave engineering: John Wiley & Sons; 2009.

sideband (LSB). A double-sideband (DSB) signal contains both upper and lower sidebands, while a single-sideband (SSB) signal can be produced by filtering or using a SSB mixer.

Conversely, Fig. 1.35B shows the process of frequency down-conversion, as used in a receiver. In this case, an RF input signal is applied to the input of the mixer $v_{RF}(t) = \cos(2\pi f_{RF}t)$. The output of the mixer is:

$$v_{IF}(t) = K \cdot v_{RF}(t) \cdot v_{LO}(t) = K \cdot \cos(2\pi f_{RF}t) \cdot \cos(2\pi f_{LO}t)$$
$$= \frac{K}{2}[\cos(2\pi(f_{RF} - f_{LO})t) + \cos(2\pi(f_{RF} + f_{LO})t)] \quad . \tag{1.44}$$

In practice, the RF and LO frequencies are relatively close together, so the sum frequency is approximately twice the RF frequency, while the difference is much smaller than f_{RF}. The desired IF output in a receiver is the difference frequency, $|f_{LO}-f_{RF}|$, which is easily selected by low-pass filtering $f_{IF} = f_{RF} - f_{LO}$. Note that in a realistic mixer, many more frequency tones (products) will be generated due to the more complicated nonlinear behavior of the diode(s) or transistor(s) used in the mixer. The products are usually undesirable, and can be removed by filtering.

1.7.2.1 Image frequency

A receiver may receive RF signals over a relatively wide band of frequencies. The desired RF input frequency that will be down-converted to the IF frequency is $f_{RF} = f_{LO} + f_{IF}$. However, there is another high frequency signal that could be down-converted into the same IF frequency. This undesired RF input has a frequency of $f_{IM} = f_{LO} - f_{IF}$. Inserting this RF signal into the down-converter yields $-f_{IF}$. Actually this frequency is identical to f_{IF} because the Fourier spectrum of any real signal is symmetric around zero frequency, and thus contains negative frequencies as well as positive. This undesired RF frequency is called the *image response*. The image response is important in receiver design because a received RF signal at the image frequency is indistinguishable at the IF stage from the desired RF frequency signal, unless steps are taken in the RF stages of the receiver to preselect signals only within the desired RF frequency band.

1.7.2.2 Conversion loss

The conversion loss is defined as the ratio of available RF input power to the available IF output power, expressed in dB:

$$L_C = 10 \log \frac{\text{available RF input power}}{\text{available IF output power}}. \tag{1.45}$$

It should be noted that for some kinds of mixers, such as the Gilbert active mixer, it is possible to have a conversion loss < 0 dB, which means the mixer has a conversion gain. Practical diode mixers typically have conversion losses between 4 and 7 dB in the 1–10 GHz range. Transistor mixers have lower conversion loss, and, as mentioned, they may even have conversion gain depending on the architecture of the mixer.

1.7.2.3 Noise figure

Noise is generated in mixers by diode/transistor components and resistive losses. It should be noted that the noise figure of a mixer depends on whether its input is an SSB or a DSB. This is because the mixer will down-convert noise at both sideband frequencies (since they have the same IF), but the power of an SSB signal is one-half that of a DSB signal (for the same amplitude). To derive the relation between the noise figure for these two cases, first consider a DSB input signal of the form: $v_{DSB}(t) = A[\cos(\omega_{LO} - \omega_{IF})t + \cos(\omega_{LO} + \omega_{IF})t]$. Upon mixing with an LO signal and low-pass filtering, the down-converted IF signal will be $v_{IF}(t) = AK\cos(\omega_{IF}t)/2 + AK\cos(-\omega_{IF}t)/2 = AK\cos(\omega_{IF}t)$, where K is a constant accounting for the conversion loss. The power of the DSB input signal is $S_i = A^2/2 + A^2/2 = A^2$, and the power of the output IF signal is $S_o = A^2K^2/2$. Similar to the amplifier noise analysis in the last section, with an input noise power of $N_i = kT_0B$, the output noise power is equal to the input noise plus N_{added}, the noise power added by the mixer, and divided by the conversion loss L_C: $N_o = (kT_0B + N_{added})/L_C$. Then the DSB noise figure of the mixer can be found as [1]:

$$F_{DSB} = \frac{S_i/N_i}{S_o/N_o} = \frac{2}{K^2 L_C}\left(1 + \frac{N_{added}}{kT_0B}\right). \tag{1.46}$$

Therefore, to improve the noise performance of a passive mixer, which will be used in the monitoring systems of later chapters, it is important to minimize the insertion loss. Furthermore, since the SSB signal power is half of the DSB signal power, the noise figure of the SSB case is twice that of the DSB case: $F_{SSB} = 2F_{DSB}$. In dB scale, this means the SSB noise figure will normally be 3 dB higher than the DSB noise figure. The DSB noise figure should be used for direct-conversion receivers. When selecting mixers for a microwave healthcare/ biosensing system, the designer should pay attention to whether a DSB or SSB noise figure is specified.

1.7.2.4 Other mixer characteristics

Since mixers involve nonlinearity, they will produce intermodulation products. Therefore, IP3 is an important metric for mixers. Another important characteristic of a mixer is the isolation between the RF and LO ports. Ideally, the LO and RF ports would be decoupled, but in reality some LO power will be coupled to the RF port. In a receiver, the LO power coupled to the RF port will be radiated by the antenna. As such signals will likely interfere with other services or users, the Federal Communications Commission (FCC) sets stringent limits on the power radiated by receivers.

1.7.3 OSCILLATORS: OSCILLATOR STRUCTURES, PHASE NOISE

In the most general sense, an oscillator is a nonlinear circuit that converts DC power to an AC waveform. Most RF oscillators provide sinusoidal outputs, which minimize undesired harmonics and noise sidebands. The basic conceptual operation of a sinusoidal oscillator can be described with the linear feedback circuit as shown in Fig. 1.36. The closed-loop voltage gain is given by $V_o(\omega)/V_i(\omega) = A/[1 - AH(\omega)]$. If the denominator becomes zero at a particular frequency, it is possible to achieve a nonzero output voltage for a zero input voltage, thus forming an oscillator. This is known as the Barkhausen criterion. In contrast to the design of an amplifier, where we design to achieve maximum stability, oscillator design depends on an unstable circuit.

There are two basic and well-known types of oscillator feedback networks: Colpitts oscillators and Hartley oscillators. They are distinguished by the formation of the tuned circuits. They have L–C resonant circuits for frequency selection. The Colpitts oscillator uses a capacitor voltage divider in the tuned circuit to provide the correct feedback. The Hartley network uses a tapped inductor tuned circuit. Because of this, a Colpitts network normally (but not always) has two capacitors and one inductor, while a Hartley network has two inductors and one capacitor. They can be implemented in various transistor configurations, as shown in Fig. 1.37. The Clapp network is similar to the Colpitts network but has an extra capacitor in series with the inductor to improve the frequency stability [17]. A Clapp circuit is often preferred over a Colpitts circuit for constructing a variable frequency oscillator. In a Colpitts variable frequency oscillator, the voltage divider

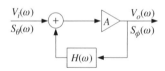

FIGURE 1.36

Block diagram of a feedback-based oscillator.

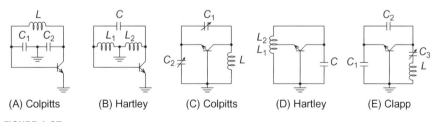

(A) Colpitts   (B) Hartley   (C) Colpitts   (D) Hartley   (E) Clapp

FIGURE 1.37

Small signal models of oscillator feedback networks: (A) Common-emitter Colpitts. (B) Common-emitter Hartley. (C) Common-base Colpitts. (D) Common-base Hartley, and (E) Common-base Clapp oscillators.

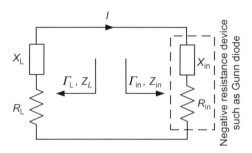

FIGURE 1.38

Model of a negative resistance oscillator.

Based on Pozar DM. Microwave engineering: John Wiley & Sons; 2009.

contains a variable capacitor (either C_1 or C_2 in Fig. 1.37). This causes the feed-back voltage to be a variable as well, sometimes making the Colpitts circuit less likely to achieve oscillation over a portion of the desired frequency range. This problem is avoided in the Clapp circuit by using fixed capacitors in the voltage divider and a variable capacitor (C_3 in Fig. 1.37) in series with the inductor.

An important design concept for microwave oscillators is negative resistance [18,19]. Fig. 1.38 shows the model of a one-port negative-resistance oscillator. In this circuit, $Z_{in} = R_{in} + jX_{in}$ is the input impedance of the active device (e.g., biased transistor), and is a function of bias current and frequency. The device is terminated with a passive load $Z_L = R_L + jX_L$. Kirchhoff's voltage law gives $(Z_L + Z_{in}) \cdot I = 0$. If oscillation occurs such that the RF current I is nonzero, then the following conditions must be satisfied: $R_L = -R_{in}$, and $X_L = -X_{in}$. For a passive load with $R_L > 0$, this leads to $R_{in} < 0$. Intuitively, a positive resistance implies energy dissipation and a negative resistance implies an energy source. It can be shown that the reflection coefficients of the source and load are correlated as $\Gamma_L = 1/\Gamma_{in}$ [1]. Therefore, the oscillation condition can be described as either $Z_L = -Z_{in}$ or $\Gamma_L \cdot \Gamma_{in} = 1$. Since Z_{in} is a function of both current and frequency, the above equation does not guarantee a stable state of oscillation. In fact, a high-Q circuit will result in maximum oscillation stability [1]. Cavity and dielectric resonators are often used for this purpose. It can be proven that the standard Smith chart can be used for negative resistances by plotting $1/\Gamma^*$ (instead of Γ). Then the resistance circle values are read as negative, while the reactance circles are unchanged [1,3].

Negative resistance oscillators are often used at higher microwave frequencies. A simple realization of negative resistance oscillators is based on the Gunn diode, also known as a transferred electron device [20,21]. With two or more transmission lines, a 50 Ω resistance can be easily matched to the negative resistance generated by a Gunn diode, achieving microwave oscillators that are frequently used in radar speed guns and microwave relay data link transmitters [22,23]. Similarly, a transistor can be used to implement a two-port negative resistance oscillator. In such a case, a transistor will provide two ports, a terminating port

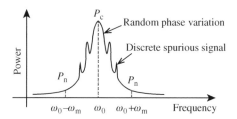

FIGURE 1.39

Spectrum of a signal generated by a realistic oscillator.

and an input port. A load-matching network and a terminating network will be designed to match the impedance. When designing a two-port negative resistance oscillator, the designer needs to select a potentially unstable transistor at the frequency of oscillation and design the terminating network to generate a negative impedance, sometimes with the help of series or shunt feedback to increase the magnitude of the reflection coefficient. Then a load network will be designed to resonate with the negative impedance and satisfy the oscillation condition.

An ideal oscillator would have a frequency spectrum consisting of a single delta function at its operating frequency, but a realistic oscillator will have a spectrum similar to Fig. 1.39. In general, the output voltage of an oscillator can be written as: $v_0(t) = V_0[1 + A(t)]\cos[\omega_0 t + \theta(t)]$, where $A(t)$ represents the amplitude fluctuations of the output waveform, and $\theta(t)$ represents the phase variation of the output waveform. The amplitude variations are usually well controlled, and have less impact on system performance. The phase variations are more important for the performance. The phase noise produced by a noise modulation of the carrier at an offset frequency of $\omega_m = 2\pi f_m$ can be modeled as $\theta(t) = \theta_p \cdot \sin\omega_m t$, where θ_p is the peak phase deviation, or the modulation index. Substituting this into the equation for oscillator output voltage and ignoring amplitude noise by setting $A(t)$ to 0, the oscillator output can be represented as $v_0(t) = V_0[\cos\omega_0 t \cdot \cos(\theta_p \cdot \sin\omega_m t) - \sin\omega_0 t \cdot \sin(\theta_p \cdot \sin\omega_m t)]$. Assuming the phase deviations are small so that $\theta_p \ll 1$, the small-angle approximation to $v_0(t)$ suggests that small phase or frequency deviations in the output of an oscillator result in modulation sidebands at $\omega_0 \pm \omega_m$, located on both sides of the carrier. Phase noise is defined as the ratio of noise power at a given offset frequency to the carrier power. It should be noted that the two-sided power spectral density associated with phase noise includes power in both sidebands, which means the noise power at a given offset frequency included in the two sidebands is twice as high as that in a single sideband. Therefore, the SSB phase noise is 3 dB lower than the DSB phase noise. It should also be noted that, as discussed in the last section, the SSB noise figure of a mixer is normally 3 dB higher than its DSB noise figure.

The above paragraph only discusses the general meaning of phase noise. To describe the power spectral density of oscillator phase noise, Leeson's model is frequently used [24]. Starting from the oscillator linear feedback loop shown in

Fig. 1.36, if the gain of the amplifier is included in the feedback transfer function $H(\omega)$, the closed-loop voltage transfer function of the oscillator is $V_o(\omega)/V_i(\omega) = 1/[1-H(\omega)]$. If we consider oscillators with a high-Q resonant circuit in the feedback loop (e.g., Colpitts, Hartley, and Clapp oscillators), then $H(\omega)$ can be represented as the voltage transfer function of a parallel RLC resonator:

$$H(\omega) = \frac{1}{1 + jQ\left(\dfrac{\omega}{\omega_0} - \dfrac{\omega_0}{\omega}\right)} = \frac{1}{1 + 2jQ\Delta\omega/\omega_0}, \qquad (1.47)$$

where ω_0 is the resonant frequency of the oscillator, Q is the quality factor, and $\Delta\omega = \omega - \omega_0$ is the frequency offset relative to the resonant frequency. Besides the kTB thermal noise, transistors generate a flicker noise, or the so-called $1/f$ noise, that is inversely proportional to frequency at frequencies below a corner frequency of f_α. Therefore, the noise power spectral density applied to the input of the oscillator in Fig. 1.36 can be modeled as $S_\theta(\omega) = kTF(1 + K\omega_\alpha/\Delta\omega)/P_0$, where K is a constant accounting for the strength of the $1/f$ noise, and $\omega_\alpha = 2\pi f_\alpha$ is the flicker noise corner frequency. The corner frequency depends primarily on the type of transistors used in the oscillator [25]. As a result, the oscillator output phase noise becomes:

$$S_\phi(\omega) = \left|\frac{1}{1-H(\omega)}\right|^2 S_\theta(\omega) = \frac{kT_0F}{P_0}\left(\frac{K\omega_\alpha\omega_h^2}{\Delta\omega^3} + \frac{\omega_h^2}{\Delta\omega^2} + \frac{K\omega_\alpha}{\Delta\omega} + 1\right). \qquad (1.48)$$

where $\omega_h = \omega_0/2Q$ is the half-power (3dB) bandwidth of the resonator. There are two possible cases of the results, depending on which of the middle two terms is more significant. In most of today's systems, the $f_h > f_\alpha$ case holds true and people use SSB phase noise more often. Additionally, it is traditional to normalize the mean-square noise density (i.e., the noise power per unit bandwidth) to the mean-square carrier power and then report the ratio in decibels. Therefore, the most popular definition of phase noise is commonly expressed as "decibels below the carrier per hertz," or "dBc/Hz," specified at a particular offset frequency $\Delta\omega$ from the carrier frequency ω_0 [25,26]. For example, a 5.35-GHz CMOS voltage controlled oscillator's phase noise could be " $-$ 116.5 dBc/Hz at 1-MHz offset" [27]. Following the above analysis on phase noise, Leeson's model predicts a SSB phase noise as [24]:

$$L\{\Delta\omega\} = 10\log\left[\frac{2FkT}{P_{\text{sig}}}\left\{1 + \left(\frac{\omega_0}{2Q\Delta\omega}\right)^2\right\}\left(1 + \frac{\Delta\omega_{1/f^3}}{|\Delta\omega|}\right)\right], \qquad (1.49)$$

where F is an empirical fitting parameter determined from measurements, P_{sig} is the signal power, Q is the quality factor of the tank, $\Delta\omega_{1/f^3}$ is the flicker noise corner frequency, and ω_0 is the carrier frequency. The phase noise curve predicted by the Lesson's model is shown in Fig. 1.40. When the offset frequency is below $\Delta\omega_{1/f^3}$, the phase noise rolls off at a speed of -30 dB/decade (or -9 dB/octave), whereas when the offset frequency is between $\Delta\omega_{1/f^3}$ and $\omega_0/2Q$, the phase noise rolls off at a rate of -20 dB/decade (or -6 dB/octave) [10].

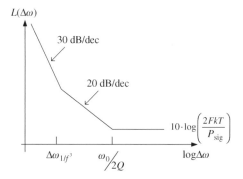

FIGURE 1.40

Phase noise curve predicted by Leeson's model.

1.8 SYSTEM ARCHITECTURES

The front-end architecture plays an important role in many biomedical systems such as implantable devices, physiological radars, and integrated biosensors. In order to maximize the detection sensitivity, extend the detection range, and reject interferences, various RF front-end architectures have been reported. In addition to the widely used homodyne, heterodyne, and DSB architectures, direct IF sampling and self-injection locking were recently adopted in biomedical systems. The properties and design considerations of these front-end architectures will be discussed in this section.

1.8.1 HETERODYNE ARCHITECTURE

The principle of heterodyne architecture was developed in 1918 to overcome the deficiencies of early vacuum tube triodes used as high-frequency amplifiers in radio direction finding [28]. The solution was later adapted to many broadcasting/receiving systems and communication systems including the frequency modulation (FM) radio. In a heterodyne architecture, there are normally two LOs that generate reference signals at the RF and the IF, respectively. The receiver down-converts the received signal using the RF LO to an IF first, then further down-converts the IF signal to the baseband using the IF LO. This has the advantage of being robust against DC offset, making it easier to handle low-frequency flicker noise, and offering improvements in IF channel selection for communication systems. For biomedical applications such as radar vital sign detection, however, a single-channel heterodyne system has the null detection point problem [29]. In order to overcome the null detection point problem, quadrature IF down-conversion or frequency-tuning on a single channel has to be used.

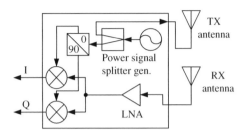

FIGURE 1.41

Simplified block diagram of homodyne transceiver without a circulator.

1.8.2 HOMODYNE ARCHITECTURE

The receiver of a homodyne solution is also known as the direct-conversion receiver architecture, or zero-IF receiver. For a biomedical transceiver based on homodyne architecture, its transmitter transmits a signal at a given carrier frequency, and its receiver demodulates the incoming radio signal using synchronous detection driven by an LO whose frequency is identical to, or very close to, the carrier frequency. Sometimes a single LO would serve for both the transmitter and receiver to achieve coherent detection. In many applications such as biomedical radar vital signs detection, a single channel transceiver will inevitably have the null detection problem [29], which means the detection sensitivity depends on the residual phase that varies as the detection distance changes. A popular way to eliminate the null detection point problem is the quadrature solution for homodyne architecture. A simplified block diagram of a homodyne transceiver without a circulator is shown in Fig. 1.41. If a circulator is used, the number of antennas can be reduced to one for simultaneous transmitting and receiving. Since there are in-phase and quadrature (I/Q) baseband outputs, there is always one channel not at the null detection point, thus realizing reliable biomedical signal detection. Moreover, since most of the biomedical vital sign signals to be detected for healthcare applications have a low bandwidth, the two output channels can be easily combined in software to perform further signal preconditioning, such as complex signal demodulation [30,31], arctangent demodulation [32], and the extended differentiate and cross-multiply algorithm [33], achieving low-cost baseband solutions.

1.8.3 DOUBLE SIDEBAND ARCHITECTURE

In order to eliminate the need for generating quadrature LO and several filtering requirements in traditional heterodyne architecture, a DSB transmission heterodyne architecture was proposed [34]. The DSB architecture is a type of heterodyne architecture without image band rejection. After mixing RF LO and IF LO, both USB and LSB are transmitted, and the receiver down-converts the reflected signals of both sidebands by mixing them with the same RF LO and the same IF LO. Thus, the biomedical signals carried by both the USB and the LSB are automatically combined in baseband during the down-conversion procedure. This is useful when designing biomedical transceivers operating at high frequencies such as millimeter-wave or above, since it eliminates the need for generating accurate quadrature RF LO for homodyne architecture or using image rejection filters for heterodyne architecture at high frequencies. Because USB and LSB have different wavelengths, the null detection points for USB and LSB are not collocated till every 1/8 of the IF wavelength, which is relatively large because the wavelength of the IF is much larger than the wavelength of the RF. Furthermore, by tuning the IF frequency, the locations of overlapping USB and LSB null points can be adjusted, providing an IF frequency tuning solution to avoid the null detection point problem. As there is no need for I/Q generation or an image rejection filter, the double sideband architecture has the advantage of being easy for system-on-chip integration.

1.8.4 DIRECT IF SAMPLING ARCHITECTURES

For the conventional quadrature homodyne architecture, I/Q imbalance and DC offset are two inevitable challenges that degrade the demodulation accuracy and the output SNR [35]. It is difficult to achieve satisfactory I/Q amplitude and phase balance without using complicated circuit designs or calibration procedures. Using a digital I/Q demodulation technique, the two output channels can have a perfect quadrature phase relationship in the digital domain, significantly reducing the complexity of analog/RF circuits or calibration [36]. Direct IF sampling architecture also has advantages in overcoming the DC offset problem because the signal after analog down-conversion is far from DC, while the digital down-conversion does not suffer from the DC offset problem. Fig. 1.42 shows the block diagram of direct IF sampling architecture. In the receiver chain, the mixer after low-noise amplifier down-converts the received signal into an IF signal. Then high speed analog-to-digital converter is used to convert the analog IF signal to digital. After that, quadrature demodulation is performed using a high speed digital signal processor.

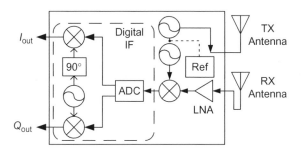

FIGURE 1.42

Simplified block diagram of direct IF sampling architecture.

1.8.5 OTHER ARCHITECTURES

There are other biomedical transceiver architectures such as self-injection locking [37], mutual-injection locking [38], time-domain ultra-wideband [39,40], frequency-modulated continuous-wave [41,42], and stepped frequency continuous-wave [43,44]. They are advanced solutions for various biomedical applications, which will be used and discussed in detail within other chapters of the book.

ACKNOWLEDGMENTS

The authors would like to acknowledge the contributions of the many researchers and educators to the field of RF/microwave engineering who made it possible to apply related technologies to human healthcare and biosensing. As fundamentals of RF/microwave engineering have been systematically presented by many excellent books, the contents of this chapter are largely based on the works of [1] and [3], and can be viewed as a brief guideline to the knowledge necessary for understanding the topics discussed in other chapters of this book. The contributions of Dr. Guillermo Gonzalez and Dr. David Pozar to microwave engineering education are deeply honored.

REFERENCES

[1] Pozar DM. Microwave engineering. John Wiley & Sons; 2009.

[2] Sorrentino R, Bianchi G. Microwave and RF engineering, vol. 1. Chichester: John Wiley & Sons; 2010.

[3] Gonzalez G. Microwave transistor amplifiers: analysis and design, vol. 2. New Jersey: Prentice Hall; 1997.

[4] Ulaby FT, Michielssen E, Ravaioli U. Fundamentals of applied Electromagnetics 6edition. Prentice Hall; 2001.

[5] Kuo F. Network analysis and synthesis. John Wiley & Sons; 2006.

[6] Matthaei GL, Young L, Jones E. Microwave filters, impedance-matching networks, and coupling structures, vol. 1. Artech House; 1964.

[7] Wilkinson EJ. An N-way hybrid power divider. IRE Trans Microwave Theory Tech 1960;8:116−18.

[8] Wilkinson EJ. Power divider. Google Patents; 1963.

[9] Sedra AS, Smith KC. Microelectronic circuits, vol. 1. New York: Oxford University Press; 1998.

[10] Lee TH. The design of CMOS radio-frequency integrated circuits. Cambridge University Press; 2003.

[11] Liao SY. 2nd ed. Microwave devices and circuits, vol. 1. Englewood Cliffs: Prentice Hall; 1985.

[12] Caverly RH. Microwave and RF semiconductor control device modeling. Artech House; 2016.

[13] Pozar DM. Microwave and RF design of wireless systems. John Wiley & Sons, Inc; 2000.

[14] Razavi B, Behzad R. RF microelectronics, vol. 2. New Jersey: Prentice Hall; 1998.

[15] Haus H, Atkinson W, Branch G, Davenport Jr W, Fonger W, Harris W, et al. Representation of noise in linear twoports. Proc IRE 1960;48:69–74.

[16] Cripps SC. RF power amplifiers for wireless communications. (Artech House Microwave Library) Artech House; 2006.

[17] Clapp J. An inductance-capacitance oscillator of unusual frequency stability. Proc IRE 1948;36:356–8.

[18] Negative resistance oscillator, Google Patents; 1961.

[19] Kurokawa K. Some basic characteristics of broadband negative resistance oscillator circuits. Bell Syst Tech J 1969;48:1937–55.

[20] Copeland JA. Theoretical study of a Gunn diode in a resonant circuit. IEEE Trans Electron Devices 1967;14:55–8.

[21] Strangeway RA, Ishii T, Hyde JS. Low-phase-noise Gunn diode oscillator design. IEEE Trans Microw Theory Tech 1988;36:792–4.

[22] Ho K, Mavrokoukoulakis N, Cole R. Propagation studies on a line-of-sight microwave link at 36 GHz and 110 GHz. IEE J Microwaves, Opt Acoust 1979;3:93–8.

[23] Meinel HH. Commercial applications of millimeterwaves: history, present status, and future trends. IEEE Trans Microw Theory Tech 1995;43:1639–53.

[24] Leeson D. A simple model of feedback oscillator noise spectrum. Proc IEEE 1966; 54:329–30.

[25] Hajimiri A, Lee TH. A general theory of phase noise in electrical oscillators. IEEE J Solid-State Circuits 1998;33:179–94.

[26] Rael J, Abidi AA. Physical processes of phase noise in differential LC oscillators, in Custom Integrated Circuits Conference, 2000. CICC. Proceedings of the IEEE 2000, 2000, pp. 569–72.

[27] Hung C-M, Floyd BA, Park N, Kenneth K. Fully integrated 5.35-GHz CMOS VCOs and prescalers. IEEE Trans Microwave Theory Tech 2001;49:17–22.

[28] Armstrong EH. The super-heterodyne-its origin, development, and some recent improvements. Proc Inst Radio Eng 1924;12:539–52.

[29] Droitcour AD, Boric-Lubecke O, Lubecke VM, Lin J, Kovacs GTA. Range correlation and I/Q performance benefits in single-chip silicon Doppler radars for noncontact cardiopulmonary monitoring. IEEE Trans Microwave Theory Tech 2004; 52:838–48.

[30] Li C, Lin J. Complex signal demodulation and random body movement cancellation techniques for non-contact vital sign detection. Microwave symposium digest, 2008. IEEE MTT-S International; 2008. p. 567–70.

[31] Li C, Lin J. Random body movement cancellation in doppler radar vital sign detection. IEEE Trans Microwave Theory Tech 2008;56:3143–52.

[32] Park B-K, Boric-Lubecke O, Lubecke VM. Arctangent demodulation with DC offset compensation in quadrature Doppler radar receiver systems. IEEE Trans Microwave Theory Tech 2007;55:1073–9.

[33] Wang J, Wang X, Chen L, Huangfu J, Li C, Ran L. Non-contact distance and amplitude independent vibration measurement based on an extended DACM algorithm. IEEE Trans Instrum Meas 2014;63:145–53.

[34] Xiao Y, Lin J, Boric-Lubecke O, Lubecke VM. Frequency-tuning technique for remote detection of heartbeat and respiration using low-power double-sideband transmission in the ka-band. IEEE Trans Microwave Theory Tech 2006;54:2023—32.

[35] Abidi AA. Direct-conversion radio transceivers for digital communications. IEEE J Solid-State Circuits 1995;30:1399—410.

[36] Gu C, Li C, Lin J, Long J, Huangfu J, Ran L. Instrument-based noncontact doppler radar vital sign detection system using heterodyne digital quadrature demodulation architecture. IEEE Trans Instrum Meas 2010;59:1580—8.

[37] Wang F-K, Horng T-S, Peng K-C, Jau J-K, Li J-Y, Chen C-C. Detection of concealed individuals based on their vital signs by using a see-through-wall imaging system with a self-injection-locked radar. IEEE Trans Microwave Theory Tech 2013;61:696—704.

[38] Wang F-K, Horng T-S, Peng K-C, Jau J-K, Li J-Y, Chen C-C. Single-antenna Doppler radars using self and mutual injection locking for vital sign detection with random body movement cancellation. IEEE Trans Microwave Theory Tech 2011;59:3577—87.

[39] Wang Y, Fathy AE. Advanced system level simulation platform for three-dimensional UWB through-wall imaging SAR using time-domain approach. IEEE Trans Geosci Remote Sens 2012;50:1986—2000.

[40] Zhang C, Kuhn MJ, Merkl BC, Fathy AE, Mahfouz MR. Real-time noncoherent UWB positioning radar with millimeter range accuracy: theory and experiment. IEEE Trans Microwave Theory Tech 2010;58:9—20.

[41] Wang G, Gu C, Inoue T, Li C. A hybrid FMCW-interferometry radar for indoor precise positioning and versatile life activity monitoring. IEEE Trans Microwave Theory Tech 2014;62:2812—22.

[42] Wang G, Munoz-Ferreras J, Gu C, Li C, Gomez-Garcia R. Application of linear-frequency-modulated continuous-wave (LFMCW) radars for tracking of vital signs. IEEE Trans Microwave Theory Tech 2014;62:1387—99.

[43] Mercuri M, Schreurs D, Leroux P. SFCW microwave radar for in-door fall detection, in Biomedical Wireless Technologies, Networks, and Sensing Systems (BioWireleSS), 2012 IEEE Topical Conference on, 2012, pp. 53—56.

[44] Mercuri M, Soh PJ, Pandey G, Karsmakers P, Vandenbosch GA, Leroux P, et al. Analysis of an indoor biomedical radar-based system for health monitoring. IEEE Trans Microwave Theory Tech 2013;61:2061—8.

Interaction between electromagnetic waves and biological materials

2

M.-R. Tofighi

Pennsylvania State University, Harrisburg, PA, United States

CHAPTER OUTLINE

2.1 INTRODUCTION

Interaction of electromagnetic (EM) energy at low through microwave and millimeter wave frequencies with biological materials has received a great deal of interest since the mid-20th century [1–9]. Some of this interest stems from the potential health hazards of significant (in accordance to regulatory limits) heat generation due to the body's exposure to wireless communication equipment [6,10], as well as therapeutic and diagnostic applications of microwaves [7–9]. Apart from these, other potential applications are evolving, including (label-free) characterization and viability studies of living cells using EM biosensors at microwave frequencies [11–13].

C. Li, M. Tofighi, D. Schreurs and T-Z. J. Horng (Eds): Principles and Applications of RF/Microwave
in Healthcare and Biosensing.

The history and an overview of the biological effects and medical applications of radio frequency (RF) and microwave can be found in review papers such as those by Guy [5] and Rosen et al. [9], as well as books on the subject such as those by Rosen and Rosen [7] and Vander Vorst et al. [8]. The field has experienced a tremendous growth within the past two decades, and advanced and emerging applications, some of which are highlighted in other chapters of this book, have been pursued by researchers and engineers worldwide. This growth has been to a great extent due to the advances in telecommunication technology and the microelectronics industry, and the availability of cost-effective processing technology for miniaturized sensors and devices, as well as sensing and testing equipment at RF through millimeter wave frequencies. Emerging technological trends and growth in the field of RF and microwave in medicine can also be traced in various publications cited in this chapter and dedicated journals and magazine special issues on the topic, such as those in *IEEE Transaction on Microwave Theory and Techniques* [14−18] and *IEEE Microwave Magazine* [19,20], along with many other journals, magazines, and conference proceedings.

Among the medical and biological applications of RF and microwave discussed in this book and elsewhere are microwave hyperthermia treatment of cancer, RF and microwave catheter ablation, microwave radiometry (thermography) of body tissues, EM sensors for label-free cell identification, wireless implantable sensors, vital sign monitoring using radar, microwave imaging of tissues, and high-field magnetic resonance imaging. What distinguishes these applications from the traditional applications of RF and microwave in communication, military, and space technologies is the fact that the interaction of RF and microwave with biological tissues and materials is a main aspect of the application and must be understood and accounted for. A common aspect of all these applications is that the EM waves generated or detected by antennas or near-field sensors at the presence of a body may propagate through, reflect, or scatter from biological substances. Analyzing such interactions between EM waves and biological materials requires knowledge of the complex permittivity of these materials.

Biological materials and tissues are complex structures and their complex permittivity is quite dispersive. In what follows, the relaxation theory of dielectrics, which explains the frequency variation of the complex permittivity at DC through millimeter wave frequencies, is discussed. Moreover, various relaxation mechanisms and complex permittivity dispersion modeling for biological materials are explained. In addition, some methods of measurement of complex permittivity using resonant, reflection, and transmission (two-port) techniques are presented.

2.2 COMPLEX PERMITTIVITY OF BIOLOGICAL MATERIALS AND RELAXATION THEORY

Quantifying the interaction of EM fields with biological materials requires knowledge of the dielectric parameters of those materials. Dielectric theory is discussed thoroughly in classical books such as those by Frohlich [21] and Daniel [22]. The basic macroscopic theory and static and dynamic properties of dielectrics, as well as topics such as dipolar interaction and dipolar molecules in gases and dilute solutions, can be found in Frohlich's book [21]. Classical relaxation theory is presented comprehensively in Daniel's book [22]. More recent review of the state of the subject can be found in a review article by Jonscher [23] and a book chapter by Feldman [24]. As observed by Jonscher [23], dielectric science is "remarkably multidisciplinary," falling within the realm of physics, chemistry, and electrical engineering. On the other hand, biological substances are quite complex and heterogeneous, and different mechanisms contribute to their dielectric properties, namely the complex permittivity of those substances, from DC to microwave frequencies. Complete understanding of the interaction of EM fields and biological materials requires sufficient knowledge in biochemistry, biology, electrical engineering, and physics [1–3,25,26]. Pethig [25,26] presents relevant issues from those disciplines and reviews the dielectric properties of biopolymers (i.e., amino acids, polypeptides, side chains, and DNA), the role of water in biological systems, heterogeneous material (Maxwell–Wagner and counterion theory), biological electrolytes, and cells. From an electrical engineering perspective, significant contributions to the field from Schwan and his coworkers are well-documented in their manuscripts, which appeared from the 1950s to the 1980s, and provide the macroscopic theory, modeling, and interpretation of different relaxation mechanisms, from DC up to the microwave region [1–3].

Because the focus of this chapter is on the fundamentals of dielectric theory as related to RF and microwave applications in medicine, it attempts to approach the topic from a perspective that would be of interest to RF and microwave engineers. Therefore, the focus of the discussion that follows will be on the macroscopic property (bulk complex permittivity) modeling. The underlying physics, biochemistry, and microscopic theory of dielectrics can be found elsewhere, such as in some of the above-mentioned references.

2.2.1 COMPLEX PERMITTIVITY AND LOSS

Since for all practical purposes, the magnetic permeability of biological materials is that of free space [3], conductivity and permittivity are the two constitutive parameters that need to be discussed. Recall the Maxwell's curl equation for the magnetic field ($\mathcal{H}$) in time domain

$$\nabla \times \mathcal{H}(t) = \sigma_f \mathcal{E}(t) + \partial \mathcal{D}(t)/\partial t \qquad (2.1)$$

where $\mathcal{E}$ and $\mathcal{D}$ are the electric field intensity and electric displacement, respectively. Moreover, σ_f is the conductivity associated with free charges, which could be in general free electrons in conductors, electrons and holes in semiconductors, or free ions in electrolytes and biological media. For this presentation, the two terms at the right side of the equation then can be viewed as the conduction and displacement current densities, respectively. The presence of an electric field $\mathcal{E}$

causes the polarization of atoms, polarization of nonpolar molecules, reorientation of dipolar moment of polar molecules (e.g., water), or polarization of charges due to the build-up of charge at the interface between two dissimilar dielectrics in heterogeneous media such as biological tissues [1–3,25,26]. This leads to a net electric moment and thus the electric polarization $\mathcal{P}$ that plays the main role in establishing the relationship between $\mathcal{D}$ and $\mathcal{E}$:

$$\mathcal{D}(t) = \varepsilon_0 \mathcal{E}(t) + \mathcal{P}(t) \tag{2.2}$$

where $\varepsilon_0 = 8.854 \times 10^{-12}$ (F/m) is the free space permittivity. This equation maintains its form for the time-harmonic case (after replacing the time domain fields and polarization with their phasors), where the steady-state situation for single frequency sine waves is considered. For a linear medium (which is a realistic assumption for biological tissues and the electric field intensity less than 10^5 V/m [25]), the electric polarization is linearly proportional to the electric field intensity, i.e.:

$$\mathbf{P}(\omega) = \varepsilon_0 \chi(\omega) \mathbf{E}(\omega) \tag{2.3}$$

where ω $(=2\pi f)$ is the radian frequency, and $\chi(\omega)$ is the electric susceptibility. The susceptibility is denoted as a function of frequency to emphasize its frequency dependency (i.e., its dispersive nature) for tissues and biological materials. Thus Eq. (2.2) leads to the following relation in phasor domain:

$$\mathbf{D}(\omega) = \varepsilon_0 [1 + \chi(\omega)] \mathbf{E}(\omega) \tag{2.4}$$

The parameter in the bracket is the relative complex permittivity attributed to the electric polarization

$$\begin{aligned} \varepsilon_p(\omega) &= 1 + \chi(\omega) \\ &= \varepsilon_p'(\omega) - j\varepsilon_p''(\omega) \end{aligned} \tag{2.5}$$

where ε_p' and ε_p'' are the real and imaginary permittivity, respectively. Rewriting Eq. (2.1) in frequency domain yields

$$\begin{aligned} \nabla \times \mathbf{H} &= \sigma_f \mathbf{E}(\omega) + j\omega\varepsilon_0 [1 + \chi(\omega)] \mathbf{E}(\omega) \\ &= j\omega\varepsilon_0 \left(-j\frac{\sigma_f}{\omega\varepsilon_0} + \varepsilon_p' - j\varepsilon_p'' \right) \mathbf{E}(\omega) \end{aligned} \tag{2.6}$$

The quantity inside the parenthesis is the overall relative complex permittivity including the conductivity for free charges, i.e.:

$$\begin{aligned} \varepsilon(\omega) &= \varepsilon_p'(\omega) - j \left[\varepsilon_p''(\omega) + \frac{\sigma_f}{\omega\varepsilon_0} \right] \\ &= \varepsilon'(\omega) - j\varepsilon''(\omega) \end{aligned} \tag{2.7}$$

with $\varepsilon'(\omega) = \varepsilon_p'(\omega)$ and $\varepsilon''(\omega) = \varepsilon_p''(\omega) + \sigma_f/\omega\varepsilon_0$, which are denoted as a function of frequency to emphasize the dispersive nature of the complex permittivity in biological media. The real and imaginary permittivity (ε' and ε'') are also called the dielectric constant and the loss factor in the literature. Moreover, they (along with ε) represent relative quantities at most biological tissue parameter references. Note also that, as far as the loss of EM energy in biological media is concerned, ε_p'' and $\sigma_f/\omega\varepsilon_0$, i.e., the loss factors related to the polarization and free charges' movement, are indistinguishable. Nonetheless, as the frequency increases, the contribution of σ_f to the total loss factor decays as the inverse of frequency. In addition to the total loss factor, an accompanying total conductivity can be also considered, which is

$$
\begin{aligned}
\sigma(\omega) &= \omega\varepsilon_0\varepsilon''(\omega) \\
&= \omega\varepsilon_0\varepsilon_p''(\omega) + \sigma_f
\end{aligned}
\tag{2.8}
$$

Besides the total conductivity and loss factor, the extent of the loss of EM energy in a material can be understood by another parameter, i.e., the loss tangent, which is the ratio of the imaginary to the real permittivity:

$$
\begin{aligned}
\tan\delta &= \frac{\varepsilon''}{\varepsilon'} \\
&= \frac{\sigma}{\omega\varepsilon'\varepsilon_0}
\end{aligned}
\tag{2.9}
$$

In a linear system, a field can be represented by the summation of plane waves. Moreover, the plane wave parameters of a dielectric medium also represent the wave parameters for the transverse electromagnetic (TEM) wave propagation in transmission lines homogenously filled by the same dielectric. Thus, to gain some insight about the wave propagation in a dielectric medium, it would be meaningful to discuss the plane wave propagation parameters of that medium. Two important wave parameters are the phase constant (β) and attenuation constant (α), which are the real and imaginary parts of propagation constant (γ)

$$
\begin{aligned}
\gamma &= \alpha + j\beta \\
&= j\omega\sqrt{\mu_0\varepsilon_0\varepsilon} \\
&= j\omega\sqrt{\mu_0\varepsilon_0(\varepsilon' - j\varepsilon'')}
\end{aligned}
\tag{2.10}
$$

which yields

$$
\alpha = \frac{\omega}{c}\sqrt{\frac{\varepsilon'}{2}}\sqrt{\sqrt{1 + \tan\delta^2} - 1}
\tag{2.11}
$$

$$
\beta = \frac{\omega}{c}\sqrt{\frac{\varepsilon'}{2}}\sqrt{\sqrt{1 + \tan\delta^2} + 1}
\tag{2.12}
$$

Plane wave wavelength (λ) is then

$$
\begin{aligned}
\lambda &= \frac{2\pi}{\beta} \\
&= \lambda_0\left[\sqrt{\frac{\varepsilon'}{2}}\sqrt{\sqrt{1 + \tan\delta^2} + 1}\right]^{-1}
\end{aligned}
\tag{2.13}
$$

where c is the speed of the light in vacuum, and λ_0 ($=c/f$) is the wavelength in free space. The inverse of the attenuation constant is called the penetration depth or skin depth δ ($=1/\alpha$), and describes the extent of the plane wave field penetration in the medium, which is of the exponential form $e^{-z/\delta}$. A sufficiently large block of lossy material (allowing plane wave propagation along z direction) presents a decay of e^{-1} (36.8%) for the field intensity at a depth δ for an incoming plane wave.

The penetration of EM fields in a lossy medium results in the absorption of electric energy. This leads to the transformation of EM energy into heat, and an increase in the medium's temperature. Specific absorption rate (SAR) is a quantity generally used to assess EM wave absorption in biological materials. SAR is defined as an incremental EM power (dP) dissipated in an incremental mass (dm) contained in an incremental volume (dV) with a given mass density ρ.

$$\text{SAR} = (1/\rho)(dP/dV) \quad [\text{W/kg}] \tag{2.14}$$

For the time-harmonic case, SAR relates to the magnitude of the electric field intensity as

$$\text{SAR} = (\sigma/2\rho)|\mathbf{E}|^2 \quad [\text{W/kg}] \tag{2.15}$$

Note that finding $\mathbf{E}$ in biological media which are dispersive, inhomogeneous, or irregularly shaped is not a trivial task and generally requires solving Maxwell's equations analytically or numerically. The scenarios for solving these equations vastly differ depending on the nature of the applications. Some scenarios are biological objects exposed to the plane wave irradiation or the irradiation by the far-field of antennas for dosimetry purposes or bio-radars. Others are irradiating tissues placed in the near-field of antennas in hyperthermia, or EM imaging and sensing applications, or wave propagation from inside or into the body for wireless implants' communication.

2.2.2 DIELECTRIC POLARIZATION MECHANISMS

The electrical properties of a material exposed to EM fields are, in general, frequency-dependent. A material that demonstrates significant permittivity changes in the frequency range of interest is referred to as a dispersive one in that range. In order to understand the physics of dielectrics, the first step is developing a theory for the static properties of dielectric materials, i.e., when the electric field has no time variation or its variation is very slow compared to the time that the electric polarization needs to achieve its steady state in response to the field. Note that electric polarization in materials in general may be due to distinctly different mechanisms.

The two types of polarization mechanisms that are responsible for dielectric response and dispersion in infrared through ultraviolet range are atomic and electronic polarizations. In the case of electronic polarization, the applied electric field causes a displacement of the electron cloud relative to the nuclei in each atom, whereas in atomic polarization, the displacement is for the atomic nuclei relative to one another. In ionic crystals, in addition to the ion being polarized

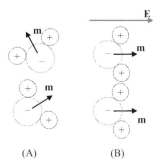

FIGURE 2.1

(A) Random orientation of dipoles with moment **m** (for water the dipole moment is the vector sum of two dipoles at 104.45 degrees due to bonds between oxygen and hydrogen atoms) and (B) Their alignment upon the application of an electric field.

individually, the positive and negative ions will be displaced with respect to each other. Atomic and electronic polarizations are determined by the internal properties of the molecules or atoms, and are characterized by resonance absorption, manifested by a drastic change of the real and imaginary permittivity around a resonance (characteristic) frequency. While the atomic polarization exhibits changes in the infrared range (10^{11}–10^{14} Hz), those for the electronic polarization fall within visible light to ultraviolet (10^{14}–10^{17} Hz).

Up to about millimeter wave frequencies, the atomic and electronic polarizability remain unchanged [25], meaning that those polarizations are formed almost instantly in response to the field at those frequencies. On the other hand, the dominant mechanism of dispersion in biological tissue from few tens of MHz to millimeter wave is due to the dipolar orientation process. Dipolar orientation polarization occurs when a polar molecule (a molecule with a permanent dipole) such as water is subjected to an electric field. The permanent dipole rotates to align itself with the direction of the field (Fig. 2.1). The timescale that this rotation occurs on depends on the exerted torque on the dipole and the viscosity of the material. The timescale of the response depends on the size of the molecule, and ranges from microseconds for large globular proteins to picoseconds for smaller polar molecules such as water. Consequently, the center frequency of the dispersion will be in MHz to GHz region [2]. At the microwave frequency range, orientation polarization of free water is the main mechanism of dispersion of tissues and biological material and is called γ-dispersion [2,26,27]. A weaker dispersion effect at the range of about 100 MHz to low GHz, called δ-dispersion, due to the protein bound water is also recognized in the literature. According to Schwan [27], this dispersion likely overlaps with some small contributions due to amino acids and polar subgroups of proteins. Thus bound water in biological material appears to display a broad spectrum of dispersions extending from about 100 MHz to some GHz. The quantitative details of the γ-dispersions are fairly well understood. Measured dielectric data indicates that the tissue water behaves identical to the normal water except for the small fraction near proteins.

Below the MHz range, the main mechanism for dispersion of biological materials is the interfacial polarization, or Maxwell−Wagner effect [24−27]. As previously mentioned, free charges in electrolytes and biological media, i.e., free ions, can freely move in those materials. This would lead to an ionic (or DC) conductivity (σ_f in Eq. 2.7). Biological materials and tissues are also electrically heterogeneous and composed of different entities. Besides exhibiting DC conductivity due to the ionic conductivity of the electrolyte medium, biological materials also exhibit dispersion due to this heterogeneity. In a heterogeneous material, in order to satisfy the boundary conditions for the internal electric fields at interfaces between different media, charge accumulation would build up between the media interfaces. Theoretical models have long been presented to explain the Maxwell−Wagner dispersion, such as the suspension of cells, i.e., spheres covered by shells with different permittivity values, suspended in a third medium, as analyzed by Schwan [1−3].

The conducting membranes separating cytoplasm and the extracellular region present a membrane capacitance C_m in the order of 1 μF/cm^2. Upon the application of an electric field, the accumulation of charges over the membrane occurs in a time scale related to the charging time constant of the membrane. This time constant depends on the membrane capacitance, and intra- and extra-cellular conductivities [2,26]. The membrane charging (Maxwell−Wagner) polarization is responsible for complex permittivity dispersion in the range of 10^4−10^8 Hz, and is referred to as β-dispersion [3].

Schwan [27] suggests that β-dispersion caused by the cell membrane polarization (charging) is superimposed by the β-dispersion caused by other effects at the higher frequency range of this dispersion. Maxwell−Wagner contributions caused by organelles inside the cell (cell nuclei and mitochondria) and dipolar effects due to proteins' amino acid residues are mentioned.

Besides the membrane charging process, another cause of dispersion in biological tissues is the Maxwell−Wagner polarization associated with electrical double layers occurring at membrane surfaces or around solvated macromolecules [26]. As pointed out by Pethig [26], due to acidic phospholipids that are main components of the cell membrane lipid bilayer, a net negative charge exists over the membrane. Therefore, a charged double-layer is formed due to the presence of mobile counterions on the membrane surface [2,26]. The counterion effect has been speculated to be responsible for the dispersion in the Hz to KHz frequency range of the tissue, and in solutions of biological particles and long chain macromolecules such as DNA. The low frequency dispersion in the Hz to kHz range is called α-dispersion and is manifested by a very large dielectric constants (in the order of 10^4 below 1 KHz) in tissue and biological solutions. Besides the counterion effect, α-dispersion could also be due to the frequency dependency of protein channels in cell membranes [27].

Fig. 2.2 illustrates an idealized description of frequency variation (dispersion) of real permittivity (dielectric constant) and conductivity of tissues and cell

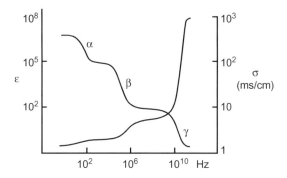

FIGURE 2.2

An idealized description of frequency variation (dispersion) of real permittivity (dielectric constant) and conductivity of tissues and cell suspension from Schwan [27]. Weaker δ-dispersion may be observed around 10^8 Hz due to the bound water.

With permission from IEEE.

suspension from Schwan [27]. The weaker δ-dispersion may be observed around 10^8 Hz due to bound water. Note also that, as opposed to what is depicted in Fig. 2.2, α and β dispersions may not be as well separated [26].

Charge accumulation at the interface of an electrode and the biological sample leads to a Maxwell–Wagner type polarization as well. In dielectric spectroscopy in the Hz to MHz range, it is common that a tissue or conductive electrolyte sample is placed in contact with electrodes. The dispersion due to this "electrode polarization" effect would then superimpose on the sample dispersion and may mask the frequency response of the sample. Different techniques have been utilized to correct for this electrode polarization effect [28–30].

So far, different polarization mechanisms in biological materials and the fact that they may contribute to the permittivity dispersion (frequency response) at various frequency ranges have been discussed. Each of these polarization effects would be associated with a relaxation (characteristic) frequency related to the timescale at which they respond to the change of an applied electric field. It is then necessary to express such frequency responses in terms of models that quantify the dispersion. Besides the frequency domain representation, it is also instructive to discuss the dielectric response in time domain. These discussions are presented next.

2.2.3 RELAXATION THEORY; DEBYE MODEL

Recalling Eq. (2.2), one may now recognize that the electric polarization $\mathcal{P}(t)$ is a combination of electronic ($\mathcal{P}_e$), atomic ($\mathcal{P}_a$), orientation ($\mathcal{P}_{or}$), and interfacial/Maxwell–Wagner ($\mathcal{P}_{mw}$) polarizations.

$$\mathcal{P} = \mathcal{P}_e + \mathcal{P}_a + \mathcal{P}_{or} + \mathcal{P}_{mw} \qquad (2.16)$$

In addition, each polarization type may have different contributors. For instance, the interfacial effect of charging the cell membrane capacitance is manifested by β-dispersion (kHz to MHz range), while that of the counterion effect is the source of α-dispersion (Hz to kHz). Up to millimeter wave frequencies, atomic and electronic polarization responses are almost instant. However, those of the Maxwell−Wagner interfacial effect and dipolar orientation are known to be exponential in nature [3] and can be characterized by a time constant, often referred to as the relaxation time. If a material is linear (no dependency of the constitutive parameter on the field level), the dielectric response of the material to a field stimulus can be obtained by the superposition of responses from each individual relaxation specie.

Empirical relations have been developed over time to characterize the (macroscopic) complex permittivity response with frequency. Debye theory [21−23,25] is a phenomenological approach for the mathematical modeling of dielectric relaxation. With the assumption of only a single relaxation mechanism, the theory suggests a first-order differential equation system, similar to the charging of a linear resistor−capacitor (RC) circuit. The polarization is assumed to be composed of two parts: one arises from the atomic and electronic displacements ($\mathcal{P}_1 = \mathcal{P}_e + \mathcal{P}_a$), which respond instantly to the applied field, and the other is due to the slower polarization process ($\mathcal{P}_0$), which could be indeed an interfacial or orientational effect

$$\frac{d\mathcal{P}_0(t)}{dt} + \frac{\mathcal{P}_0(t)}{\tau} = \frac{\chi_0 \varepsilon_0 \mathcal{E}(t)}{\tau} \tag{2.17}$$

where τ (i.e., the relaxation time) is a time constant of this first order system and χ_0 is the steady state susceptibility corresponding to the slower process. This model suggests that the polarization response to a step change of the electric field has the exponential form of $e^{-t/\tau}$. Similar to the responses of first order linear systems in electrical circuits and systems, for an initial condition of zero ($\mathcal{P}_0(0^-) = 0$), one can readily find the response to a step change of the field from 0 to $\mathcal{E}_0$ (for $\mathcal{E}(t) = \mathcal{E}_0 u(t)$, where u(t) is the unit step function) as

$$\mathcal{P}_0(t) = \chi_0 \varepsilon_0 \mathcal{E}_0 (1 - e^{-t/\tau}) \quad t > 0 \tag{2.18}$$

Note that it is assumed that the fast atomic and electronic processes instantly change with the applied field at the slower process's timescale. This means that $\mathcal{P}_1 = \chi_1 \varepsilon_0 \mathcal{E}_0$, with χ_1 being the steadystate susceptibility corresponding to the faster process. The corresponding equation for the electric displacement is then

$$\mathcal{D}(t) = \varepsilon_0 \mathcal{E}_0 + \chi_1 \varepsilon_0 \mathcal{E}_0 + \chi_0 \varepsilon_0 \mathcal{E}_0 (1 - e^{-t/\tau}) \quad t > 0 \tag{2.19}$$

Eq. (2.19) reveals that $\mathcal{D}(t)$, which is directly related to the arrangement of charges in the medium, has an exponential form in response to the step electric field stimulus. The steady state response is

$$\begin{aligned}\mathcal{D}(\infty) &= \varepsilon_0 \mathcal{E}_0 + \chi_1 \varepsilon_0 \mathcal{E}_0 + \chi_0 \varepsilon_0 \mathcal{E}_0 \\ &= \varepsilon_0 (1 + \chi_1 + \chi_0) \mathcal{E}_0 \\ &= \varepsilon_0 \varepsilon_s \mathcal{E}_0 \end{aligned} \tag{2.20}$$

This is indeed the response to a DC (static) field $\mathcal{E}_0$, and ε_s $(=1+\chi_1+\chi_0)$ is then called the static permittivity (a relative permittivity quantity). For the electrical range from DC to millimeter wave frequencies, one can combine the first two terms in Eq. (2.20) and allocate to it what is called the infinite (or optical) permittivity, ε_∞ $(=1+\chi_1)$.

$$\varepsilon_0\varepsilon_\infty = \varepsilon_0(1+\chi_1) \tag{2.21}$$

Note that then $\chi_0 = \varepsilon_s - \varepsilon_\infty$ signifies the permittivity change from DC to very high frequencies, at which the relaxation mechanism has no longer any influence. In addition, Eq. (2.2) is often written as [31]:

$$\mathcal{D}(t) = \varepsilon_0\varepsilon_\infty \mathcal{E}(t) + \mathcal{P}_0(t) \tag{2.22}$$

From the theory of linear systems, the response to an impulse stimulus $\mathcal{E}(t) = \delta(t)\hat{u}$ (with $\hat{u}$ being the unit vector indicating the direction of the field and $\delta(t)$ being the Dirac delta function) is obtained by taking the derivative of the step response. By doing this, the impulse responses for $\mathcal{P}_0(t)$ and $\mathcal{D}(t)$ are obtained by taking derivatives in Eqs. (2.18) and (2.19) respectively, and they are ($|\mathcal{E}_0| = 1$):

$$\mathcal{P}_0(t) = h_0(t)\hat{u} = \frac{(\varepsilon_s - \varepsilon_\infty)\varepsilon_0}{\tau}e^{-t/\tau}\hat{u} \quad t>0 \tag{2.23a}$$

$$\mathcal{D}(t) = [\varepsilon_0\varepsilon_\infty\,\delta(t) + h_0(t)]\hat{u} \quad t>0 \tag{2.23b}$$

where $h_0(t)$ is the impulse response of the polarization. Note that, generally, $h_0(t)$ could be any slower process's impulse response, not just that of the simple Debye differential equation (Eq. 2.17). Furthermore, recalling the linear system theory again, for any arbitrary electric field $\mathcal{E}(t)$ applied to a medium with the impulse response of the electric polarization $h_0(t)$, the electric displacement response is obtained from a convolution integral. Thus, from Eq. (2.23b)

$$\mathcal{D}(t) = [\varepsilon_0\varepsilon_\infty\,\delta(t) + h_0(t)]^*\mathcal{E}(t)$$
$$= \varepsilon_0\varepsilon_\infty\mathcal{E}(t) + \varepsilon_0\int_{\Lambda=0}^{t}h_0(t-\Lambda)\mathcal{E}(\Lambda)d\Lambda \tag{2.24}$$

Debye exponential response (Eq. 2.18), is then a special case, for which

$$\mathcal{D}(t) = \varepsilon_0\varepsilon_\infty\mathcal{E}(t) + \int_{\Lambda=0}^{t}\varepsilon_0\left(\frac{\varepsilon_s-\varepsilon_\infty}{\tau}\right)e^{-(t-\Lambda)/\tau}\mathcal{E}(\Lambda)d\Lambda \tag{2.25}$$

In addition, the complex permittivity is obtained by taking the Fourier transform of Eq. (2.25):

$$\varepsilon_0\varepsilon(\omega) = H(\omega) = \frac{\mathbf{D}(\omega)}{\mathbf{E}(\omega)}$$

$$= \varepsilon_0 + \varepsilon_0\chi_1 + \frac{\chi_0\varepsilon_0}{\tau}\frac{1}{1/\tau+j\omega} \tag{2.26}$$

$$= \varepsilon_0\varepsilon_\infty + \frac{\varepsilon_0(\varepsilon_s-\varepsilon_\infty)}{1+j\omega\tau}$$

The complex permittivity is then expressed by what is known as the Debye relation

$$\varepsilon = \varepsilon' - j\varepsilon'' = \varepsilon_\infty + \frac{\varepsilon_s - \varepsilon_\infty}{1 + j\omega\tau} \tag{2.27}$$

For small and relatively simple molecular structures (e.g., water), there is often only a single relaxation process, and the Debye relation is sufficient to express the dispersion. ε_∞ is considered as the permittivity value at sufficiently high frequencies, at which the orientational effects are disappeared. In contrast, for polymers and biological tissues, the dielectric dispersion can consist of several components, associated with small side chain movements and the whole macromolecular movement, or interfacial polarization due to the heterogeneity of the material as discussed in the previous subsection. From Eq. (2.27), the real and imaginary parts of the complex permittivity, ε, are

$$\varepsilon' = \varepsilon_\infty + \frac{\varepsilon_s - \varepsilon_\infty}{1 + (\omega\tau)^2} \tag{2.28a}$$

$$\varepsilon'' = \frac{(\varepsilon_s - \varepsilon_\infty)\omega\tau}{1 + (\omega\tau)^2} \tag{2.28b}$$

Moreover, the conductivity is

$$\sigma = \omega\varepsilon_0\varepsilon'' = \frac{\varepsilon_0(\varepsilon_s - \varepsilon_\infty)\omega^2\tau}{1 + (\omega\tau)^2} \tag{2.29}$$

For Debye relaxation, the characteristic frequency is defined as $f_c = 1/(2\pi\tau)$. At $f = f_c$, $\varepsilon' = (\varepsilon_s + \varepsilon_\infty)/2$. Eq. (2.28a) indicates a monotonous decrease of ε' from ε_s at zero frequency to ε_∞ at high frequencies ($f \gg f_c$). On the other hand, ε'' (Eq. 2.28b) exhibits a peak value of $(\varepsilon_s - \varepsilon_\infty)/2$, with zero value at low and high frequencies. To verify if the complex permittivity follows Debye dispersion, it is often useful to plot ε'' versus ε'. This plot is called the Cole–Cole diagram, which can be shown to be a semicircle within the first quadrant in the complex ε plane [1,22]:

$$\left(\varepsilon' - \frac{\varepsilon_s + \varepsilon_\infty}{2}\right)^2 + (\varepsilon'')^2 = \left(\frac{\varepsilon_s - \varepsilon_\infty}{2}\right)^2 \tag{2.30}$$

This semicircle is centered at $[(\varepsilon_s + \varepsilon_\infty)/2, 0]$ (Fig. 2.3). Debye theory is the basis of relaxation models proposed for the interpretation of the observed dispersion of real materials.

For liquid, it can be assumed that the molecular dipoles are continuously changing direction as a result of thermal agitation. A simple model can be used to show that for a sphere whose rotation is opposed by the viscosity of the surrounding medium, the relaxation time is [25,26]

$$\tau = \frac{4\pi\eta a^3}{kT} \tag{2.31}$$

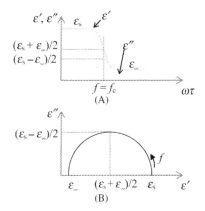

FIGURE 2.3

Dielectric properties for Debye model (single relaxation time) and no DC (ionic) conductivity; (A) Graphs of ε' and ε'', and (B) Cole–Cole diagram.

where a is the radius of the rigid sphere, and η is the macroscopic viscosity of the medium. For water, considering a radius one-half of the interoxygen distance, or 0.14 nm, and a viscosity of 10^{-3} kg/ms, Pethig [25] obtains a value of $\tau = 8.5$ ps at room temperature ($T = 293°$K), which he suggests to be close to the measured value of 9.3 ps.

As mentioned earlier, free water relaxation is the main contributor to the dielectric dispersion in biological material at microwave range, with a dielectric loss peak (characteristic frequency) of around 20 GHz. However, the tissue permittivity is decreased with respect to that of water by about 20, due to the fraction of volume of water that is displaced by proteins, as mentioned by Schwan [2]. The Debye parameters of water are well-known and can be found in papers by Von Hippel [32,33] and others [34–36], provided as a function of temperature. The model provided by Von Hippel suggests $\varepsilon_\infty = 5$, $\varepsilon_s = 78.3$, and $\tau = 8.3$ ps at $T = 25°$C. Graphs of ε', ε'', and σ versus frequency, as well as the Cole–Cole diagram of water, are plotted in Fig. 2.3.

2.2.4 EFFECT OF DC (IONIC) CONDUCTIVITY

Free movement of charges (ions) in tissue electrolytes is manifested by an ionic conductivity (σ_I) that adds to the complex permittivity according to Eq. (2.7) (subscript "I" is used to emphasis the ionic nature). This can be also accounted for by adding a term—$j\sigma_I/\omega\varepsilon_0$—to Eq. (2.27). The real permittivity would not exhibit a different frequency variation than what is expressed in Eq. (2.28b), and Figs. 2.3 and 2.4. However, the imaginary permittivity is modified by a term $\sigma_I/\omega\varepsilon_0$. For a single relaxation frequency, this would result in an indefinite increase of the imaginary permittivity and the conductivity approaching the ionic conductivity at low

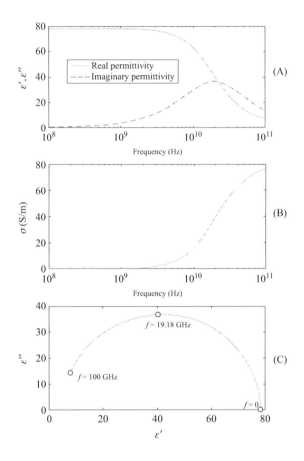

FIGURE 2.4

Dielectric properties of water at 25°C; (A) ε' and ε'', (B) σ, and (C) Cole–Cole diagram.

frequencies. For instance, for the physiological (0.9%) saline with normality $N = 0.154$, according to the relations provided by Stogryn [37], the model parameters (see Subsection 2.2.5 for the Cole–Cole model) are $\varepsilon_\infty = 4.5$, $\varepsilon_s = 75.3$, $\tau = 8.10$ ps, $\sigma_I = 1.55$ S/m, and $\alpha = 0.02$ at $T = 25°C$. α is an additional parameter that will be discussed later. Graphs of ε', ε'', and σ versus frequency, as well as the Cole–Cole diagram of the physiological saline (Fig. 2.5) reveal an imaginary permittivity local minimum around 2.87 GHz, and a corresponding deviation of the Cole–Cole diagram at its right side (low frequencies) from a circle.

2.2.5 DISTRIBUTION OF RELAXATION TIME; COLE–COLE MODEL

In most real materials, particularly biological ones, the electrical interaction between the relaxing species would usually lead to a distribution of relaxation

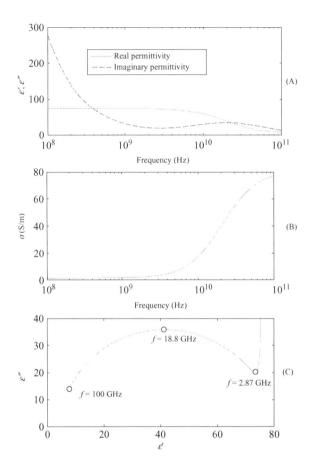

FIGURE 2.5

Dielectric properties of physiological saline (Normality: $N = 0.154$) at 25°C; (A) ε' and ε'', (B) σ, and (C) Cole–Cole diagram.

time, rather than a single relaxation time [1,22,25]. This can be thought of as the superposition of parallel combinations of linear series RC circuits with different time constants. In concentrated systems, to account for this distribution, the following relation is used:

$$\varepsilon = \varepsilon_\infty + (\varepsilon_s - \varepsilon_\infty) \int_0^\infty \frac{F(\tau)}{1 + j\omega\tau} d\tau - \frac{j\sigma_I}{\omega\varepsilon_0} \qquad (2.32)$$

where $F(\tau)$ is a function which indicates the distribution of relaxation times. Note that with equating the two sides of Eq. (2.32) for the real permittivity at $\omega = 0$, $F(\tau)$ should have the following constraint

$$\int_0^\infty F(\tau) d\tau = 1 \qquad (2.33)$$

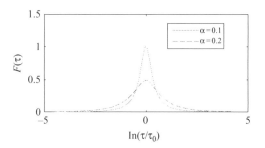

FIGURE 2.6

Plot of $F(\tau)$ for $\alpha = 0.1$ and 0.2.

The Cole−Cole model is a relaxation model that was originally suggested by Cole and Cole [38], and was verified to closely fit experimental data, particularly for tissue dispersion [1]. For the Cole−Cole distribution,

$$F(\tau) = (1/2\pi)[\sin(\beta\pi)/(\cosh(\beta s) + \cos(\beta\pi))] \tag{2.34}$$

where $\beta = 1 - \alpha$ and $s = \ln(\tau/\tau_0)$, $0 \le \alpha \le 1$. Larger α values correspond to wider distributions and broader imaginary permittivity peaks. τ_0 is the relaxation time where $F(\tau)$ peaks, and represents the center of the dispersion, i.e., the characteristic frequency ($f_c = 1/(2\pi\tau_0)$). α is often small and less than 0.2. A plot of $F(\tau)$ for $\alpha = 0.1$ and 0.2 is provided in Fig. 2.6. Debye distribution (single relaxation time τ_0) can then be mathematically represented by $F(\tau) = \delta(\tau - \tau_0)$.

Eqs. (2.32) and (2.34) lead to the well-known Cole−Cole relation for the complex permittivity:

$$\varepsilon = \varepsilon_\infty + \frac{(\varepsilon_s - \varepsilon_\infty)}{1 + (j\omega\tau_0)^{1-\alpha}} - \frac{j\sigma_I}{\omega\varepsilon_0} \tag{2.35}$$

Note that in this relation the ionic conductivity σ_I is not ignored. The real and imaginary parts of the complex permittivity then are

$$\varepsilon' = \varepsilon_\infty + \frac{(\varepsilon_s - \varepsilon_\infty)\left[1 + (\omega\tau_0)^{1-\alpha}\sin\frac{\alpha\pi}{2}\right]}{1 + 2(\omega\tau_0)^{1-\alpha}\sin\frac{\alpha\pi}{2} + (\omega\tau_0)^{2(1-\alpha)}} \tag{2.36}$$

$$\varepsilon'' = \frac{\sigma_I}{\omega\varepsilon_0} + \frac{(\varepsilon_s - \varepsilon_\infty)(\omega\tau_0)^{1-\alpha}\cos\frac{\alpha\pi}{2}}{1 + 2(\omega\tau_0)^{1-\alpha}\sin\frac{\alpha\pi}{2} + (\omega\tau_0)^{2(1-\alpha)}} \tag{2.37}$$

It should be mentioned that for $\alpha = 0$, these relations would reduce to those for the Debye dispersion. Moreover, a plot of ε' versus $\varepsilon'' - \sigma_I/\omega\varepsilon_0$ (replacing ε'' with $\varepsilon'' - \sigma_I/\omega\varepsilon_0$ in Eq. 2.30) will be part of a circle depicted in the first quadrant of the complex plane, with a center in the fourth quadrant (referred to as a depressed center by Schwan [1]). The subtended angle is $(1 - \alpha)\pi$ as depicted in Fig. 2.7.

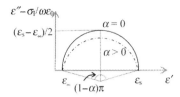

FIGURE 2.7

Cole–Cole diagram for Debye ($\alpha = 0$) and Cole–Cole ($\alpha > 0$) dispersion models.

Physically realizable frequency responses must satisfy causality, which can be described by the Kramers–Kronig relations for the complex permittivity [22,25].

$$\varepsilon'(f) - \varepsilon_\infty = \frac{2}{\pi} \int_0^\infty \frac{x\varepsilon''(x)}{x^2 - f^2} dx \tag{2.38}$$

$$\varepsilon''(f) = \frac{-2f}{\pi} \int_0^\infty \frac{\varepsilon'(f) - \varepsilon_\infty}{x^2 - f^2} dx \tag{2.39}$$

In other words, the real permittivity and imaginary permittivity (or conductivity) are mathematically interrelated and can be obtained by one another. One simple observation that can be made for the Debye dispersion is that the conductance and real permittivity changes between two frequencies ω_1 and ω_2 are interrelated through [2]

$$\sigma_1 - \sigma_2 = \frac{(\varepsilon_2' - \varepsilon_1')\varepsilon_0}{\tau} \tag{2.40}$$

Causality of the Cole–Cole relation is proven and discussed in [39].

2.2.6 COMPLEX PERMITTIVITY OF BIOLOGICAL TISSUES

Each of the α, β, or γ relaxation regions is in its simplest form characterized by a Cole–Cole relation. The Maxwell–Wagner effect is responsible for β-dispersion, around several tens of KHz. Different possible mechanisms for α-dispersion (about several tens of Hz) have been speculated as mentioned earlier. As Schwan suggests [2], the temperature coefficients for the characteristic (relaxation) frequency for α and β are the same as the temperature coefficient for the conductivity of the electrolyte medium, or 2% per °C. On the other hand, dipolar relaxations of bound and free water are major contributors to the permittivity of tissues at MHz to GHz region. However, in contrast to tissues, water as a pure liquid exhibits a single relaxation frequency close to 20 GHz at room temperature and 25 GHz at 37°C. Thus, γ-dispersion has a relaxation frequency near 20 GHz at room temperature, which is more pronounced for higher water content tissues. The temperature dependence of the relaxation frequency for γ-dispersion in tissues is thus equal to that of water and is about 2% per °C [2]. δ-Dispersion is due to the proteins' bound water, is less pronounced, and is observed in a broad

Table 2.1 Range of Characteristic Frequencies Observed for Biological Materials for α, β, γ, and δ dispersions [3]

Dispersion	α	β	δ	γ
Frequency range (Hz)	$1-10^4$	10^4-10^8	10^8-10^9	2×10^{10}

frequency range from some 200 to 3000 MHz [2,3]. The ranges of the characteristic frequencies for the various tissue dispersions are given at Table 2.1 [3].

Since early reports in the 1950s (Cook [40] and Schwan [1]), many articles have appeared that provide the complex permittivity of biological tissues and biological solutions such as [41–44]. Examples of in vivo data for some tissues can be found at [45,46]. Tabulation and listing of complex permittivity data can be found in manuscripts by Stuchly and Stuchly [47], Schwan and Foster [2], Duck [48], and Gabriel et al. [49,50].

The 1996 articles by Gabriel et al. [49–51], are among the well-recognized and frequently cited reports in the field of the complex permittivity of biological tissues. In their survey of prior reported complex permittivity data published by that time, the authors included the human tissue and in vivo measurement results at temperatures as low as 20°C. They also measured the tissue's dielectric parameters [50] from 10 Hz to 20 GHz at 37°C, and then provided parametric models for 17 types of tissues [51] according to a four-term Cole–Cole relation:

$$\varepsilon(\omega) = \varepsilon' - j\varepsilon'' = \varepsilon_\infty + \sum_{n=1}^{4} \frac{\Delta\varepsilon_n}{1 + (j\omega\tau_n)^{1-\alpha_n}} - \frac{j\sigma_\mathrm{I}}{\omega\varepsilon_0} \tag{2.41}$$

In this equation, τ_n, $\Delta\varepsilon_n$, and α_n ($n = 1, 2, 3, 4$) represent Cole–Cole parameters corresponding to four different dispersion regions. The corresponding parameters were obtained by fitting the measurement results to the above equation. Parameters of Eq. (2.41) and tissue dispersion graphs (of ε' and σ) for up to 100 GHz for the tissues studied can be found in [51].

Fig. 2.8 depicts the real and imaginary permittivity, as well as the conductivity, for muscle and fat (infiltrated). The graphs are generated using the parameters for the four-term Cole–Cole model (Eq. 2.41) in Gabriel's paper [51]. For 1 Hz–100 GHz, the overall range of variation is about seven orders of magnitude for ε', eight for ε'', and three for σ. However, for frequencies of 100 MHz to 100 GHz the complex permittivity variation remains roughly below two orders of magnitude.

In situations for which the interaction of EM waves with tissue at microwave frequency range is analyzed, the four-term model in Eq. (2.41) contains unnecessarily redundant parameters. Above a few hundred MHz the dipolar relaxation of water is the dominant polarization mechanism and a single-term relaxation (Eq. 2.35) could be sufficient. σ_I in that equation then represents the conductivity due to lower frequency polarization mechanisms, for which $\omega\tau \ll 1$. A least

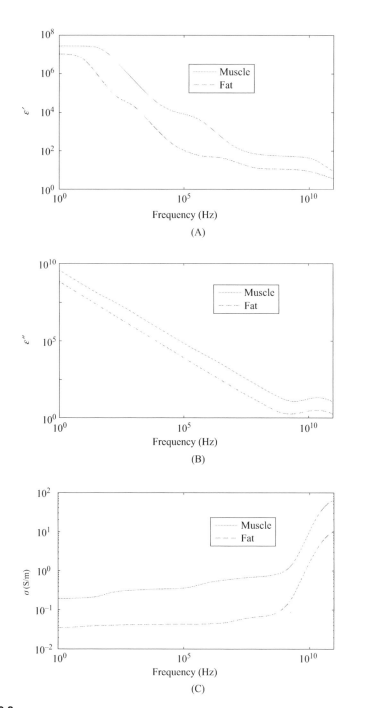

FIGURE 2.8

Complex permittivity versus frequency for muscle and fat (infiltrated); (A) Real permittivity,
(B) Imaginary permittivity, and (C) Conductivity. The graphs are generated using the
parameters for the four-term Cole–Cole model in Gabriel's paper [51].

Table 2.2 Fitting Parameters for the Complex Permittivity Values of Muscle and Fat From [52] Fitted to a Single-Term Cole–Cole Relation, From 0.5 to 30 GHz [52]

Tissue	τ_0 (ps)	ε_∞	ε_s	α	σ (s/m)	E_ε	Max Error (%)	
							$\Delta\varepsilon'/\varepsilon'$ (@f, GHz)	$\Delta\varepsilon''/\varepsilon''$ (@f, GHz)
Muscle	7.23	2.93	55.32	0.125	0.777	0.0049	−2.7(0.5)	−1.7(1.4)
Fat-Inf	7.98	2.41	11.62	0.212	0.073	0.0016	−0.9(0.5)	−0.8(1.1)

With permission from IEEE.

squares minimization procedure was applied by Gabriel et al. [51] at the frequency range above 400 MHz to provide a one-term Cole–Cole model for γ-dispersion of several tissue types, along with the 95% confidence interval for the model parameters.

As another example discussed in [52], single-term Cole–Cole model parameters for muscle and fat are presented from 0.5 to 30 GHz (Table 2.2). They are obtained by fitting the complex permittivity values from the four-term Cole–Cole model (Eq. 2.41) to a one-term model (Eq. 2.35), by minimizing the sum of the square of the fitting error magnitudes for 100 frequency points from 0.5 to 30 GHz [52]. The maximum fitting errors for the real and imaginary permittivity ($\Delta\varepsilon'/\varepsilon'$ and $\Delta\varepsilon''/\varepsilon''$), and $|\Delta\varepsilon/\varepsilon|$, are listed in Table 2.2 are −2.7% for muscle and −0.9% for fat, both at 0.5 GHz [52]. In addition, the ratio of the root-mean-square (rms) value of the fitting error to the rms value of the complex permittivity given by

$$E_\varepsilon = \left[\left(\int_{f_1}^{f_2} |\varepsilon_{\text{fit}} - \varepsilon|^2 df \right) / \left(\int_{f_1}^{f_2} |\varepsilon|^2 df \right) \right]^{1/2} \qquad (2.42)$$

is also obtained, where $f_1 = 0.5$ GHz and $f_2 = 30$ GHz.

Gabriel's Cole–Cole models for various tissues are inclusive of various reported tissue data in the literature available at the time. However, depending on applications that may need more specific tissue differentiation (e.g., healthy versus tumor tissue) and the investigated frequency range, a variety of tissue model parameters (Debye or Cole–Cole) for a tissue type are found in the literature. Some examples include two-term Debye models for muscle from 20 MHz to 20 GHz [53], two-term Cole–Cole models of rat brain gray and white matter (25°C and 37°C from 45 MHz to 26.5 GHz) [54], single-term Debye models of rat brain gray and white matter (at 27°C from 15 to 50 GHz) [55], single-term Debye models of normal, malignant, and cirrhotic human liver tissues from 0.5 to 20 GHz (in vivo and ex vivo) [56], and various models for normal and cancerous breast tissues [57–63]. The wealth of data on the breast permittivity models for normal and cancerous breast tissue is because of the great deal of interest for detecting breast cancer through microwave imaging [64,65]. Reports include normal and malignant one- and two-term Cole–Cole models from 0.5 to 20 GHz [58,59] and a one-term model from 3.1 to 10.6 GHz [59], two-term Cole–Cole model for tumors from 0.5 to 20 GHz [61], and one-term Cole–Cole model for tumors from 0.5 to 50 GHz [62].

Because of their higher water content, tumors exhibit higher permittivity than normal tissues. Sugitani et al. [61] observe that the dielectric constant of cancerous breast tissues at 6 GHz is approximately four times more than that of the adipose tissues, and find insignificant differences between cancer and stroma tissues. They conclude that a correlation between the volume fraction of cancer cells in the tumor tissue to the measured conductivity and dielectric constant exists. In studying the ultra-wideband microwave dielectric property of normal tissue, Lazebnik et al. [58] observe that the dielectric properties of breast tissue are primarily determined by the adipose content of the tissue, with negligible effects from factors such as patient age, tissue temperature, and time between excision and measurement. They also mention substantial tissue heterogeneity, causing a large variation in the dielectric properties of the adipose content of the tissue in normal breast tissue.

2.2.7 CELL AND CELL CONSTITUENTS' COMPLEX PERMITTIVITY

There are some data on the complex permittivity of cell solutions in microwave and millimeter wave frequencies [66–71]. Bao et al. have reported the microwave dielectric property of erythrocyte suspensions [66]. Dielectric measurements of liposome solutions have been reported for a frequency range of 1 kHz–60 GHz by Pottel et al. [67] and for 10 kHz–100 MHz by Stuchly et al. [68]. Merla et al. [71] studied the complex permittivity of liposome (100 MHz–2 GHz) as well as erythrocyte (100 MH–1 GHz) solutions [72], in order to obtain the membrane permittivity. They considered a Maxwell–Wagner three-layered model to obtain the membrane permittivity from the mixture permittivity. A Debye dispersion model was used to fit the complex permittivity data, where relaxation frequencies of $f = 325.6$ and $f = 179.85$ MHz were obtained for the erythrocyte membrane and the liposome membrane lipid bilayer. Another work was conducted by Kolsgen et al. [69], who measured phospholipid samples from 1 MHz to 2 GHz, where a two-term Debye model for membrane permittivity was suggested within that range. The two relaxation frequencies increase with temperature and are estimated to be in ranges of 15–100 and 100–500 MHz for temperature changes from 286 to 323°K. Ebara et al. [70] have introduced a method of measurement of the complex permittivity of phospholipid membranes using dielectric mixture theory. Well-known relations for the complex permittivity of composite materials were used to estimate the complex permittivity of the phospholipid, in a mixture of phospholipid and Ringer's solution. The measurements were performed from 0.8 to 6 GHz.

2.3 COMPLEX PERMITTIVITY MEASUREMENT

A survey of the available literature on permittivity measurement and complex permittivity of lossy material and biological tissues reveals a rich and long history. Some earlier reviews of permittivity measurement techniques are those of Westphal [73], Fox and Sucher [74], and Bussey [75]. Methods developed by 1980s, particularly those for biological tissues and liquids, can be found in a book by Grant et al. [76] and an article by Stuchly and Stuchly [77], as well as a review article by Afsar et al. [78]. Some useful and more recent reviews and topics related to permittivity measurement techniques can be found in a book by Chen et al. [79], a book chapter by Tofighi and Daryoush [80], and articles by Gregory and Clarke [81] and Kaatze [82]. The latter, in particular, provides a comprehensive historical review of methods for measuring the dielectric property of materials known by 2012.

Fig. 2.9 illustrates a conceptually simple and straightforward method of measuring the complex permittivity. $C_0 = \varepsilon_0 A/d$ is the capacitance of the empty capacitor. When the capacitor is filled with the permittivity ε, the circuit representation is a parallel combination of a capacitance C with a conductance G, whose admittance is

$$\begin{aligned} Y &= j\omega\varepsilon C_0 = j\omega(\varepsilon' - j\varepsilon'')C_0 \\ &= jB + G = j\omega\varepsilon'C_0 + \omega\varepsilon''C_0 \end{aligned} \tag{2.43}$$

with the conductance $G = \omega\varepsilon''C_0$ and the susceptance $B = \omega\varepsilon'C_0$ ($C = \varepsilon'C_0$). The capacitor fringing field is neglected. The complex permittivity can thus be obtained from the measured admittance through

$$\varepsilon = Y/j\omega C_0 = -jdY/(\omega\varepsilon_0 A) \tag{2.44}$$

This method becomes erroneous and ineffective when the size of the capacitor, e.g., dimensions of A or d, become comparable to the wavelength. Thus, a parallel plate capacitor or similar implementations of the capacitance are useful up to frequencies in the order of a few tens of MHz. On the other hand, the dielectric

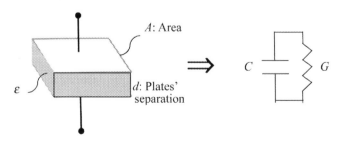

FIGURE 2.9

Conceptual representation of an ideal parallel plate capacitor, with an air-filled capacitance C_0, filled with a material with the complex permittivity ε.

property measurements from impedance (admittance) measurement to these frequency ranges are also aided by available equipment for direct measurement of impedance (admittance) in the MHz range [50].

Although Fig. 2.9 depicts a parallel plate capacitance, this concept can be applied to any other measurement probe with a known empty capacitance C_0. The most well-known example is an open-ended coaxial probe with terminating empty capacitance C_0. Nonetheless, Y may deviate from $j\omega\varepsilon C_0$, particularly at higher frequencies, in a manner such that Y would be no longer a linear function of ε, due to the fringing fields, frequency dependency of field distribution in the capacitor, and antenna (radiation) effects.

An overview of older as well as more recent permittivity measurement methods can be found in [80]. In general, at microwave frequency ranges and beyond, the methods that are often described for the permittivity measurements of biological material can be categorized as transmission line methods versus resonance methods. Due to the versatility of these methods, probing structures, permittivity extraction procedures, and measurement calibration techniques, an exhaustive literature review would require much more space than would be permitted for this chapter. Thus, only highlights of the most important aspects of the permittivity measurement of biological tissue at microwave and millimeter wave frequencies are provided.

2.3.1 RESONANT METHODS

The idea behind the resonant methods for permittivity measurement is that for a resonator loaded with a sample, the resonance frequency (f_0) and the quality factor (Q) of the cavity will change due to the loading by the sample [83–94]. The real permittivity primarily affects the resonance frequency, and the imaginary permittivity (or conductivity) is the primary factor changing the quality factor from that of the unloaded resonator. The advantages of the resonant methods are that they may not require expensive equipment such as vector network analyzers (VNAs) (a swept generator and a power detector may be sufficient), can provide accurate results without complex calibration or postprocessing of the measured data, and can be used to measure very low loss material. On the other hand, they only provide single frequency information and are not suitable for broadband measurements. Moreover, for lossy material, it may be hard to accurately evaluate the resonance frequency and quality factor due to the response peak broadening.

Resonant cavities have been frequently used for permittivity measurement using resonant methods [84–92]. Fig. 2.10 illustrates the idea behind the resonant cavity methods. For a sample whose volume V_1 is small compared to the cavity volume V_0, the perturbation theory [83,91] can be used to describe the shift in the resonance frequency after the cavity is loaded. For nonmagnetic materials including biological substances,

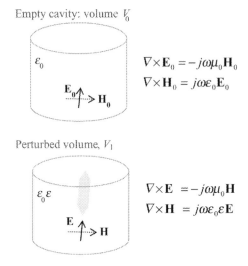

FIGURE 2.10

Perturbation of an empty cavity by a sample with the relative permittivity ε.

$$\frac{\Delta\Omega}{\omega_0} = \frac{\iiint_{V_1}[\varepsilon_0(1-\varepsilon)\mathbf{E}_0.\mathbf{E}]dv}{\iiint_V(\varepsilon_0|\mathbf{E}_0|^2 - \mu_0|\mathbf{H}_0|^2)dv} \qquad (2.45)$$

where $\Delta\Omega = \Omega - \Omega_0$, and Ω and Ω_0 are the complex resonance frequencies for the sample loaded and empty cavity.

$$\Omega = \omega\left(1 + \frac{j}{2Q}\right)$$

$$\Omega_0 = \omega_0\left(1 + \frac{j}{2Q_0}\right) \qquad (2.46)$$

In addition, ω and ω_0 and Q and Q_0 are the actual resonance frequencies and quality factors of the loaded and empty cavities, respectively. The denominator in Eq. (2.45) is $4W_0$, where W_0 is the stored energy in the empty resonator. Certain assumptions are made for obtaining Eq. (2.45), i.e., the perturber should be of small size compared to the cavity volume, and the $\mathbf{E}$ and $\mathbf{H}$ fields outside it will remain the same as for the empty cavity case. Furthermore, if the sample is placed in a region of the cavity with no field variation across it, Eq. (2.45) can be further simplified as

$$\frac{\Delta\Omega}{\omega_0} = \frac{\varepsilon_0(1-\varepsilon)(\mathbf{E}_0.\mathbf{E})V_1}{4W_0} \qquad (2.47)$$

For certain sample shapes, the field inside the sample ($\mathbf{E}$) can be related to the field outside it ($\mathbf{E_0}$) by approximation relations found in the literature [83].

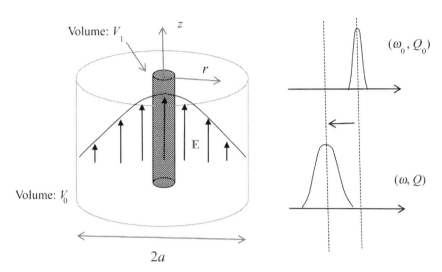

FIGURE 2.11

Perturbation of a cylindrical cavity resonator resonating at TM_{010} by a sample with relative permittivity ε.

Cylindrical cavity resonators resonating at TM_{010} mode are frequently used in permittivity measurements [90,91] (Fig. 2.11). The **E** field is circularly symmetric and has no variation with height. The sample, with a uniform cross section and small cross-sectional dimensions, fills the entire central region of the cavity from the top to the bottom. Given the fact that the z-directional **E** field inside and immediately outside the sample is the same under this condition [83], Eq. (2.47) then reduces to

$$\frac{\Delta\Omega}{\omega_0} = \frac{\varepsilon_0(1-\varepsilon)|\mathbf{E}_0|^2 V_1}{4W_0} \tag{2.48}$$

where

$$W_0 = \frac{1}{2}\varepsilon_0 V_0 J_1^2(2.405)|\mathbf{E_0}|^2 \tag{2.49}$$

and J_1 is the first-order Bessel function of the first kind. The real and imaginary permittivity are then obtained from

$$\varepsilon' = 1 + 2C\left(\frac{f_0-f}{f}\right)$$

$$\varepsilon'' = C\left(\frac{1}{Q} - \frac{1}{Q_0}\right) \tag{2.50}$$

with $C = J_1^2(2.405)\frac{V_0}{V_1}$.

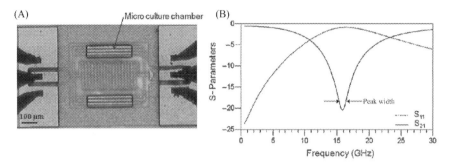

FIGURE 2.12

(A) Photograph of a microchip resonator sensor for the measurement of biological cells and (B) Its typical frequency response [99].

With permission from Elsevier.

The broadening of the resonance peak causes some difficulty for evaluating Q for lossy samples, and methods such as that suggested in [92] might be used to evaluate Q. Among other types of resonant cavities in the literature are TE_{101} rectangular resonant cavities [89,92], stripline cavities [84−86], sample terminated coaxial cavities [93,94], and open resonator cavities, which are particularly useful for millimeter wave measurements [95−98].

Recent interest in resonant methods has also been directed toward biosensor chips for the measurement of biological cells [99−101]. The motivation behind the microwave biosensors for biological cell measurements has been to discriminate between different cell types, for instance between stem and differentiated cells [99] or aggressive cancer cells [100]. Proposed microchips include LC resonator metallization, with an interdigitated coupling capacitor. The capacitor is exposed to biological cells through small microculture chambers ($\sim 200\,\mu m^2$) [99−101] (Fig. 2.12). In order to capture the resonance graph, two-port (transmission) measurements (S_{21}) are done using coplanar waveguide (CPW) probes by a VNA. Some reported graphs illustrate frequency ranges of 11−21 GHz [99], 5−15 GHz [100], and 3−10 GHz [101]. The shift of resonance frequency due to the samples is in the order of few tens of MHz.

2.3.2 OPEN-ENDED COAXIAL LINE

The majority of the published works on microwave permittivity measurement discuss methods that utilize an open-ended coaxial probe [45,77,102−105], where the change in the terminating impedance of the line causes a change in its input reflection coefficient. In other words, different complex permittivity values of the tissue under test cause variation in the capacitance and conductance of the line termination, hence impacting the amplitude and the phase of the reflected wave. The open-ended coaxial method is closely related to the lumped capacitance method (Fig. 2.9), where

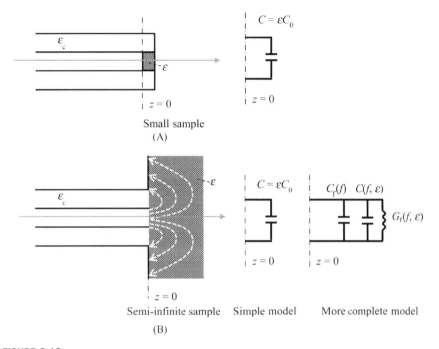

FIGURE 2.13

The lumped element approach: (A) The sample is placed within the gap between the inner conductor of the coaxial line and the short end wall; and (B) The sample is in contact with the open-ended coaxial probe. In its simplest form, the circuit model is an empty capacitor C_0, filled with the material. A more complex model incorporates frequency and permittivity dependent radiation resistance and end capacitance. z axis is along the central axis of the line.

the terminating impedance of a transmission line (e.g., a coaxial cable) is a capacitance that is filled with the sample under test. Fig. 2.13A represents the earlier version [106,107], where the sample is completely placed inside a gap between the inner conductor of the line and the end wall. On the other hand, for an open-ended coaxial line [108–121] (Fig. 2.13B), the open transmission line is in contact with the sample, which ideally should be infinite in extent. The equivalent circuit model of the line termination could be as simple as a lumped capacitor C_0 (empty sensor) filled with the sample permittivity, or it could involve a fringing capacitance C_f, due to the higher-order modes inside the line, and two external components that are a capacitance $C(f, \varepsilon)$ and a radiation conductance $G_r(f, \varepsilon)$. These two components could be a function of frequency and the sample's permittivity [113–117]. Some early reviews (until the 1980s) of the terminated coaxial line methods, discussing equivalent circuits, measurement calibration considerations, and uncertainty analysis, can be found in the papers Stuchly and Stuchly [77] and Marsland and Evans [105]. A review of the method up to 2004 can be found in [80].

The great deal of interest in the open-ended coaxial lines for complex permittivity measurement since 1970s coincided with the availability and evolution of VNAs and advances in their calibration techniques. VNAs are also being used for permittivity measurement using planar transmission line, such as CPWs [99–101]. Planar transmission lines may lend themselves better than the coaxial cables for measuring small sample sizes, microfluidics, and biological cells, as already mentioned [99–101]. Nonetheless, the methods and processes highlighted in this section for the complex permittivity measurement using open-ended coaxial cables can find applications where other types of transmission lines are used. This would be particularly the case where a similar terminating impedance model for the open end could be established regardless of the type of the transmission line.

The lumped element approach (Fig. 2.13), in its simplest form, assumes an empty capacitor C_0, which is filled with the material complex permittivity, such that the terminating impedance is [105]

$$Y_S(f, \varepsilon) = G + jB = j\omega(\varepsilon C_0 + C_f) = j\omega(\varepsilon' - j\varepsilon'')C_0 + j\omega C_f \qquad (2.51)$$

with

$$B = j\omega(\varepsilon' C_0 + C_f), \quad G = \omega\varepsilon'' C_0 \qquad (2.52)$$

In this simplified treatment, the radiation resistance $G_r(f, \varepsilon)$ is neglected, and $C(f, \varepsilon) = \varepsilon C_0$ in Fig. 2.13B is assumed, denoting an air-filled (empty) external capacitor that is not a function of ε. These approximations are often used when the dimension of the open-ended probe is much smaller than the wavelength, and the radiation effect can be neglected. Since in frequency domain measurements C_0 is measured (or as seen later, it is often taken care of by a proper calibration process) or analytically evaluated at each frequency, its changes with frequency from its static value would not affect this analysis, as long as the dimension of the open-ended probe is much smaller than the wavelength. C_f in Eq. (2.51) is sometimes neglected. However, as described in [105], its impact would also be taken into account by the calibration process and would not further add to the complexity of the measurement process as discussed later.

In order to find ε, if the calibration process incorporates known samples, an explicit knowledge of C_0 and C_f values is not required. Nonetheless, it is instructive to discuss relations for finding ε when the reflection coefficient of the sample Γ is known at the plane of sample ($z = 0$ in Fig. 2.13):

$$\Gamma = \frac{1 - Z_0 Y_S}{1 - Z_0 Y_S} \qquad (2.53)$$

Using Eq. (2.51), the complex permittivity of the sample is then obtained by

$$\varepsilon = \frac{1 - \Gamma}{j\omega Z_0(1 + \Gamma)} - \frac{C_f}{C_0} \qquad (2.54)$$

Cautions should be advised when interpreting Eq. (2.54). First, open-ended probes may not lend themselves to being suitably calibrated at the plane of the sample using standard network analyzer calibration processes such as open-short-matched load. In other words, probe tips may not be compatible with the existing calibration standards, and designing them for compatibility with those standards would add to the complexity of the probe structure and the circuit model (Fig. 2.13). Second, although C_f and C_0 may be obtainable analytically, numerically, or empirically, any deviations from the ideal shape of the probe tip such as those caused by grinding, filing, or any other fabrication process of the tip may result in sizable error in the extracted permittivity. Because of these issues, it is not surprising that the literature after the 1980s overwhelmingly uses calibrations using known sample materials that simultaneously set the measurement reference plane, and take into account circuit parameters (C_f and/or C_0) in order to find the complex permittivity of an unknown sample.

Fig. 2.14 illustrates a process by which the systematic error [122] due to the VNA's internal circuitry, the adapter connecting the probe to the VNA's circuitry, and the length of the probe are modeled as an error network. Furthermore, since C_f is due to the fringing fields within the line, it can be also included in the error

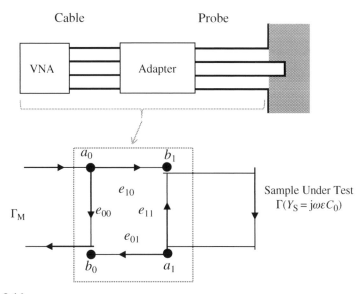

FIGURE 2.14

The systematic error due to the VNA's internal circuitry, the adapter connecting the probe to the VNA's circuitry, the internal fringing capacitance (C_f), and the line length of the probe are modeled as an error network, to be determined with a calibration using three known samples.

network. A one-port calibration process using three known standards can be used to find the error network parameters e_{11}, e_{00}, and $e_{10}e_{01}$ [116]. Resolving e_{10} and e_{01} individually is unnecessary in this calibration process. The error parameters are obtained from the following expressions [122]:

$$e_{00} = \frac{\Gamma_1\Gamma_2\Gamma_{M3}(\Gamma_{M1} - \Gamma_{M2}) + \Gamma_2\Gamma_3\Gamma_{M1}(\Gamma_{M2} - \Gamma_{M3}) + \Gamma_3\Gamma_1\Gamma_{M2}(\Gamma_{M3} - \Gamma_{M1})}{\Gamma_1\Gamma_2(\Gamma_{M1} - \Gamma_{M2}) + \Gamma_2\Gamma_3(\Gamma_{M2} - \Gamma_{M3}) + \Gamma_3\Gamma_1(\Gamma_{M3} - \Gamma_{M1})}$$ (2.55)

$$e_{11} = \frac{\Gamma_3(e_{00} - \Gamma_{M2}) + \Gamma_2(\Gamma_{M3} - e_{00})}{\Gamma_2\Gamma_3(\Gamma_{M3} - \Gamma_{M2})}$$ (2.56)

$$e_{01}e_{10} = \frac{(\Gamma_{M2} - e_{00})(1 - e_{11}\Gamma_2)}{\Gamma_2}$$ (2.57)

where

$$\Gamma_{Mi} = e_{00} + \frac{e_{01}e_{10}\Gamma_i}{1 - e_{11}\Gamma_i} \quad i = 1, 2, 3$$ (2.58)

is the measured reflection coefficient of the known sample i, and $\Gamma_i = (1 - Z_0Y_{Si})/(1 + Z_0Y_{Si})$ is the reflection coefficient of the terminating end capacitor for the known sample i. Note that if C_f is already included in the error network,

$$Y_{Si} = j\omega\varepsilon_i C_0$$ (2.59)

Finally, the true reflection coefficient of the sample is obtained from

$$\Gamma = \frac{\Gamma_M - e_{00}}{e_{11}(\Gamma_M - e_{00}) + e_{01}e_{10}}$$ (2.60)

and its complex permittivity is obtained from Eq. (2.54) (with the term with C_f dropped since it is already included in the error network).

Note that in the above process, the error network parameters are explicitly obtained. Thus, C_0 needs to be known for that purpose. However, an alternative approach, as recognized by some researchers and articulated by Marsland and Evans [105], can be used to find ε, without finding C_0 and the error network parameters. Note that the relation between the true and measured reflection coefficients (Eq. 2.60) is a bilinear transform. Marsland and Evans [105] point out that the cross-ratio invariance property of the bilinear transform implies the existence of the following relation between the measured reflection coefficients and the terminating admittances:

$$\frac{(Y_S - Y_{S1})(Y_{S3} - Y_{S2})}{(Y_S - Y_{S2})(Y_{S1} - Y_{S3})} = \frac{(\Gamma_M - \Gamma_{M1})(\Gamma_{M3} - \Gamma_{M2})}{(\Gamma_M - \Gamma_{M2})(\Gamma_{M1} - \Gamma_{M3})}$$ (2.61)

where Γ_M (and Γ_{Mi}, $i = 1,2,3$) and Y_S (and Y_{Si}, $i = 1,2,3$) are the measured reflection coefficient and terminating admittance of the sample under test (and known standards), respectively. The relation for Y_S (and Y_{Si}) would be a complex model in general, with the one in Eq. (2.51) the simplest form in particular. Furthermore, the cross-ratio invariance property implies that Eq. (2.61) also holds for any arbitrary bilinear transformation of the admittance. A useful such transformation is [105]

$$Y'_S(f, \varepsilon) = Y_S(f, \varepsilon)/j\omega C_0 - C_f/C_0 \tag{2.62}$$

with Eq. (2.61) rewritten as

$$\frac{(Y'_S - Y_{S1})(Y'_{S3} - Y'_{S2})}{(Y'_S - Y'_{S2})(Y'_{S1} - Y'_{S3})} = \frac{(\Gamma_M - \Gamma_{M1})(\Gamma_{M3} - \Gamma_{M2})}{(\Gamma_M - \Gamma_{M2})(\Gamma_{M1} - \Gamma_{M3})} \tag{2.63}$$

If the simple model in Eq. (2.51) is used, then from Eq. (2.62)

$$Y'_S(f, \varepsilon) = \varepsilon \tag{2.64}$$

An explicit relation for the complex permittivity of the sample is then

$$\varepsilon = -\frac{\Delta_{M1}\Delta_{32}Y'_{S3}Y'_{S2} + \Delta_{M2}\Delta_{13}Y'_{S1}Y'_{S3} + \Delta_{M3}\Delta_{21}Y'_{S2}Y'_{S1}}{\Delta_{M1}\Delta_{32}Y'_{S1} + \Delta_{M2}\Delta_{13}Y'_{S2} + \Delta_{M3}\Delta_{21}Y'_{S3}} \tag{2.65}$$

with $\Delta_{Mi} = \Gamma_M - \Gamma_{Mi}$ and $\Delta_{ij} = \Gamma_{Mi} - \Gamma_{Mj}$. The calibration standards can be open, short, distilled water, methanol, saline, or any other known liquids [116,123,124]. Generally, two known liquids with their permittivity values close to the upper and lower bounds of the expected permittivity values of the unknown samples are desirable [123,124]. The third sample can be chosen as a liquid within the expected range of the unknown sample, a short [105], or an open. The latter is often incorporated as an integral part of the process when measuring the complex permittivity using time domain reflectometry [123].

It has been long known that the terminating capacitance of the probe should be optimized in order to achieve the least degree of uncertainty in determining the permittivity due to the uncertainty in the magnitude and the phase of the reflection coefficient of the probe [77,102,106,107]. A relation suggested in [106] for the optimum choice of C_0 to minimize $\frac{\Delta\varepsilon'}{\varepsilon'}$ and $\frac{\Delta\varepsilon''}{\varepsilon''}$ is

$$C_0 = \frac{1}{\omega Z_0 \sqrt{\varepsilon'^2 + \varepsilon''^2}} \tag{2.66}$$

This optimum value is under the condition that the uncertainty in the phase of the reflection coefficient (Eq. 2.53) is the same as the uncertainty in its magnitude, i.e., $\Delta\theta = \Delta|\Gamma|/|\Gamma|$. A plot of C_0 versus frequency for various $\sqrt{\varepsilon'^2 + \varepsilon''^2}$ from [106] is provided in Fig. 2.15. For instance, for $\sqrt{\varepsilon'^2 + \varepsilon''^2} = 1$, $Z_0 = 50\ \Omega$, and $f = 2$ GHz, the optimum C_0 is 1.6 pF. Eq. (2.66) and Fig. 2.15 suggest that for higher frequencies and permittivity values, a smaller C_0 is desired, while a larger C_0 is preferable for lower frequencies and permittivity values. For coaxial lines with a fixed characteristic impedance Z_0, C_0 increases with the outer conductor's diameter. In general, larger diameter probes having a larger C_0 are preferable at lower frequencies, while smaller ones are preferable at higher frequencies [123]. One method of controlling the C_0 is by extending the inner conductor's tip. Fig. 2.16 illustrates the graphs for the C_0, C_f, and $C_{total} = C_0 + C_f$ versus the probe tip length (d), for a 3.5 mm Teflon-filled

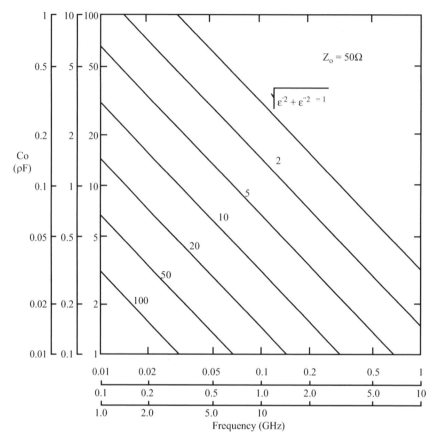

FIGURE 2.15

Optimum empty capacitance C_0 versus frequency for various $\sqrt{\varepsilon'^2 + \varepsilon''^2}$ [106].

With permission from IEEE.

coaxial cable ($\varepsilon_r = 2.1$) with $Z_0 = 50$ Ω. The graphs are obtained from simulation using High Frequency Structures Simulator (HFSS, ANSYS, Canonsburg, PA). To find C_0 and C_f, using least square fitting, the simulated admittances for the empty probe and water (at 25°C) are fitted to Eq. (2.51). For evaluating C_0 and C_f, only frequencies below a certain frequency at which the probe admittance starts deviating from the linear model (Eq. 2.51) are considered. Fig. 2.17 illustrates the change of the magnitude of the simulated probe admittance (solid line), along with the one obtained from the linear model (dashed line), versus frequency, and for the end tip lengths of $d = 0$, 0.4, 1, and 1.5 mm. The small fluctuations in the graph are due to the reflection from imperfect radiation boundaries that are placed in the model to limit the problem size. It is observed that the probe admittance deviates from the linear model (Eq. 2.51) at a certain frequency, above which the linear model becomes

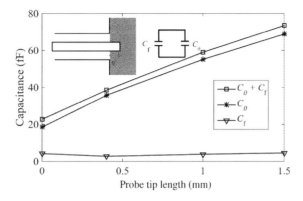

FIGURE 2.16

C_0, C_f, and $C_{total} = C_0 + C_f$ versus the probe tip length (d), for a 3.5 mm Teflon-filled coaxial cable with $Z_0 = 50\ \Omega$, obtained using High Frequency Structures Simulator (HFSS, ANSYS, Canonsburg, PA).

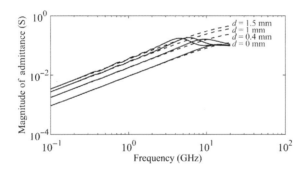

FIGURE 2.17

Magnitude of the simulated probe admittance for water versus frequency (solid line) compared to the one obtained from the linear model (dashed line, from Eq. 2.51) for various probe tip lengths (d).

no longer valid. At that frequency, the radiation effect (Fig. 2.13B) must be considered. In addition, for longer tips, the onset of this deviation moves toward lower frequencies. For water, this frequency is roughly 8, 6, 4, or 3 GHz for $d = 0$, 0.4, 1, or 1.5 mm, respectively (Fig. 2.17). This indicates that an extended tip that increases the empty capacitance C_0, more appropriate for lower frequency and permittivity, would also decrease the frequency range of the measurement or at least the range of validity of the simple capacitance model of Eq. 2.51.

Analytical studies of the aperture admittance that incorporate the radiation effect as well as the frequency dependency of the probe conductance and susceptance (Fig. 2.13B) can be found in references such as those by Misra and his colleagues [113–115] and Xu et al. [94]. In those approaches, the admittance of the probe is represented by a stationary integral that leads to series expansions of G and B as polynomial functions of frequency. Polynomial coefficients include constant multipliers and multipliers in the form of a power of ε. From the calibration perspective, each term of the order of $(f)^n$ (n being an integer) added to Eq. (2.51), which is basically what the series expansion of the aperture admittance suggests, requires that one additional calibration standard be measured.

One model that has been used by some [105,116] to expand the linear model of Eq. (2.51) and extend its validity to the higher frequencies (radiation region), is

$$Y_S(f, \varepsilon) = G + jB = G_0(f)\varepsilon^{2.5} + j\omega(\varepsilon C_0 + C_f) \tag{2.67}$$

where G_0 is in general a function of f, and the coefficient of the lowest-order term that can added to account for the radiation effect [94,113−115]. Consequently, to incorporate the radiation effect, Eq. (2.64) is replaced by [105]

$$Y'_S(f, \varepsilon) = \varepsilon + G_n\varepsilon^{2.5} \tag{2.68}$$

where $G_n = G_0/(j\omega\varepsilon C_0)$. In addition, Eq. (2.65) is consequently replaced by

$$G_n\varepsilon^{2.5} + \varepsilon + \frac{\Delta_{M1}\Delta_{32}Y'_{S3}Y'_{S2} + \Delta_{M2}\Delta_{13}Y'_{S1}Y'_{S3} + \Delta_{M3}\Delta_{21}Y'_{S2}Y'_{S1}}{\Delta_{M1}\Delta_{32}Y'_{S1} + \Delta_{M2}\Delta_{13}Y'_{S2} + \Delta_{M3}\Delta_{21}Y'_{S3}} = 0 \tag{2.69}$$

Since G_n is unknown, Marsland and Evans [105] suggest using a fourth standard ($\varepsilon = \varepsilon_4$) for finding G_n, that can be found by using Eq. (2.69) by replacing subscript "M" by 4. Eq. (2.69) can then be used again to find the permittivity of the unknown sample ε from its measured reflection coefficient Γ_M.

2.3.3 TWO-PORT METHODS

For higher frequency measurements up to millimeter wave, for thin samples, or for liquids in microfluidic structures, using the open-ended coaxial line may not be the best approach. Coaxial transmission lines with typical sizes suffer from the presence of higher-order modes at millimeter wave frequencies. Furthermore, there are a few sources of error when establishing a good interface with the tissue, such as the existence of random air gaps, improper positioning of the probe, variations of the probe's contact pressure, or accumulation of dried tissue at the probe tip. On the other hand, in recent years, researchers have turned to planar type transmission lines, often designed for two-port measurements, particularly at the higher microwave and millimeter wave frequencies. Moreover, since powerful two-port calibration techniques such as Through-Reflect-Line (TRL) [125,126] have emerged, two-port measurements have lent themselves very well to accurate permittivity measurement. Besides providing reflection coefficients (S_{11} and S_{22}), two-port measurements also provide transmission measurements (S_{21} and S_{12}, where $S_{21} = S_{12}$ because of the reciprocity).

A two-port structure that is designed for permittivity measurement may consist of either a continuing section of transmission line that can be partly loaded with a sample (Fig. 2.18A), or two pieces (of the same type) of transmission lines that are

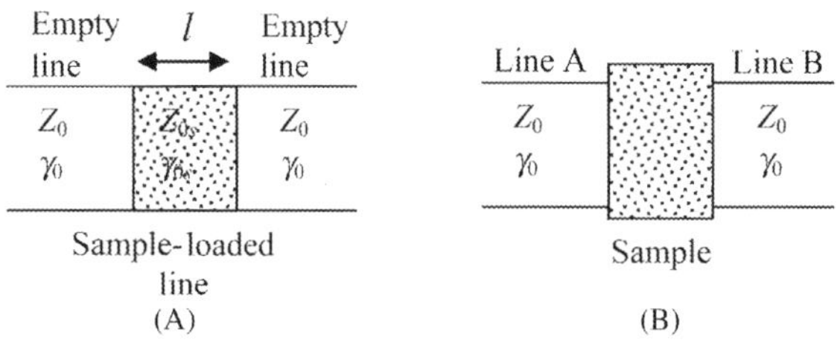

FIGURE 2.18

(A) A section of a transmission line with the characteristic impedance Z_0 and the propagation constant γ_0 is loaded with a sample, causing the filled line's characteristic impedance Z_{0s} and propagation constant γ_{0s} to change from their values for the unloaded case. (B) Transmission lines A and B are coupled through a sample.

electrically (electromagnetically) coupled to one another, with the sample placed in the coupling region between the two (Fig. 2.18B). The two lines can be coupled electromagnetically or connected through a passive circuit. The circuit may include a capacitor whose capacitance would be changed when loaded with tissue. The scattering parameters in both cases would change through loading with the sample. Although any of the scattering parameters can be used for complex permittivity extraction, transmission coefficient S_{21} is likely more appropriate than the reflection coefficients S_{11} and S_{22}. This is because the transmission coefficient is more sensitive to the permittivity variation than the reflection coefficient, and this leads to the more accuracy (less uncertainty) of the measured permittivity [80,127].

Examples of methods proposed for two-port permittivity measurements include that of Belhadj-Tahar et al. [128] for frequencies from 45 MHz to 18 GHz (a gap built in a coaxial cable is filled with the sample), sample loaded discontinuities in microstrip (for $1-7$ GHz) and various waveguides (for $8-26$ GHz) reported by Abdulnour et al. [129], and a microstrip line with the sample laying over it for measurements from 45 MHz to 14 GHz (Queffelec and Gelin [130]). None of the above-mentioned examples was reported for biological material measurements. However, these, along with examples of two-port measurements for biological materials discussed later in this chapter, exemplify various approaches for the two-port measurement.

Because of the complexity of the calibration procedure and permittivity extraction, methods that could offer the least degree of complexity are desirable. One such method is to extract the complex permittivity directly from the measured propagation constant of a transmission line filled with a sample (Fig. 2.18A) [131−136]. The sample-loaded transmission line should be uniform across its length. Moreover, it should only operate with its dominant mode at the operating frequency. This method follows the same steps taken in a TRL calibration procedure in order to find the propagation constant of a transmission line from the scattering parameter measurement of the sample-loaded transmission line with two different lengths. The lengths of the two lines are denoted by $l = l_1$ and $l = l_2$ ($l_1 > l_2$), with $\Delta l = l_1 - l_2$ being the difference in the lengths. The two-port scattering parameters of the two sample-filled lines are then measured. It is customary to utilize transmission wave matrices ($[\mathbf{T}]$) rather than $[\mathbf{S}]$ matrices, where

$$[\mathbf{T}] = \begin{bmatrix} T_{11} & T_{12} \\ T_{21} & T_{22} \end{bmatrix} = \frac{1}{S_{21}} \begin{bmatrix} S_{21}S_{12} - S_{11}S_{22} & S_{11} \\ -S_{22} & 1 \end{bmatrix} \tag{2.70}$$

In addition, for the two lines [135]

$$[\mathbf{T}_{l1}] = \begin{bmatrix} e^{-\gamma l_1} & 0 \\ 0 & e^{\gamma l_1} \end{bmatrix} \tag{2.71a}$$

$$[\mathbf{T}_{l2}] = \begin{bmatrix} e^{-\gamma l_2} & 0 \\ 0 & e^{\gamma l_2} \end{bmatrix} \tag{2.71b}$$

It can be shown that [134,135], the propagation constant can be obtained from

$$\gamma = \frac{\ln(\Lambda)}{\Delta l} + \frac{j2n\pi}{\Delta l} \tag{2.72}$$

with

$$\Lambda = \frac{1}{2}\left(\lambda_1 + \frac{1}{\lambda_2}\right) \tag{2.73}$$

where λ_1 and λ_2 are the eigenvalues of the matrix $[\mathbf{T}_M]$, which are obtained from the measured transmission wave matrices $[\mathbf{T}_{M1}]$ and $[\mathbf{T}_{M2}]$ for lengths l_1 and l_2 respectively.

$$[\mathbf{T}_M] = [\mathbf{T}_{M1}][\mathbf{T}_{M2}^{-1}] \tag{2.74}$$

For the most reliable results, the length difference at the middle of the frequency band should be a quarter of a wavelength (wavelength of the transmission line loaded with the sample). $n > 0$ (n being an integer) in Eq. (2.72) is necessary to account for Δl values greater than half a wavelength. Zero n ($n = 0$) would correspond to Δl less than half a wavelength, and a larger nonzero n must be properly chosen to correctly identify β, as discussed in [134].

The complex permittivity would then be extracted from the propagation constant obtained from this procedure, also referred to as the line–line method by some authors [131–133]. The extraction process would be straightforward if close form relations are available for γ versus ε. This would be the case for simple transmission lines such as coaxial cables or rectangular waveguides. The latter (Fig. 2.19A–C) were used by Huynen and her colleagues [131–133] to measure the complex permittivity of lossy materials, liquids, and biological substances [131,132]. Other examples of reports using this method include articles by Janezic and Jargon [135], Wan et al. [136], and Roelvink et al. [134].

Roelvink et al. [134] propose a planar CPW section with length l_1 or l_2 that would be loaded with a sample under test (Fig. 2.19D). However, to avoid variations due to the imprecision of the sample length placed on the line, the feed lines are placed on the other side of the board, and the side facing the sample

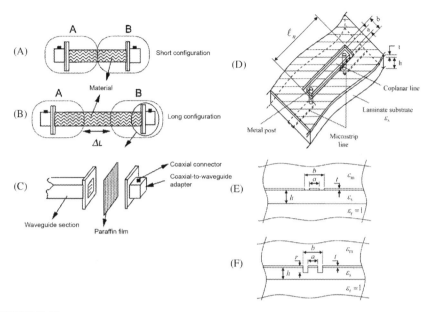

FIGURE 2.19

Examples of measurement fixtures for using the line—line method of complex permittivity measurement. Left: two waveguide lines with different lengths are used for measuring the complex permittivity of material filling (A) A shorter and (B) A longer line [133].
(C) Description of the interface with the sample-filled waveguide [133]. Right: (D) A planar coplanar waveguide (CPW) section with length l_1 or l_2 would be loaded by a sample under test [134]. Cross section of the loaded CPW (E) Without and (F) With recess [134].

With permission from IEEE.

is completely metalized except for the CPW region. This implies necessitating large sample sizes that extend well beyond the CPW region, as is the case for samples measured by the authors in [134]. Moreover, because the transmission lines were fabricated using a milling machine, not through microfabrication processing, recess as depicted in Fig. 2.19F is introduced. To be able to extract the complex permittivity of the sample from the measured γ (Eq. 2.72), close form quasi-static expressions are provided for relating the effective permittivity of the sample-loaded CPW (ε_{eff}) versus the sample's permittivity (ε_m), as well as the substrate's permittivity and other parameters in Fig. 3.19E and F. Such quasi-static expressions for a CPW loaded with dielectric layers are obtained from conformal mapping, involve terms with the complete elliptic integral, and are well- known from past literature [137,138]. The limitation of a quasi-static formulation's ability to fully account for the frequency variations and the lack of miniaturization (microfabrication) cause errors in higher frequencies, which may explain the limited frequency range of the reported data up to 5 GHz in [134].

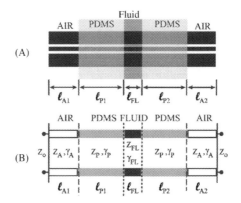

FIGURE 2.20

(A) Schematic and (B) Transmission line models of the microfluidic structure reported by Booth et al. [140].

With permission from IEEE.

Microfabricated CPW structures for two-port measurements of biological cells and small sample solutions have gained popularity in recent years. Seo et al. [139] use CPW on Pyrex substrate for measuring the living cell suspension from 1 to 32 GHz. Facer et al. [11] use CPW on glass for measuring solutions of protein and nucleic acid from 40 Hz to 26.5 GHz. Greiner et al. [12] report CPW on quartz for measuring human umbilical vein endothelial cell suspension. One of the main motivations for such studies, which have been receiving a great deal of attention in recent years, has been the label-free detection of biological substances. Some studies using similar sensors and for similar purposes, but utilizing resonating structures, have been mentioned earlier in this chapter.

One advantage of a microfabricated CPW is the capability for integration with microfluidic structures. Booth and his colleagues [140] report a CPW structure with microfluidic channels for measuring liquids of nanoliter quantities up to 40 GHz (Fig. 2.20). Microfluidic channels are based on polydimethylsiloxane (PDMS). The CPW substrate is quartz. The equivalent transmission line model (Fig. 2.20B) would consequently be a cascade of five transmission line sections. The overall transmission wave matrix is [140]

$$\mathbf{T}_M = \left[\mathbf{Q}_{Z_A}^{Z_0}\right]\left[\mathbf{T}_{l_{A1}}\right]\left[\mathbf{Q}_{Z_P}^{Z_A}\right]\left[\mathbf{T}_{lP1}\right]\left[\mathbf{Q}_{Z_{FL}}^{Z_P}\right]\left[\mathbf{T}_{l_{FL}}\right]\left[\mathbf{Q}_{Z_P}^{Z_{FL}}\right]\left[\mathbf{T}_{lP2}\right]\left[\mathbf{Q}_{Z_A}^{Z_P}\right]\left[\mathbf{T}_{l_{A2}}\right]\left[\mathbf{Q}_{Z_0}^{Z_A}\right] \tag{2.75}$$

where $\left[\mathbf{Q}_{Z_m}^{Z_n}\right]$ is an impedance transformer matrix

$$\left[\mathbf{Q}_{Z_m}^{Z_n}\right] = \frac{1}{2Z_m}\left|\frac{Z_m}{Z_n}\right|\sqrt{\frac{\mathrm{Re}(Z_n)}{\mathrm{Re}(Z_m)}}\begin{bmatrix} Z_m + Z_n & Z_m - Z_n \\ Z_m - Z_n & Z_m + Z_n \end{bmatrix} \tag{2.76}$$

$Z_0 = 50\ \Omega$ is the reference impedance, as well as the impedance of the CPW probes, used for measuring the S-parameters.

Authors in [140] use a wafer probe station and TRL calibration to set the reference plane on the device substrate. They also use the well-known lumped element model for the characteristic impedance and the propagation constant of each section.

$$\gamma = \sqrt{(R + j\omega L)(G + j\omega C)} \tag{2.77}$$

$$Z_c = \sqrt{\frac{R + j\omega L}{G + j\omega C}} \tag{2.78}$$

where R, G, L, and C are per unit length quantities and may also be frequency dependent. The process of determining the complex permittivity from the measurement and the model of Fig. 2.20B is lengthy and is only highlighted here. Using the quasi-static assumption, the authors independently calculate the resistance, inductance, capacitance, and conductance per unit length. In addition, it is assumed that the conductance and capacitance per unit length of the fluid-filled section change with its permittivity [140]. Two-dimensional simulations are then performed to establish the relationship between these circuit parameters and the complex permittivity of the fluid. The circuit parameters of other regions are obtained through a combination of measurement and simulation as further articulated in [140]. To assist this, they also measure two additional transmission lines, one with no loading (air), and the other one being a combination of air and the PDMS-loaded line.

The method just mentioned utilizes the straightforward relation between the circuit elements of the sample-loaded line and its permittivity, obtained through simulation. The line—line method discussed earlier benefits from the direct relationship that exists between the propagation constant of the sample-loaded line and its permittivity, obtained by close form relations or simulation.

The permittivity extraction procedure may not be as straightforward if two separate transmission lines are coupled through the tissue (Fig. 2.18B), where a close form relation or circuit model for expressing the relationship between the permittivity and measured parameters such as S_{21} or γ may not exist. In such cases, numerical simulations are needed to create lookup tables or empirical relations that relate the measured scattering parameters with the complex permittivity of the sample under test. Tofighi and Daryoush report a two-port microstrip test fixture (Fig. 2.21) to measure the complex permittivity of brain gray and white matter [55,141] and neurological cell solutions [142] from 10 to 50 GHz. Two open-ended microstrip transmission lines on fused silica substrates are coupled to one another through two small apertures. The sample, sandwiched between two glass plates, is inserted between the microstrip ground planes. Using a TRL calibration, reference measurement planes (planes A-A' and B-B') are set at the middle of the apertures. Rational functions are obtained to fit the simulated S_{21} versus the complex permittivity data at each frequency. Details of the calibration and extraction procedures are discussed in [55,127,141]. One advantage of this arrangement is that it allows the top and bottom microstrip lines to simultaneously

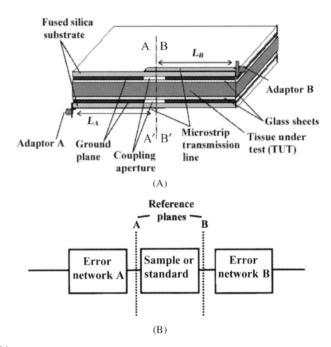

FIGURE 2.21

(A) Schematic of the microstrip test fixture utilized by Tofighi and Daryoush for measuring the complex permittivity of brain gray and white matter [55,127,141] and neurological cell solutions [142], as well as near-filed brain imaging [143]. (B) Error networks would be de-embedded using a TRL calibration, with reference planes set at the middle of the two apertures.

With permission from IEEE.

move along x and y directions, while the sample container remains fixed. This property has been exploited to perform imaging of rat brain slices with mm resolution at millimeter wave frequencies up to 50 GHz, as explained in [143]. This achievement would be at the expense of additional complexity for the extraction procedure. Since no straightforward model of the coupling between the two lines could be established, extensive 3-dimensional full-wave simulation followed by curve fitting for simulated S_{21} versus complex permittivity data is required.

2.4 CONCLUSION

Applications of RF and microwave in medicine and biology are diverse, as exemplified by what will be covered in the future chapters of this book or found in many other articles and books. Many of those applications require that the interaction between EM waves and biological materials be characterized. This often necessitates having proper knowledge of the complex permittivity of these materials. Complex permittivity of biological substances is dispersive and is often modeled by Debye or Cole—Cole models as described in this chapter. Due to the lossy and dispersive nature of biological materials, many methods for complex permittivity measurement, some of which are discussed in this chapter, have been developed. Methods of measurement of biological substances at microwave and millimeter wave frequencies also require accurate calibration procedures and modeling of the pertinent EM problems. Lessons learned by using appropriate sensing structures, measurement procedures, and EM field modeling for complex permittivity measurement would be also applicable in many other cases in which an EM probe, applicator, or antenna is in the proximity of or in contact with tissue and should interact with it. Thus, the understanding of tissue dispersion and issues related to interfacing with tissue through sensing probes, such as the coaxial, microstrip, and CPW transmission lines discussed in this chapter, should provide a useful perspective, beneficial in many medical and biological applications of RF and microwave.

REFERENCES

[1] Schwan HP. Electrical properties of tissues and cell suspensions. Adv Phys Med Biol 1957;5:147—209.
[2] Schwan HP, Foster KR. RF-field interactions with biological systems: electrical properties and biophysical mechanism. Proc IEEE 1980;68:104—13.
[3] Schwan HP. Dielectric properties of biological tissues and biophysical mechanisms of electromagnetic field interaction. In: Illinger KH, editor. Biological effects of non-ionizing radiation. Washington DC: ACS Symposium Series; 1981. p. 109—31.
[4] Johnson CC, Guy AW. Nonionizing electromagnetic wave effects in biological materials and systems. Proc IEEE 1972;60:692—718.
[5] Guy AW. History of biological effects and medical applications of microwave energy. IEEE Trans Microw Theory Tech 1984;32(9):1182—200.
[6] Polk C, Postow E, editors. CRC handbook of biological effects of electromagnetic fields. Boca Raton, FL: CRC Press; 1986.
[7] Rosen A, Rosen H, editors. New frontiers in medical device technologies. New York: John Wiley; 1995.
[8] Vander Vorst A, Rosen A, Kotsuka Y. RF/microwave interaction with biological tissues. Hoboken, NJ: Wiley; 2006.
[9] Rosen A, Stuchly MA, Vorst AV. Applications of RF/microwave in medicine. IEEE Trans Microw Theory Tech 2002;50(3):963—74.
[10] Gandhi OP, Lazzi G, Furse CM. Electromagnetic absorption in the human head and neck for mobile telephones at 835 and 1900 MHz. IEEE Trans Microw Theory Techn 1996;44(10):1884—97.
[11] Facer GR, Notterman DA, Sohn LS. Dielectric spectroscopy for bioanalysis: from 40 Hz to 26.5 GHz in a microfabricated waveguide. Appl Phys Lett 2001;8(7): 996—8.

[12] Grenier K, Dubuc D, Poleni PE. New broadband and contactless RF/microfluidic sensor dedicated to bioengineering. Digest of 2009 IEEE Int Microw Symp 2009;1329−32.

[13] Kim YI, Park TS, Kang JH, Lee MC, Kim JT, Park JH, et al. Biosensors for label free detection based on RF and MEMS technology. Sens Actuators 2006;592−9.

[14] Rosen A, Vander Vorst A. Special issue on medical application and biological effects of RF/microwaves. IEEE Trans Microw Theory Tech 1996;44(10): Part II, pp. 1753−1973.

[15] Rosen A, Vander Vorst A. Special issue on medical applications and biological effects of RF/microwaves. IEEE Trans Microw Theory Tech 2000; 48(11): Parts I & II, pp. 1781−2198.

[16] Lazzi G, Gandhi OP, Ueno S. Special issue on medical applications and biological effects of RF/microwaves. IEEE Trans Microw Theory Tech 2004; 52(8): Parts II, pp. 1853−2084.

[17] Rosen A, Liu WLG, Tofighi MR. "Editorial," special issue on RF and microwave techniques in wireless implants and biomedical applications. IEEE Trans Microw Theory Tech 2009;57(10):2477−9.

[18] Tofighi MR, Chiao J-C. "Editorial," special issue on biomedical applications of RF/microwave technologies. IEEE Trans Microw Theory Tech 2013;61(5):1977−9.

[19] Chiao J-C, Kissinger D. Medical applications of RF and microwaves—therapy and safety. IEEE Microw Mag 2015;16(2):12−84.

[20] Chiao J-C, Kissinger D. Medical applications of radio-frequency and microwaves—sensing, monitoring, and diagnostics. IEEE Microw Mag 2015;16(4):28−113.

[21] Frohlich H. Theory of dielectrics, dielectric constants, and dielectric loss. Amen House, London: Oxford University Press; 1958.

[22] Daniel VV. Dielectric relaxation. London: Academic Press; 1967.

[23] Jonscher AK. Dielectric relaxation in solids. J. Phys D: Appl Phys 1999;32:R57−70.

[24] Feldman Y, Puzenko A, Ryabov Y. Dielectric relaxation phenomena in complex materials. In: Coffey WT, Kalmykov YP, editors. Fractals, diffusion, and relaxation in disordered complex systems: advances in chemical physics, part A, 133. Hoboken, NJ: Wiley; 2006.

[25] Pethig R. Dielectric and electronic properties of biological materials. New York: John Wiley & Sons; 1979.

[26] Pethig R, Kell DB. The passive electrical properties of biological systems: their significance in physiology. Biophys Biotechnol, Phys Med Biol 1987;32(8):933−70.

[27] Schwan HP. Electrical properties of tissues and cell suspensions: mechanisms and models. Proc of 16th annual international conference of the IEEE engineering in medicine and biology society 1994 1, pp. A70−A71.

[28] Schwan HP. Electrode polarization impedance and measurements in biological materials. Ann N Y Acad Sci 1968;148:191−209.

[29] Ben Ishai P, Sobol Z, Nickels JD, Agapov AL, Sokolov AP. An assessment of comparative methods for approaching electrode polarization in dielectric permittivity measurements. Rev Sci Instrum 2012;83. pp. 083118-1−083118-1.

[30] Pizzitutti F, Bruni F. Electrode and interfacial polarization in broadband dielectric spectroscopy measurements. Rev Sci Instrum 2001;72(5):2502−4.

[31] Vaughan WE. Dielectric relaxation. Annu Rev Phys Chem 1979;30:103−24.

[32] Von Hippel A. The dielectric relaxation spectra of water, ice, and aqueous solutions, and their interpretation: 1. Critical survey of the status-quo for water. IEEE Trans Electr Insul 1988;23(5):801−16.

[33] Von Hippel A. The dielectric relaxation spectra of water, ice, and aqueous solutions, and their interpretation: 2. Tentative interpretation of the relaxation spectrum of water in the time and frequency domain. IEEE Trans Electr Insul 1988;23(5):817–23.

[34] Malmberg CG, Maryott AA. Dielectric constant of water from 0 to 100° C. J Res Natl Bur Stand (1934) 1956;56(1):1–7.

[35] Tables of Dielectric Dispersion Data for Pure Liquids and Dilute Solutions, National Bureau of Standards Circular 589, November 1958.

[36] Grant EH, Buchanan TJ, Cook HF. Dielectric behavior of water at microwave frequencies. J Chem Phys 1957;26:156–61.

[37] Stogryn A. Equations for calculating the dielectric constant of saline water. IEEE Trans Microw Theory Tech 1971;19(8):733–6.

[38] Cole KS, Cole RH. Dispersion in dielectrics; I. Alternating current characteristics. J Chem Phys 1941;9:341–51.

[39] Van Gemert MJC. A note on the Cole—Cole dielectric permittivity equation in connection with causality. Chem Phys Lett 1972;14(5):606–8.

[40] Cook HF. The dielectric behavior of some types of human tissues at microwave frequencies. Br J Appl Phys 1951;2:295–300.

[41] Foster KR, Schepps JL, Schwan HP. Microwave dielectric relaxation in tissue: a second look. Biophys J 1980;29:271–81.

[42] Grant EH, Keefe SE, Takashima S. The dielectric behavior of aqueous solutions of bovine serum albumin from radiowave to microwave frequencies. J Phys Chem 1968;72:4373–80.

[43] Grant EH, Szwarnowski S, Sheppard RJ. Dielectric properties of water in microwave and far-infrared regions. In: Illinger KH, editor. Biological effects of nonionizing radiation. Washington D.C.: ACS Symposium Series; 1981. p. 47–56.

[44] Grant EH, Nightingale RV, Sheppard RJ. Dielectric properties of water in myoglobin solution. In: Illinger KH, editor. Biological effects of nonionizing radiation. Washington D.C.: ACS Symposium Series; 1981. p. 57–62.

[45] Burdette EC, Cain FL, Seals J. *In vivo* probe measurement technique for determining dielectric properties at VHF through microwave frequencies. IEEE Trans Microw Theory Tech 1980;28(4):414–24.

[46] Kraszewski A, Stuchly MA, Stuchly SS, Smith M. *In vivo* and *in vitro* dielectric properties of animal tissues at radio frequencies. Bioelectromagnetics 1982;3:421–32.

[47] Stuchly MA, Stuchly SS. Dielectric properties of biological substances-tabulated. J Microw Power 1980;1(15):19–26.

[48] Duck FA. Physical properties of tissue: a comprehensive reference book. London: Academic Press; 1990.

[49] Gabriel C, Gabriel S, Corthout E. The dielectric properties of biological tissues: I. Literature survey. Phys Med Biol 1996;41:2231–49.

[50] Gabriel S, Lau RW, Gabriel C. The dielectric properties of biological tissues: II. Measurements in the frequency range 10 Hz to 20 GHz. Phys Med Biol 1996;41:2251–69.

[51] Gabriel S, Lau RW, Gabriel C. The dielectric properties of biological tissues: III. Parametric models for the dielectric spectrum of tissues. Phys Med Biol 1996;41:2271–93.

[52] Tofighi MR. FDTD modeling of biological tissues Cole—Cole dispersion for 0.5 to 30 GHz using relaxation time distribution samples—novel and improved implementations. IEEE Trans Microw Theory Tech 2009;57(10):2588–96.

[53] Gandhi OP, Gao B-Q, Chen J-Y. A frequency-dependent finite difference time-domain formulation for general dispersive media. IEEE Trans Microw Theory Tech 1993;41(4):658–65.

[54] Bao J, Lu S, Hurt WD. Complex dielectric measurements and analysis of brain tissues in the radio and microwave frequencies. IEEE Trans Microw Theory Tech 1997;45(10):1730–40.

[55] Tofighi MR, Daryoush AS. Characterization of the complex permittivity of brain tissues up to 50 GHz utilizing a two-port microstrip test fixture. IEEE Trans Microw Theory Tech 2002;50(10):2217–25.

[56] O'Rourke AP, Lazebnik M, Bertram JM, Converse MC, Hagness SC, Webster JG, et al. Dielectric properties of human normal, malignant and cirrhotic liver tissue: *in vivo* and *ex vivo* measurements from 0.5 to 20 GHz using a precision open-ended coaxial probe. Phys Med Biol 2007;52:4707–19.

[57] Li X, Hagness SC. A confocal microwave imaging algorithm for breast cancer detection. IEEE Microw Wireless Compon Lett 2001;11:130–2.

[58] Lazebnik M, McCartney M, Popovic D, Watkins CB, Lindstrom MJ, Harter J, et al. A large-scale study of the ultrawideband microwave dielectric properties of normal breast tissue obtained from reduction surgeries. Phys Med Biol 2007;52:2637–56.

[59] Lazebnik M, Okoniewski M, Booske JH, Hagness SC. Highly accurate debye models for normal and malignant breast tissue dielectric properties at microwave frequencies. IEEE Microw Wireless Compon Lett 2007;17(12):822–4.

[60] Said T, Varadan VV. The role of the concentration and distribution of water in the complex permittivity of breast fat tissue. Bioelectromagnetics 2009;30:669–77.

[61] Sugitani T, Kubota S-I, Kuroki S-I, Sogo K, Arihiro K, Okada M, et al. Complex permittivities of breast tumor tissues obtained from cancer surgeries. Appl Phys Lett 2014;104. pp. 253702-1–253702-1.

[62] Martellosio A, Pasian M, Bozzi M, Perregrini L, Mazzanti A, Svelto F, et al. 0.5–50 GHz dielectric characterization of breast cancer tissues. Electron Lett 2015;51(13):974–5.

[63] Romeo S, Di Donato L, Bucci OM, Catapano I, Crocco L, Scarfi MR, et al. Dielectric characterization study of liquid-based materials for mimicking breast tissues. Microw Opt Technol Lett 2011;53(6):1276–80.

[64] Fear EC, Sill J, Stuchly MA. Experimental feasibility study of confocal microwave imaging for breast tumor detection. IEEE Trans Microw Theory Tech 2003;51(3): 887–92.

[65] Fear EC, Li X, Hagness SC, Stuchly MA. Confocal microwave imaging for breast tumor detection: localization of tumors in three dimensions. IEEE Trans Biomed Eng 2002;49:812–22.

[66] Bao J-Z, Davis CC, Swicord ML. Microwave dielectric measurements of erythrocyte suspensions. Biophys J 1994;66:2172–80.

[67] Pottel R, Gopel KD, Henze R, Kaatze U, Uhlendorf. The dielectric permittivity spectrum of aqueous colloidal phospholipid solutions between 1 KHz and 60 GHz. Biophys Chem 1984;19:233–44.

[68] Stuchly MA, Stuchly SS, Liburdy RP, Rousseau DA. Dielectric properties of liposome vesicles at the phase transition. Phys Med Biol 1988;33:1309–23.

[69] Klosgen B, Reichle C, Kohlsmann S, Kramer KD. Dielectric spectroscopy as a sensor of membrane headgroup mobility and hydration. Biophys J 1996;71:3251–60.

[70] Ebara H, Tani K, Onishi T, Uebayashi S, Hashimoto O. Method for estimating complex permittivity based on measuring effective permittivity of dielectric mixtures in radio frequency band. IEICE Trans Commun 2005;E88-B(8):3269–74.

[71] Merla C, Liberti M, Apollonio F, d'Inzeo G. Quantitative assessment of dielectric parameters for membrane lipid bi-layers from RF permittivity measurements. Biolectromagnetics 2009;30:286–98.

[72] Merla C, Liberti M, Apollonio F, Nervi C, D'Inzeo G. A microwave microdosimetric study on blood cells: estimation of cell membrane permittivity and parametric EM analysis. Digest of 2009 IEEE International Microwave Symposium 2009, pp. 601–4, Boston, USA.

[73] Westphal WB. Dielectric measuring techniques. In: Von Hippel AR, editor. Dielectric material and applications. New York: Wiley; 1954. p. 63–122.

[74] Fox J, Sucher M. Handbook of microwave measurements. New York: Polytechnique Institute of Brooklyn; 1954.

[75] Bussy HE. Measurement of RF properties of materials, a survey. Proc IEEE 1967;55(6):1046–53.

[76] Grant EH, Sheppard RJ, South GP. Dielectric behavior of biological molecules in solution. Oxford: Oxford University Press; 1978.

[77] Stuchly MM, Stuchly SS. Coaxial line reflection methods for measuring dielectric properties of biological substances at radio and microwave frequencies—a review. IEEE Trans Instrum Meas 1980;29(3):176–83.

[78] Afsar MN, Birch JR, Clarke RN. The measurement of the properties of materials. Proc IEEE 1986;74(1):183–99.

[79] Chen LF, Ong CK, Neo CP, Varadan VV, Varadan VK. Microwave electronics: measurement and materials characterization. New York: Wiley; 2004.

[80] Tofighi MR, Daryoush AS. Measurement techniques for electromagnetic characterization of biological materials. In: Bansal R, editor. Chapter 18 of Handbook of engineering electromagnetics. New York: Marcel Dekker; 2004. p. 631–76.

[81] Gregory AP, Clarke RN. A review of RF and microwave techniques for dielectric measurements on polar liquids. IEEE Trans Dielectr Electr Insul 2006;13(4):727–42.

[82] Kaatze U. Measuring the dielectric properties of materials. Ninety-year development from low-frequency techniques to broadband spectroscopy and high-frequency imaging. Meas Sci Technol 2013;24 012005 (31pp).

[83] Harrington RF. Time-harmonic electromagnetic fields. New York: McGraw-Hill; 1961.

[84] Waldron RA. Theory of strip-line cavity for measurement of dielectric constants and gyromagnetic-resonance line-width. IEEE Trans Microw Theory Tech 1964;12(1):123–31.

[85] Jones CA, Kantor Y, Grosvenor JH, Janezic MD. Striple Resonator for Electromagnetic Measurements of Materials, NIST Technical Note 1505, National Institute of Standard and Technology, Boulder, Colorado, 1998.

[86] Jones CA. Permittivity and permeability measurements using stripline resonator cavities—a comparison. IEEE Trans Instrum Meas 1999;40(4):843–8.

[87] Lakshminarayana MR, Partain LD, Cook WA. Simple microwave technique for independent measurement of sample size and dielectric constant with results for a gunn oscillator system. IEEE Trans Microw Theory Tech 1979;27(7):661–5.

[88] Parkash A, Vaid JK, Mansingh A. Measurement of dielectric parameters at micro-wave frequencies by cavity perturbation technique. IEEE Trans Microw Theory Tech 1979;27(9):791−5.

[89] Fenske K, Misra D. Dielectric materials at microwave frequencies. Appl Microw Wireless 2000;12(10):92−100.

[90] Land DV, Campbell AM. A quick accurate method for measuring the microwave dielectric properties of small tissue samples. Phys Med Biol 1992;37(1):183−92.

[91] Carter RG. Accuracy of microwave cavity perturbation measurements. IEEE Trans Microw Theory Tech 2001;49(5):918−23.

[92] Wang Z, Che W, Chang Y. Permittivity measurement of biological materials with improved microwave cavity perturbation technique. Microw Opt Technol Lett 2008;50(7):1800−4.

[93] Tanabe E, Joines WT. A nondestructive method for measuring the complex permit-tivity of dielectric materials at microwave frequencies using an open transmission line resonator. IEEE Trans Instrum Meas 1978;25(3):222−6.

[94] Xu D, Liu L, Jiang Z. Measurement of the dielectric properties of biological sub-stances using and improved open-ended coaxial line resonator method. IEEE Trans Microw Theory Tech 1987;35(12):82−6.

[95] Jones RG. Precise dielectric measurements at 35 GHz using an open microwave res-onator. Proc IEE 1976;123(4):285−90.

[96] Clarke RN, Rosenberg CB. Fabry-Perot and Open Resonators at Microwave and Millimeter Wave Frequencies, 2−300 GHz. J Phys E: Sci Instrum 1982;15:9−24.

[97] Hirvonen TM, Vainikainen P, Lozowski A, Raisanen A. Measurement of dielectrics at 100 GHz with an open resonator connected to a network analyzer. IEEE Trans Microw Theory Tech 1996;45(4):780−6.

[98] Afsar MN, Huachi X. An automated 60 GHz open resonator system for precision dielectric measurement. IEEE Trans Microw Theory Tech 1990;38(12):1845−53.

[99] Dalmay C, Cheray M, Pothier A, Lalloué F, Jauberteau MO, Blondy P. Ultra sensi-tive biosensor based on impedance spectroscopy at microwave frequencies for cell scale analysis. Sens Actuators A: Phys 2010;162(2):189−97.

[100] Zhang LY, Bounaix Morand du Puch C, Lacroix A, Dalmay C, Pothier A, Lautrette C, et al. Microwave biosensors for identifying cancer cell aggressiveness grade. Digest of 2012 IEEE International Microwave Symposium 2012; Montreal, Canada.

[101] Landoulsi A, Zhang LY, Dalmay C, Lacroix A, Pothier A, Bessaudoul A, et al. Tunable frequency resonant biosensors dedicated to dielectric permittivity analysis of biological cell cytoplasm. Digest of 2013 IEEE International Microwave Symposium 2013, Seattle, WA.

[102] Athey TW, Stuchly MA, Stuchly SS. Measurement of radio frequency permittivity of biological tissues with an open-ended coaxial line: part I. IEEE Trans Microw Theory Tech 1982;30:82−6.

[103] Wei YZ, Sridhar S. Technique for measuring the frequency dependent complex dielectric constant of liquids up to 20 GHz. Rev Sci Instrum 1989;60:3041−6.

[104] Wei Y, Sridhar S. Radiation-corrected open-ended coaxial line technique for dielec-tric measurements of liquids up to 20 GHz. IEEE Trans Microw Theory Tech 1991;39:526−31.

[105] Marsland TP, Evans S. Dielectric measurements with an open-ended coaxial probe. Microw, Antennas Propagation, IEE Proc H 1987;134(4):341−9.

[106] Rzepecka MA, Stuchly SS. A lumped element capacitance method for the measurement of the permittivity and conductivity in the frequency and time domain—a further analysis. IEEE Trans Instrum Meas 1975;24(1):27–32.

[107] Stuchly SS, Rzepecka MA, Iskander MF. Permittivity measurements at microwave frequencies using lumped elements. IEEE Trans Instrum Meas 1974;23(1):56–62.

[108] Iskander MF, Stuchly SS. Fringing field effect in lumped-capacitance method for permittivity measurement. IEEE Trans Instrum Meas 1978;27:107–9.

[109] Kraszewski A, Stuchly SS, Stuchly MA, Symons S. On measurement accuracy of the tissue permittivity in-vivo. IEEE Trans Instrum Meas 1983;32(1):37–42.

[110] Stuchly MA, Athey TW, Samaras TW, George M, Taylor GE. Measurement of radio frequency permittivity of biological tissues with an open-ended coaxial line: part II—experimental results. IEEE Trans Microw Theory Tech 1982;30(1):87–92.

[111] Kraszewski A, Stuchly SS. Capacitance of open-ended dielectric field coaxial lines-experimental results. IEEE Trans Instrum Meas 1983;32(4):517–19.

[112] Gajda G, Stuchly SS. An equivalent circuit of an open-ended coaxial line. IEEE Trans Microw Theory Tech 1983;31(5):380–4.

[113] Misra DK. A quasi-static analysis of open coaxial lines. IEEE Trans Microw Theory Tech 1987;35:925–38.

[114] Staebell KF, Misra D. An experimental technique for *in vivo* permittivity measurement of materials at microwave frequencies. IEEE Trans Microw Theory Tech 1990;38(3):337–9.

[115] Misra DM, Chabbra M, Epstein BR, Mirotznik M, Foster KR. Noninvasive electrical characterization of materials at microwave frequencies using an open-ended coaxial line: test of an improved calibration technique. IEEE Trans Microw Theory Tech 1990;38(1):8–13.

[116] Nyshadham A, Sibbald CL, Stuchly SS. Permittivity measurements using open-ended sensors and reference liquid calibration—an uncertainty analysis. IEEE Trans Microw Theory Tech 1992;40(2):305–14.

[117] Stuchly SS, Sibbald CL, Anderson JM. A new admittance model for open-ended waveguides. IEEE Trans Microw Theory Tech 1994;42(2):192–8.

[118] Anderson JM, Sibbald CL, Stuchly SS. Dielectric measurements using a rational function model. IEEE Trans Microw Theory Tech 1994;42(2):199–200.

[119] Baker-Jarvis J, Janezic MD, Domich PD, Geyer RG. Analysis of an open-ended coaxial probe with lift-off for nondestructive testing. IEEE Trans Instrum Meas 1994;43(5):711–18.

[120] Gabriel C, Chan TYA, Grant EH. Admittance models for open ended coaxial probes and their place in dielectric spectroscopy. Phys Med Biol 1994;39:2183–99.

[121] Okoniewski O, Anderson JA, Okoniewska E, Gupta K, Stuchly SS. Further analysis of open-ended sensors. IEEE Trans Microw Theory Tech 1995;43(8):1986–9.

[122] Lane R. De-embedding device scattering parameters. Microw J 1984;149–56.

[123] Hager N, Domszy RC, Tofighi MR. Smith-chart diagnostics for multi-GHz time-domain-reflectometry dielectric spectroscopy. Rev Sci Instrum 2012;83(2), 025108 (1-11).

[124] Hager NE, Domszy RC, Tofighi MR. Multi-GHz monitoring of cement hydration using time-domain-reflectometry dielectric spectroscopy. In: 4th International Symposium on Soil Water Measurement using Capacitance, Impedance and TDT; 2014.

[125] Eagen GF, Hoer CA. Thru-reflect-line: an improved technique for calibrating the dual six-port automatic network analyzer. IEEE Trans Microw Theory Tech 1979;27(12):987−93.

[126] Soares RA, Gouzien P, Legaud P, Follot G. A unified approach to two-port calibration techniques and some applications. IEEE Trans Microw Theory Tech 1989;37 (11):1669−73.

[127] Tofighi MR, Daryoush AS. Biological tissue complex permittivity measured from S_{21}—error analysis and error reduction by reference measurements. IEEE Trans Instrum Meas 2009;58(7):2316−27.

[128] Belhadj-Tahar N, Fourrier-Lamer A, Chanterac H. Broad-band simultaneous measurement of complex permittivity and permeability using a coaxial discontinuity. IEEE Trans Microw Theory Tech 1990;38(1):1−7.

[129] Abdulnour J, Akyel C, Wu K. A generic approach for permittivity of dielectric materials using a discontinuity in a rectangular waveguide or a microstrip line. IEEE Trans Microw Theory Tech 1995;43(5):1060−6.

[130] Queffelec P, Gelin P. Influence of higher order modes on the measurements of complex permittivity and permeability of materials using a microstrip discontinuity. IEEE Trans Microw Theory Tech 1996;44(6):814−24.

[131] Duhamel F, Huynen I, Vander Vorst A. Measurements of complex permittivity of biological and organic liquids up to 110 GHz. In: 1997 IEEE International Microwave Symposium Digest 1997; 1: 107−110, Denver, CO.

[132] Fossion M, Huynen I, Vanhoenacker D, Vander Vorst A. A new and simple calibration method for measuring planar lines parameter up to 40 GHz. Proc. 22nd European Microwave Conference 1992: 180−5, Espoo, Finland.

[133] Huynen I, Steukers C, Duhamel F. A wideband line-line dielectrometric method for liquids, soils, and planar substrates. IEEE Trans Instrum Meas 2001;50 (5):1343−8.

[134] Roelvink J, Trabelsi S, Nelson SO. A planar transmission-line sensor for measuring the microwave permittivity of liquid and semisolid biological materials. IEEE Trans Instrum Meas 2013;62(11):2974−82.

[135] Janezic MD, Jargon JA. Complex permittivity determination from propagation constant measurements. IEEE Microw Guided Wave Lett 1999;9(2):76−8.

[136] Wan C, Nauwelaers B, De Raedt W, Van Rossum M. Two new measurements methods for explicit determination of complex permittivity. IEEE Trans Microw Theory Tech 1998;46(11):1614−19.

[137] Hillberg W. From approximations to exact relations for characteristic impedances. IEEE Trans Microw Theory Tech 1969;17(5):259−69.

[138] Raj A, Holmes WS, Judah SR. Wide bandwidth measurement of complex permittivity of liquids using coplanar lines. IEEE Trans Instrum Meas 2001;50 (4):905−9.

[139] Seo S, Stintzing T, Block I, Pavlidis D, Rieke M, Layer PG. High frequency wideband permittivity measurements of biological substances using coplanar waveguides and application to cell suspensions. Digest of 2008 IEEE International Microwave Symposium 2008: 915−8.

[140] Booth JC, Orloff ND, Mateu J, Janezic M, Rinehart M, Beall JA. Quantitative permittivity measurements of nanoliter liquid volumes in microfluidic channels to 40 GHz. IEEE Trans Microw Theory Tech 2010;59(12):3279−88.

[141] Tofighi MR, Daryoush AS. Comparison of two post-calibration correction methods for complex permittivity measurement of biological tissues up to 50 GHz. IEEE Trans Instrum Meas 2002;51(6):1170−6.

[142] Tofighi MR, Daryoush AS. Study of the activity of neurological cell solutions using complex permittivity measurement. Digest of 2002 IEEE International Microwave Symposium Digest 2002; 2: 1763−6, Seattle, WA.

[143] Tofighi MR, Daryoush AS. Near field microwave brain imaging. Electron Lett 2001;37(13):807−8.

Microwave cancer diagnosis

3

F. Töpfer and J. Oberhammer
KTH Royal Institute of Technology, Stockholm, Sweden

CHAPTER OUTLINE

3.1 INTRODUCTION

Microwaves have been successfully applied to the nondestructive testing and evaluation of dielectric materials and objects in a wide range of applications, including the examination of concrete and composite materials, porosity evaluation in polymers, moisture measurements of soil, and many more [1−3]. Electromagnetic measurement and imaging techniques have also been applied to biological tissue and the discovered difference in dielectric properties between healthy tissue and malignant tumor tissue [4−6] raised the hopes toward the application of microwave for medical diagnostics. Especially the potential

C. Li, M. Tofighi, D. Schreurs and T-Z. J. Horng (Eds): Principles and Applications of RF/Microwave
in Healthcare and Biosensing.

application of microwave imaging for the detection of early-stage malignant tumors in the human breast has served as a strong motivation for the development of advanced microwave imaging systems [7−9]. Besides the breast, other tumor sites such as the skin and the brain have been targeted by research toward microwave diagnostic applications. The nature of biological tissue and the complexity of the human body possess some considerable challenges for the application of microwaves, and microwave diagnostic systems have not yet entered to clinical practice. In biological tissue, microwaves are subjected to complex scattering and high absorption; signal levels fall to a fraction of their initial power after a distance of only a few wavelengths. As a consequence, the application of microwaves to the human body often comes with a trade-off between penetration depth and resolution, and the sites commonly targeted by microwave diagnostics and imaging are comparably small in size and easily accessible from the outside. High-resolution imaging of large portions of the human body using microwaves, as possible by X-ray imaging techniques, is impeded by the high absorption and the complex scattering of microwaves in tissue. Furthermore, tissue within the human body is highly heterogeneous and can exhibit a large variation in properties within an individual and between different individuals, which makes simulations and image reconstruction highly complex and computationally demanding. However, the advances in computational power and the improving performance of microwave components and devices ease some of the challenges that microwave diagnostic systems face, and the intensive work of researchers and engineers in the field has led to advanced systems for cancer diagnosis and imaging that have the potential to be used in the clinics in the future. This chapter starts with a summary of the relevant knowledge about the dielectric properties of different types of biological tissue and the differences in permittivity between healthy and malignant tumor tissue. Then, general aspects which are important for cancer diagnosis using microwaves are discussed before the most common techniques are explained in detail. Following that, several application areas are discussed, and the state of the art of microwave diagnostic systems in these areas is presented.

3.2 THE DIELECTRIC PROPERTIES OF BIOLOGICAL TISSUE IN THE MICROWAVE RANGE

Biological material, including tissue, can be described as a dielectric with losses on a macroscopic level. The dielectric properties of tissue are discussed in detail in Chapter 2; therefore only some information on the bulk electromagnetic behavior of tissue which is important for the understanding of its use for diagnostic applications is summarized here.

The response of tissue to an alternating electromagnetic field is characterized by relaxation phenomena, which are caused by different mechanisms, such as interfacial polarization, (partial) reorientation of dipoles, and counterion polarization effects. In general, these different relaxation processes result in a rather stepwise decrease of the relative permittivity of tissue with increasing frequency. In tissue, the main dispersion steps are referred to as α-, β-, and γ-dispersion. The latter is attributed to the dipolar relaxation of water, which is the dominant relaxation process in tissue at microwave frequencies. Each dispersion is characterized by a relaxation time τ. The γ-dispersion has a relaxation frequency around 25 GHz at body temperature. Furthermore, dielectric losses are associated with the different relaxation processes in tissue; e.g., energy is absorbed by the reorienting dipoles from an alternating electric field and dissipated as heat.

Therefore, the relative permittivity can be presented as a complex number, where both the real and imaginary part are highly frequency-dependent:

$$\varepsilon(\omega) = \varepsilon'(\omega) - j\varepsilon''(\omega)$$

Here, ε' is the real part, which describes the dielectric constant, and ε'' is the imaginary part, which describes the absorption and dissipation of electromagnetic energy from an external field; $\omega = 2\pi f$ is the angular frequency. The fact that both parts are often in the same order of magnitude for tissue at microwave frequencies has an influence on the choice of suitable measurement methods for tissue characterization.

Not only can the physical processes connected to the dielectric relaxation cause dissipation of energy, but the ionic (dc) conductivity of tissue can also result in electric losses. Therefore, often the losses caused by the ionic conductivity σ_I, also called static conductivity, are taken into account by an extra term:

$$\varepsilon(\omega) = \varepsilon'(\omega) - j\varepsilon''(\omega) - j\frac{\sigma_I}{\omega\varepsilon_0}$$

Here ε_0 is the permittivity of free space. Generally the losses caused by the static conductivity dominate at low frequencies and become less prominent at higher frequencies where the losses caused by the dielectric relaxation dominate.

The real and the imaginary parts of the permittivity of a dielectric cannot vary independently of each other with frequency but are connected by the Kramers–Kronig relations [10]. Hence permittivity data can be tested for consistency by applying these equations.

Generally tissue is considered nonmagnetic and hence has a relative permeability μ_r of 1.

3.3 MATHEMATICAL MODELS FOR THE COMPLEX PERMITTIVITY OF BIOLOGICAL TISSUE

A common mathematical model to describe the frequency dependency of the permittivity of dielectrics is the so-called Debye expression, which is also often used to describe the dispersion behavior of tissue permittivity:

$$\varepsilon(\omega) = \varepsilon_\infty + \frac{\varepsilon_s - \varepsilon_\infty}{1 + j\omega\tau}$$

Here ε_∞ is the permittivity at infinite frequency, ε_s is the permittivity at low frequencies, and τ is the time constant of the relaxation process. Since tissue exhibits several dispersion regions over a large frequency range, its permittivity is often described by a sum of several Debye terms. As mentioned above the static conductivity can be additionally taken into account. A general resulting model might look similar to the one introduced by Hurt for the permittivity of muscle [11]:

$$\varepsilon(\omega) = \varepsilon_\infty + \sum_{n=1}^{5} \frac{\Delta\varepsilon_n}{1 + j\omega\tau_n} - j\frac{\sigma_1}{\omega\varepsilon_0}$$

τ_n and $\Delta\varepsilon_n$ are the time constant and the change in the permittivity related to the nth dispersion region, respectively. An alternative model, which often allows for a better fitting of the biological data, is the Cole−Cole equation [12]. By introducing a distribution parameter α, a broadening of the dispersion region can be modeled. It is worth mentioning that α is an empirical parameter without theoretical justification. Again, several Cole−Cole terms might be used to describe the spectrum biological material:

$$\varepsilon(\omega) = \varepsilon_\infty + \sum_{n=1}^{N} \frac{\Delta\varepsilon_n}{1 + (j\omega\tau_n)^{(1-\alpha_n)}} - j\frac{\sigma_1}{\omega\varepsilon_0}$$

Other models have been described in literature, such as the Davidson−Cole function [13] or the Havriliak−Negami [14] function, and a best fitting of experimental data might be achieved by a combination of several of these models. However, for many applications the use of a multipole Cole−Cole model seems sufficient.

3.4 THE COMPLEX PERMITTIVITY OF MALIGNANT TUMOR TISSUE VERSUS HEALTHY TISSUE

The development of sensor systems for the diagnosis of tumors using microwaves is only possible with good knowledge of the permittivity values for healthy and malignant tissue in the respective frequency range. Exact and comprehensive permittivity data is furthermore needed in therapeutic applications, such as microwave ablation or hyperthermia for tumor treatment. Vast measurements have been done in the past and comprehensive data has been published for many tissue types, e.g., by Gabriel et al. [15]. However, there are still a lot of gaps in the available data for frequencies above 20 GHz. Furthermore, data on malignant tissue is often only available from measurements on animal tissues but not on human tissues, probably due to the practical difficulties of obtaining a satisfactory number of malignant tissue samples of human origin for these measurements. A further shortcoming of the data is that it is often obtained from ex vivo samples (excised tissue) rather than from in vivo measurements.

Even though it has been reported that differences between measurements on in vivo samples and on samples a few hours after excision are sufficiently small [16], other publications suggest that there might be differences that cannot be neglected [17]. Differences between in vivo and ex vivo data might be caused by temperature variations, loss of tissue water after excision, and the missing blood perfusion in excised tissue. Furthermore, data from different groups does not always agree. This could be due to differences in the used measurement method, in the sample preparation, and in the sample size, as well as due to the small number of samples used by single groups and the difference when it comes to the classification of the tissue type (e.g., one group might not mean the same as another one by "breast tissue").

It is generally a good idea to be aware of how the data in literature was obtained, and that published values might or might not well compare to the in vivo permittivity that is important for the diagnosis. Therefore permittivity data should always be published along with information about the measurement method used and possible sample pretreatment or preparation, as well as the size and temperature of the sample.

Even if the permittivity data was captured in vivo using a highly accurate measurement method, the spread within the data can be considerable. Biological material can be highly variable in its properties and exhibits differences between different patients as well as within patients. Furthermore, tissue is not as homogenous as it is often modeled to be for biomedical applications. Attention should be paid to the fact that there might be several different types of healthy tissue within the same organ, as well as several different tumor types. Skin consists, for instance, of several layers with quite different water content, and healthy breast tissue can have basically any ratio of adipose to glandular and fibroconnective tissue, and therefore can span over a wide range of water content. To provide only a single permittivity value for normal tissue of a certain kind is often impossible or at least an extreme simplification.

However, for many diagnostic purposes, exact values might not be needed (besides being impossible to obtain, due to the reasons given above). The fact that there is a clear difference in the permittivity and knowledge of the general range of values, along with a good estimation of the contrast between malignant and healthy tissue, might be enough to design a system that is able to identify or image a tumor within healthy host tissue.

In the following, some examples of data for the microwave dielectric properties of and the contrast between healthy and malignant tissue is given. Only data that is relevant for a promising application for microwave cancer diagnosis and that was obtained from a large number of samples using a precision measurement technique, a clear experimental protocol, and appropriate data processing is discussed. Data from in vivo measurements is given preference over data from ex vivo measurements, and data from human tissue over data obtained for animal tissue, if available.

3.4.1 DATA FOR NORMAL AND MALIGNANT BREAST TISSUE

Several research groups have published on the dielectric properties of healthy and malignant breast tissue at microwave frequencies and reported a considerable contrast between them. A summary of data from earlier studies can be found in [18] and [19]. These summaries revealed gaps and discrepancies in the existing data, as many of the published studies report on a rather limited frequency range and had measurements taken only on a small number of samples. Most of the summarized data had been measured ex vivo, and presented in vivo measurements did not concern human samples but samples from animals.

The most thorough and comprehensive study of healthy and malignant human breast tissue in the microwave range to date has been published by Lazebnik et al. [16] in 2007, and findings from this study are to be presented and discussed here briefly.

Lazebnik et al. measured the permittivity and conductivity for frequencies between 0.5 and 20 GHz using an open-ended coaxial probe on freshly excised human tissue from reduction surgery and cancer surgery. The measurements were done at room temperature and a maximum of 4 hours after excision on a total of 807 samples, 319 of these from cancer surgeries. An important difference from earlier studies lies not only in the large number of samples, but also in the data analysis. Samples were not only classified into "healthy" and "malignant," as done by most earlier studies, but also their adipose content was characterized. Based open this, three groups of healthy tissue (0−30%, 31−84%, and 85−100% adipose tissue content) were formed for further data analyses. The types of tumor in the cancer samples were also identified, but for the data analysis all samples with at least 30% tumor content were summarized into one "malignant" group. The data from each measurement was used to fit a single-pole Cole−Cole model. The corresponding Cole−Cole parameters for the model using the median data in each of the three groups of normal tissue samples and the malignant sample group obtained from cancer surgeries can be found in Table 3.1. For representations of

Table 3.1 Cole−Cole Parameters for the Median Dielectric Properties ($f = 0.5-20$ GHz) of Three Adipose-Defined Groups of Normal Breast Tissue and of Malignant Breast Tissue [16]

	Normal, 0–30% Adipose Content	Normal, 31–84% Adipose Content	Normal, 85–100% Adipose Content	Malignant, ≥30 Tumor Content
ε_∞	7.237	6.080	3.581	6.749
$\Delta\varepsilon = \varepsilon_s - \varepsilon_\infty$	46.00	19.26	3.337	50.09
τ (ps)	10.30	11.47	15.21	10.50
α	0.049	0.057	0.052	0.051
σ_1 (S/m)	0.808	0.297	0.053	0.794

the single data sets the reader is referred to the original paper [16]. The same data was also used for fitting one- and two-pole Debye models, which was published later [20]. The most important findings of the study are as follows: (1) changes due to temperature differences are small as compared to the variability among samples; the same is true for the time between excision and measurement; (2) the intra-patient variability is not significantly different from the inter-patient variability; (3) the adipose content has a major influence on the dielectric properties of healthy tissue samples and consequently the span of properties for all healthy samples is large; (4) malignant tissue properties exhibit a much smaller span than healthy tissue properties; (5) There is a statistically significant difference between normal and cancer samples if only samples with less than 10% adipose content are considered; the difference is also statistically significant when the data is not adjusted for adipose and fibroconnective tissue content; (6) if data is adjusted for both adipose and fibroconnective tissue contents, i.e., healthy glandular tissue and malignant glandular tissue are compared, no statistically significant difference in the permittivity was found.

Lazebnik et al. point out that "the contrast in the microwave-frequency dielectric properties between malignant and normal adipose-dominated tissues in the breast is considerable, as large as 10:1, while the contrast in the microwave-frequency dielectric properties between malignant and normal glandular/fibroconnective tissues in the breast is no more than about 10%" [16], p. 6093. The results therefore agree with the reported high contrast between healthy and malignant breast tissue in earlier publications, assuming that the normal samples in those cases had a high adipose content. It can be seen from this discussion that it might be necessary to take the heterogeneous nature of tissue into account in the data analysis and that a classification of one tissue type into healthy and malignant might be an oversimplification.

Data for healthy and malignant breast tissue from 102 samples taken from 35 patients has been recently published by Sugitani [21]. The complex permittivity of the samples was measured in the frequency range from 0.5 to 20 GHz using a coaxial probe, and a two-pole Cole−Cole model was used to fit the data of each sample. The authors distinguish between adipose, glandular, and tumor tissue for the comparison of the data. At 6 GHz, they revealed a clear difference in the permittivity between the different tissue types for most of the individual patients. Furthermore, they found a correlation between the dielectric properties and the volume fraction of cancer cells in the tumor samples (Fig. 3.1), and pointed out that the fraction of adipose cells in the tumor samples was negligible.

Data from measurements at a higher frequencies was published by [22] (30−900 GHz) and [23] (0.5−50 GHz). However, both groups had very few samples and only distinguished between "healthy" and "malignant" tissue; Khan et al. furthermore prepared the excised samples by fixation using formalin before measurement. The results might not be directly applicable to in vivo diagnostic applications; however, they indicate that there is a contrast also in the millimeter-wave range.

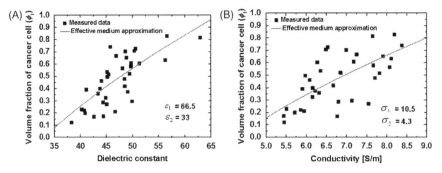

FIGURE 3.1

Correlation of the dielectric properties of tumors of the breast with the volume fraction of cancer cells. (A) Measured dielectric constant and (B) Conductivity of samples from 35 individual patients at 6 GHz [21].

3.4.2 DATA FOR NORMAL SKIN AND SKIN TUMORS

Data for the dielectric properties of skin is often published treating the skin as a single entity [24–26], even though it consists of several layers (epidermis, dermis, subcutaneous fat), which are characterized by significant differences in water content. Generally the water content gradually increases from the skin's surface toward deeper layers, and permittivity is expected to change accordingly [27]. The thickness of the different skin layers depends on the location on the body, and data for the overall skin permittivity is often categorized into different body sites to take the structural variation into account [26].

There are many different kinds of skin cancer; the three most common ones are basal cell carcinoma (BCC), squamous cell carcinoma (SCC), and malignant melanoma, of which the first two metastasize only seldom. In contrast, malignant melanoma has a high probability of developing metastases if not removed at an early stage [28]. While there are a number of publications on the dielectric properties of skin and on probes for the diagnosis of skin cancer based on the dielectric properties, there has been hardly any reliable data published on the permittivity of the different skin tumors in the microwave range.

Data for normal skin and BCC in the terahertz and millimeter-wave range (200 GHz–2 THz) has been obtained from spectroscopic measurements on ex vivo samples using a Terahertz pulsed imaging system [29,30]. The parameters for the double Debye model, fitted to the measured data, are shown in Table 3.2. Additionally, the data from ex vivo measurements on normal skin was compared to earlier published in vivo data [31], and differences were noted which are attributed to changes in the tissue after excision, such as the buildup of a layer of interstitial fluid on the sample surface.

Based on the same measurement data, Truong et al. [32] analyzed the sensitivity of the parameters of the double Debye model to the percentage of BCC tumor

Table 3.2 Double Debye Parameters as Published by [29] for the Dielectric Properties ($f = 200$ GHz–2 THz) of Normal Skin and Basal Cell Carcinoma, Obtained from Ex Vivo Measurements

	Skin Normal, Ex Vivo	Basal Cell Carcinoma, Ex Vivo
ε_α	2.58	2.65
ε_s	14.7	17.6
ε_2	4.16	4.23
τ_1 (ps)	1.45	1.55
τ_2 (ps)	0.0611	0.0614

contained in the sample, and proposed combination coefficients that exhibit a significant dependency on the tumor content and might be used for the detection of skin cancer.

Furthermore, publications which concentrated on the measurement techniques for skin cancer detection reported a difference in the dielectric properties of normal skin and skin tumors without determining explicitly the absolute permittivity levels of either [33,34].

It can be seen from these examples that data on the microwave permittivity of malignant tissue is rather incomplete, especially for skin. More comprehensive data could facilitate the further development of systems for microwave cancer diagnosis.

3.5 GENERAL CONSIDERATIONS FOR MICROWAVE SYSTEMS FOR CANCER DIAGNOSIS

The dielectric properties of a sample, e.g., tissue, influence how electromagnetic waves are reflected, absorbed, and transmitted by the sample. Therefore, both the reflected and the transmitted microwave response can be measured and used to determine the dielectric properties of the sample. Tissue is generally nonmagnetic and therefore only one of these two complex parameters, i.e., the reflection (S_{11}) or transmission (S_{21}), is enough to derive the permittivity of a sample, if the sample is isotropic and homogenous. However, in diagnostic applications the samples are often heterogeneous, and several measurements at different positions, angles, and/or frequencies are conducted in order to obtain information about the spatial distribution of the permittivity within the sample.

The choice of the experimental method and the measurement system for diagnostic purposes depends on several factors, e.g., on the expected permittivity range of the sample and the expected contrast between healthy and malignant tissue, as well as on the necessary probing depth and, in the case of imaging, the sought spatial resolution. Some general considerations will be discussed first before the most commonly used measurement techniques for microwave cancer diagnosis are presented.

A measurement can be done either in vivo (on the patient's body) or ex vivo (on an excised sample). For noninvasive tumor diagnoses, especially cancer screenings, the measurement must be done in vivo without causing pain or harm to the patient, but there are also applications where the measurements are done on excised samples, such as margin control during cancer surgery. The high attenuation of microwave signals in tissue generally makes noninvasive in vivo microwave sensing and imaging only feasible for tumors that are close to the body's surface, such as skin or breast tumors. For measurements of deeper body regions the sensor may be brought inside the body, for instance by endoscopic tools.

The frequency or frequency range has an influence on several aspects of the measurement, and must be chosen carefully considering the special characteristics and requirements of the targeted application. Generally a higher frequency means a higher spatial resolution, but at the same time a lower penetration depth means that imaging at high frequencies is limited to thin samples. Therefore transmission measurements, e.g., for breast cancer imaging, are often done at lower frequencies [35,36]. Since the permittivity of tissue is frequency dependent, the contrast between healthy and malignant tissue might be higher at certain frequencies. As water is one of the main contributors to the microwave contrast of tissue, the contrast is expected to be generally high close to the relaxation frequency of free water (around 25 GHz) and smaller for, e.g., the THz range.

Microwave measurement systems can be divided into frequency-domain (FD) systems, which apply a continuous wave to the sample, and time-domain (TD) systems, which often apply a broad band pulse. The applicable measurements instrumentation depends on the chosen domain. Prototypes operating in the FD may use commercial vector network analyzers (VNAs) for scattering parameter measurements with high accuracy and high dynamic range [37,38,39], as well as custom-built transceivers [36]. Commercially available instrumentation for TD measurements might consist of an impulse generator in combination with an oscilloscope [40]. In order to benefit from the high dynamic range of FD measurement instrumentation, TD signal can also be obtained from FD measurements, i.e., the pulse is synthetically generated from a sequence of continuous-wave measurements at different frequencies by inverse Fourier-transform [41−43].

Furthermore it can be distinguished between resonance and broadband measurement techniques; the former is based on the resonance frequency shift of an electromagnetic resonator, which is weakly coupled to the tissue to be measured, and the latter is based on the absolute values of reflected and/or transmitted power depending on the tissue properties.

Another choice concerns reflection versus transmission measurements. For in vivo measurements, samples often have a large volume as compared to the microwave wavelength and/or are only accessible from one side. In that case, reflection measurements are more feasible. However, for certain applications where the imaging site is relatively easily accessible from many angles, as in the case of the breast or the head, as well as for ex vivo measurements of small samples, microwave transmission measurements might be used.

The sample can either be in the near-field or in the far-field of the sensing antenna. Near-field measurements have the advantage that they are not diffraction-limited and allow for spatial resolution below the wavelength. However, the calculation of the near-field distribution in the sample is not trivial and generally requires sophisticated numerical methods. Therefore even the determination of the complex permittivity of a homogenous sample from nearfield measurements is not trivial [44−47]. The image reconstruction, e.g., the determination of the spatial distribution of the complex permittivity within the sample, using a set of near-field measurements obtained from different antenna positions poses an ill-defined and inverse scattering problem, and is therefore challenging and computationally demanding.

If a set of measurements with the probe at different positions in respect to the sample are needed, as in the case of imaging, either the probe can be mechanically scanned or an antenna array can be used. Advantages of mechanically scanning are that the setup can be easily adapted to different sample sizes and that the signal does not suffer from interference from other antennas. On the other hand, it requires a complex positioning system with moving cables and generally results in longer data acquisition times. Antenna arrays allow for much faster scanning times but often require a switching network which introduces additional signal losses. Furthermore, the size of a single antenna becomes the limiting factor for the distance between two antenna positions. Naturally, a combined solution, i.e., an antenna array which is mechanically scanned, is also possible.

Measurement setups can be classified into mono-, bi-, and multi-static. In a so-called monostatic configuration the same antenna is used as both transmitter and receiver. In a bistatic configuration a pair of antennas is used, with one antenna acting as transmitter and the other antenna acting as receiver, and in a multistatic configuration multiple antennas act as receivers.

Good coupling of the microwave signal into biological tissue is important for high sensitivity measurements. The tissue permittivity is generally very different from the permittivity of air, and the resulting impedance jump at the air-tissue interface results in the reflection of a large portion of the microwave energy. To avoid high reflections the tissue and the antenna can be immerged into a lossless coupling medium with a permittivity close to the that of tissue. Near-field probes that are used in direct contact should be well matched to the sample, i.e., the probes' characteristic impedance must be close to the equivalent impedance of the probes' near-field interaction with the sample at the measurement frequency.

3.6 COMMON TECHNIQUES FOR MICROWAVE CANCER DIAGNOSIS

3.6.1 PERMITTIVITY EVALUATION BY FREE-SPACE QUASI-OPTICAL SYSTEMS

The complex permittivity of a sample can be measured with a free-space quasi-optical measurement setup. For the measurement, a collimated microwave beam

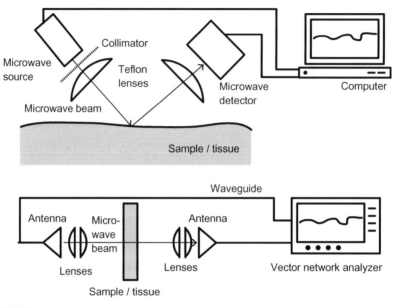

FIGURE 3.2

Examples of quasi-optical microwave systems for determining the dielectric permittivity of a tissue sample, schematic representation.

is radiated onto the sample, which is located in the far-field of the transmitting antenna. The reflected and/or transmitted beam is collected by a receiver, which often is a second antenna but in some cases may be the same as the transmitting antenna (Fig. 3.2). A common beam used is the Gaussian beam [48], which has its maximum magnitude at the axis of the beam with a radial decay that can be described by a Gaussian function. To radiate a Gaussian beam, e.g., corrugated-horn antennas can be used. From the transmitted and received radiation energy, the complex scattering matrix parameters, i.e., reflection (S_{11}) or/and transmission (S_{21}), are determined. Even though in the simplest setup the sample might be placed simply between two antennas (or in front of one), in practice, often a quasi-optical system consisting of dielectric lenses and/or mirrors focuses the beam and guides it onto the sample. The use of a focused beam has several advantages over the use of an unfocused beam. An unfocused, therefore diverging, beam might be wider than the aperture of the receiving antenna, and a correction that is prone to errors would be needed. Furthermore, a larger electromagnetic beam width also results in a larger spot size illuminated on the sample, and is therefore only practical for large homogeneous samples.

For the excitation signal there are two general options: either a continuous wave or a wideband pulse. A continuous-wave system, operating at one or several fixed or time-variable frequencies, can be built relatively inexpensively and compactly, and requires only simple data postprocessing, as compared to a wideband-pulse system.

For computation of the measurement data, the beam is assumed to follow a straight line, and is approximated by a plane wave which is attenuated while passing through the sample, but also partly reflected at the sample interface depending on its complex permittivity. This makes the data processing, e.g., the computation of S_{11} and S_{21}, simple. To obtain a two-dimensional image of the permittivity distribution, either the sample or the beam is raster-scanned so that the permittivity of the sample is determined pixel by pixel. The result is a two-dimensional projection of the sample permittivity. A three-dimensional image can be obtained if the sample is scanned at several different angles, comparable to X-ray computed tomography. Another possibility to obtain a three-dimensional image of the permittivity is to use the amplitude and phase information obtained from a transmitted or reflected pulse. However, simple image reconstructing methods, which are based on the assumption that the waves propagate along straight lines, have limited applicability to in vivo imaging as they neglect diffraction and multiple scattering [7].

Typically quasi-optical systems operate in the millimeter-wave and THz range, where a higher resolution can be achieved compared to lower frequencies. Examples are terahertz pulse spectroscopy and terahertz pulse imaging, which use THz pulses for reflection or transmission measurements.

Quasi-optical free-space techniques allow for contactless measurements and relatively easy scanning of the beam for imaging or reference measurements. However, transmission measurements are only feasible for thin samples and are hardly applicable to in vivo measurements. For quasi-optical systems, the resolution is diffraction-limited and the choice of the frequency is a trade-off between the sensitivity to the tissue's water content, which is higher at around 100 GHz than at THz frequencies [49,50], the penetration depth, which increases with decreasing frequency, and the resolution, which is higher at higher frequencies [51]. In the sub-millimeter wave range, random scattering from rough surfaces (e.g., skin) might influence the reflectivity and mask small changes in permittivity. To prevent this problem, the use of a (lossless) dielectric window to flatten the surface has been suggested [50].

Quasi-optical systems have their main application in the determination of the dielectric properties of ex vivo samples and, due to their limited penetration depth, in the diagnosis of the upper layer of the skin as well as in margin control during tumor surgeries.

3.6.2 MICROWAVE REFLECTOMETRY USING NEAR-FIELD PROBES

Microwave reflection measurements with near-field probes, commonly called "microwave reflectometry," have been used already in the 1980s to measure the dielectric properties of tissue [4,44,52]. The reflection of an incident wave from the probe-sample interface depends on the interaction of the probe's near-field with the sample, which depends on, among other factors, the probe geometry and

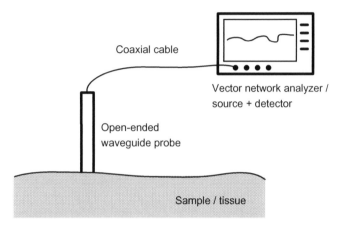

Coaxial cable

Vector network analyzer /
source + detector

Open-ended
waveguide probe

Sample / tissue

FIGURE 3.3

Example of a microwave reflectometry system with an open-ended waveguide probe,
schematic representation.

the permittivity of the penetrated sample region. Both FD and TD measurements
are possible. For the measurement the probe is put in contact with the sample or
placed in close proximity (Fig. 3.3), and the complex input reflection coefficient
(S_{11}) is measured by, e.g., a slotted line, a network analyzer, or in the case of a
TD measurement, with a TD spectrometer. An analytical expression for the terminal impedance of the probe is only known for some simplified cases, and generally there is no analytical solution to Maxwell's equations for the near-field
distribution of the probe. Therefore numerical methods are used to determine the
complex permittivity of the sample [45–47,53]. For coaxial probes an equivalent
circuit representation can be used, and its individual elements can be calculated
from measurements on reference samples with known properties [44,52,54].

Microwave reflectometry requires no sample preparation or pretreatment, and
can be used on any accessible surface tissue layer, e.g., for measurements on the
skin or for margin control during tumor removal surgery. An unwanted air gap
between the probe and the sample surface might pose a problem; however, the
soft surface of tissue makes this rather uncommon for small probes.

The simplest near-field probes are open-ended transmission lines, such as
coaxial probes, and open-ended rectangular waveguides, which are widely available and easy to operate. Coaxial probes work over a large range of frequencies,
can be used both for TD and FD measurements, and are commonly used for both
frequencies up to several GHz [33,45,55,56] and in the millimeter-wave range
[57]. Rectangular waveguide probes are used especially for higher frequencies
[34,58–62], for which coaxial probes are difficult to fabricate or have high losses.
Pyramidal horn antennas and conical horn antennas operated in near-field have
also been proposed as applicators [63]; they can achieve a larger sensing volume
than rectangular waveguide probes in the same frequency range.

If a two-dimensional image of the sample is desired, the probe can be raster-scanned over the sample, or alternatively an array of probes can be employed. Each single measurement gives a kind of average or "effective" permittivity over the probed volume, and therefore the probe size is one of the determining factors for the spatial resolution. Commercially available probes have rather large sizes and, while they might be well suited for the simple determination of the permittivity of a homogenous sample, they provide insufficient spatial resolution for most diagnostic applications. A further limitation of commercial probes comes from their poor matching to tissue. The permittivity of tissue depends on the frequency and, e.g., for the lower GHz range, the relative permittivity of tissue is around 100 and thus very high. This means that only a small portion of the electromagnetic energy penetrates the sample, and the measured reflection coefficient is close to one and very insensitive to changes in the tissue permittivity. This poses a challenge for the evaluation and limits the sensitivity of the measurement. For a meaningful measurement the probe should be designed for the intended application, taking the targeted permittivity range and the needed spatial resolution into account. In the following some possibilities for probe miniaturization are discussed; more examples of adequate probes are given in the respective application sections.

For open-ended rectangular waveguides the probe size is generally dictated by the frequency used. A smaller probe footprint can be achieved by tapering the waveguide toward the tip, but care must be taken not to compromise the signal strength that reaches the sample and therefore the sensitivity [64]. Smaller waveguide dimensions for the same frequency range can be achieved with a dielectric-core waveguide compared to an air-filled waveguide. The field at the waveguide tip can be further confined by metallizing the sidewalls of the tapered dielectric-core waveguide section [39]. For open-ended transmission line probes, like coaxial probes, the size is not rigidly tied to the frequency range. However, transmission lines with small dimensions compared to the wavelength generally suffer from high losses. Furthermore, small coaxial probes are difficult to fabricate and therefore expensive.

As an alternative to conventional coaxial probes, a group from Korea has introduced open-ended strip line probes [65−67] and planar probes with ring-shaped apertures [68,69] fabricated by micromachining. Advantages of these micromachined probes are their flexibility in size and design (to adapt to different applications), their potentially low-cost fabrication in high volumes due to batch fabrication, and the possibility for monolithic integration with the driving/evaluation circuit [70]. Furthermore, arrays of several probes can be fabricated [67,71].

Another example of a probe with miniaturized footprint is a broadband near-field antenna with partly metallized pyramidal tip [72,73], which is essentially an open-ended tapered parallel-plate transmission line with constant wave impedance to avoid reflections. Beside miniaturizing or tapering transmission lines, scattering from a small antenna, e.g., the extended inner conductor of a coaxial probe, has been used to achieve high lateral resolution [52,74−76]. Radiation into the sample through a small aperture also allows for high resolution, but only a small portion of the energy interacts with the sample, which poses a challenge for the probe's sensitivity.

An alternative design for a tissue near-field probe that focuses on improving the sensing depth compared to conventional probes has been presented by [77]. Their double-armed equiangular spiral patch probe achieved almost twice the probing depth compared to an open-ended coaxial line.

Generally near-field probes allow for high-precision measurement of the sample permittivity but should be adapted to the intended permittivity range and needed spatial resolution as well as probing depth if used for diagnostic applications. As medical diagnostic tools, they are mainly used for measurements on surface tissue, such as skin.

3.6.3 MICROWAVE TOMOGRAPHY

The term microwave tomography is generally used to refer to active, quantitative microwave imaging methods. It was initially used in connection with microwave imaging, which applied the same principles of image reconstruction algorithms as X-ray computed tomography, assuming straight-path wave transmission and calculating the permittivity distribution of two-dimensional slices of the sample from measured projections in a direct manner [78–81]. Nowadays the term microwave tomography is applied to most active microwave imaging methods that determine the spatial distribution of the complex permittivity of a sample. These methods generally employ complex algorithms for two- or three-dimensional image reconstruction that take wave scattering (multiple reflections and diffraction effects) into account [7,42,82].

A schematic representation of a microwave tomography system is shown in Fig. 3.4. For imaging the sample is illuminated by a transmitter and the scattered fields are measured at a number of locations. Measurements might be taken at a set of different frequencies. This is repeated for several different locations of the transmitter around the sample. From the measured scattered fields, the permittivity distribution of the sample is reconstructed. In practice, usually one or several antenna arrays are used where each antenna subsequently acts as transmitter while the remaining antennas measure the scattered signals. In a two-dimensional imaging approach the antennas are often positioned in a circular array around the sample, so that illumination angles from 0 to 360 degrees are achieved, and the antenna array is mechanically scanned perpendicular to this imaging plane. In this way the imaging is done slice-by-slice; i.e., the data acquisition and reconstruction of the permittivity distribution is done independently for each slice, and the three-dimensional image consists of a stack of 2-D images which were reconstructed only from the data acquired in their respective planes. Three-dimensional imaging systems use, e.g., cylindrical antenna arrays [83], or mechanically scan the transmitter and receiver pairs or arrays [84–86], in order to measure signal transmission in three dimensions. Naturally, three-dimensional tomography comes

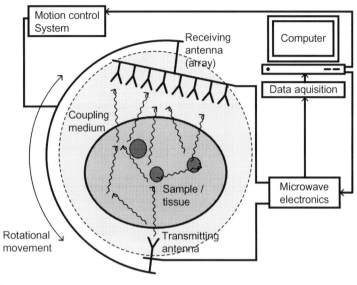

FIGURE 3.4

Example of a microwave tomography system, schematic representation.

at higher computational costs for image reconstruction than two-dimensional imaging systems.

There are several aspects that must be considered when designing a microwave tomography system for medical imaging. Generally, the requirements on the system performance are rather challenging. The receivers must be highly sensitive in order to be able to measure even extremely weak scattering fields. Imaging systems with dynamic ranges up to 135 dB have been reported [87]. Furthermore, strong reflections are to be expected at the interface of the sample, but can be avoided if the antennas and the sample are immerged in a low-loss coupling medium with dielectric properties close to the sample. However, this might be somewhat unpractical in clinical applications. Furthermore, the properties of the coupling medium might suffer from temperature drifts. As an alternative approach to avoid reflections from the air-sample interface, the use of antenna arrays on flexible substrates which are directly attached to the sample has been suggested [88]. It should also be noted that in a clinical situation, the data acquisition time cannot be excessively long. A long imaging time not only increases the discomfort for the patient and the costs for the clinic, but also makes motion artefacts more likely.

As mentioned above, the initial approach for image reconstruction, related to X-ray tomography, assumed a straight-path propagation of the microwaves between emitter and receiver and neglected diffraction effects [78–81]. However, diffraction cannot be neglected for the interaction of microwaves with the strongly heterogeneous samples that are typical for medical imaging. Therefore a second approach, microwave diffraction tomography, was soon suggested [89−91], which does not assume straight-path propagation and takes diffraction effects into account. The Born or the Rytov approximations for linearization of the problem are applied, which allows for fast and reliable image reconstruction. However, these approximations are only valid if the diffraction effects are weak [92], and therefore they are typically not applicable to biomedical imaging applications. Consequently, reconstruction of biological images with high contrast is mostly done with iterative methods that solve the inverse, nonlinear problem by optimization of a forward model and that do not rely on linearization approximations [42,82,93]. A number of different algorithms have been proposed for this iterative optimization. Newton-type iterative algorithms, such as Newton−Kantorovich [94], Gauss−Newton [95,96], Levenberg−Marquardt [97], Log-Magnitude and Phase Reconstruction [98], and the Distorted Born Iterative method [99] minimize a regularized cost function that describes the difference between computed and measured scattered fields at the receivers. Regularization, i.e., the use of a priori information, is necessary since the problem is ill-posed; the Tikhonov regularization is often used [97]. Newton-type algorithms have a high computational demand, since they require solving the forward equation at every iteration step, which is especially a challenge in the case of three-dimensional image reconstruction [82]. Alternatives such as the modified gradient method [100] and contrast source inversion [101,102] do not require solving the forward model at each iteration. Despite the resulting larger number of unknowns and iterations compared to Newton-type algorithms, they can be less computationally demanding, especially for the full three-dimensional vector case.

The development and improvement of microwave tomography systems for medical applications has been driven by several research groups and has resulted in advanced systems that are mainly, but not exclusively, aiming at complementing or even replacing X-ray imaging systems for breast cancer screening [9,103,104], as well as for brain imaging [105−108].

3.6.4 ULTRA-WIDEBAND MICROWAVE RADAR

Ultra-wideband (UWB) microwave imaging is a TD technique. It uses a pulsed-radar approach to identify the location of regions of increased scattering (scatterers) within the sample rather than aiming at the complete information of the spatial permittivity distribution [109−111]. Full-wave electromagnetic analysis is avoided, and the output of the measurement is a 2D or 3D image with the location of the scatterer, which represents the tumor within the sample.

Generally, short pulses are radiated from different positions and the scattered pulses are recorded (Fig. 3.5). The time delay between transmitted and received pulses and the shape of the pulses depend on the scattering events in the imaging volume. In order to achieve a high dynamic range and good measurement accuracy, many UWB radar systems measure in the frequency range and obtain the

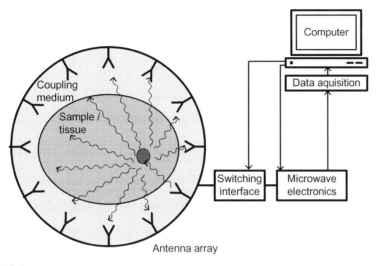

FIGURE 3.5

Example of an ultra-wideband microwave radar system for medical imaging, schematic representation.

TD waveform by inverse Fourier transform [38,41−43,111−113]. For the measurement setup, mono-, bi-, and multi-static configurations are possible. For good image reconstruction the acquisition of a high number of scattered responses at a high number of illumination angles is advantageous. This can either be achieved by mechanically scanning the antennas or by using antenna arrays, which are electronically scanned.

Different synthetic focusing algorithms can be used for image reconstruction, e.g., delay-and-sum beamforming (DAS) [43,114], delay-multiply-and-sum beamforming (DMAS) [115,116], space-time beamforming [117], and a data-adaptive algorithm termed MAMI (multistatic adaptive microwave imaging) [112]. Their ability to locate tumors correctly depends, for instance, on the heterogeneity of the surrounding tissue [118]. Time reversal algorithms have also been proposed for image reconstruction [119−121]. Research toward improved signal processing and improved image reconstruction algorithms well-adapted to specific applications is an ongoing effort [122,123].

3.6.5 PASSIVE MICROWAVE TECHNIQUES FOR CANCER DIAGNOSIS

Microwave radiometry is a passive microwave imaging method (Fig. 3.6), meaning that no external radiation is illuminating the tissue. Alternative names for this technique are microwave thermography or microwave thermometry [124,125]. This method makes use of the fact that tumors often have slightly elevated temperatures compared to the surrounding tissue due to the higher metabolic rate of

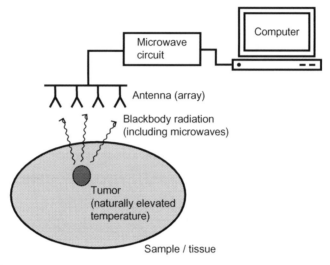

FIGURE 3.6

Schematic representation of a microwave radiometry system for medical imaging.

tumor cells and the increased vasculation (localized blood perfusion) of early tumors. It measures the thermal radiation at a certain frequency or frequency range, which is spontaneously emitted by the body. All warm bodies spontaneously emit electromagnetic radiation, and the intensity of the radiation at a certain frequency increases with the temperature of the object. The intensity is generally quite low in the microwave range, especially compared to the infrared range, but at the same time the attenuation in tissue is also comparably low. Some imaging systems therefore combine microwave radiometry, which examines the temperature of the internal tissue, with infrared measurements, which determine the temperature at the surface of the sample (e.g., breast). However, the detection of the very low power radiated by a tumor is technically challenging. Nevertheless, radiometry systems for tumor detection have been presented by several research groups in the past decades [126–128], and their performance has been and continues to be tested in clinical trials [129,130]. An approach to improve the signal strength that is radiated from the tumor is the combination of radiometry with microwave-induced heating [131]. Due to its higher conductivity, tumor tissue is heated more intensively than the surrounding tissue by high-power unfocused microwave radiation, and a better imaging contrast can be achieved.

3.6.6 HYBRID MICROWAVE TECHNIQUES FOR CANCER DIAGNOSIS

Hybrid microwave techniques do not solely rely on the interaction of microwaves with the tissue to distinguish between healthy and malignant tissue. A hybrid technique may still exploit the variations in tissue permittivity for creating an imaging contrast, but uses a different imaging modality. Another possibility is that the imaging contrast does not arise from the dielectric properties of the tissue but, e.g., from differences in mechanical properties, which can be indirectly revealed by microwave measurements.

An example for the former is a method for determining the dielectric properties based on heating kinetics. The energy that a sample absorbs from incident microwave radiation depends, among many other things, on its complex permittivity. Therefore, if a sample is illuminated by microwaves in a stepwise fashion, its permittivity can be determined from the monitored heating kinetics. This method was introduced by Chahat et al. [132]. For sample illumination, an open-ended waveguide or a horn antenna located a certain distance above the sample is used, and the temperature of the sample is monitored using a high-resolution infra-red antenna array (Fig. 3.7). Chahat et al. use a WR-15 waveguide at a distance of 17 mm from the sample with an input power of 425 mW for illumination at 60.4 GHz. As minimum features for the IR camera, a resolution of 1 mm^2 on the sample, a frame rate of 10 Hz, and a thermal resolution of 0.1°C are suggested. The temperature increase during a 10 seconds microwave exposure is measured as a function of time, and then, by fitting a theoretical model for the heating kinetics to the measurement data, the penetration depth is determined. The permittivity is subsequently determined from the obtained penetration depth by varying the expression for the permittivity in the equation for the calculation of the penetration depth (for details see Ref. [132]). This method can be used for a relatively wide range of permittivity values; however, it is only applicable for

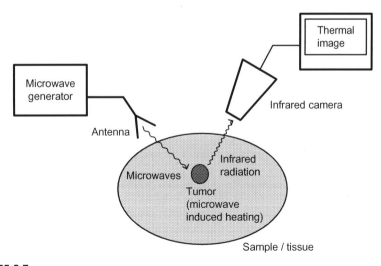

FIGURE 3.7

Schematic representation of a hybrid medical imaging system based on the kinetics of microwave-induced heating in tissue.

high-water-content samples, which include most types of tissue. As the infrared camera monitors an area on the sample, the technique can provide a two-dimensional picture of the permittivity distribution. Permittivity difference in the depth of the sample cannot be resolved. An advantage of the technique is the relatively short time needed for obtaining a measurement of an entire area. The equipment is relatively simple and inexpensive. One potential application of the technique could be screening for skin tumors.

Another method that exploits the differences in microwave-induced heating dynamics between tumors and healthy tissue is thermoacoustic imaging, also known as microwave-induced ultrasonic imaging. The basic principles of this technique are depicted in Fig. 3.8. The tissue is irradiated with microwave pulses, and expands and contracts as a consequence of the induced temperature fluctuation. This causes acoustic signals, which are collected by ultrasonic sensors. Tumor tissue absorbs more energy than the surrounding tissue and consequently produces a different (stronger) acoustic wave. Using sensor arrays and image reconstruction, thermoacoustic computed tomography can be realized, which combines the high spatial resolution of ultrasound imaging and the good imaging contrast of microwave techniques. The spatial resolution depends on the microwave pulse length, which therefore should be as short as possible. However, shorter pulses trigger higher frequency thermoacoustic signals which suffer from higher

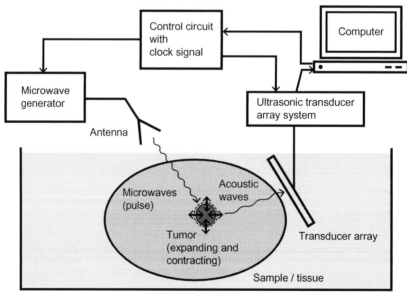

FIGURE 3.8

Schematic representation of a microwave-induced ultrasonic imaging system.

attenuation in tissue [133]. Systems with a spatial resolution of 1.5 mm [134,135] and down to 0.5 mm using a pulse length of 0.5 µs [136,137] have been reported. As with other microwave imaging techniques, image reconstruction is challenging for highly heterogeneous tissue. Here an additional uncertainty arises from the nonuniform traveling time of the ultrasonic signals in heterogeneous tissue. Another disadvantage is the high microwave radiation power required which can be at kilowatt levels [133,135].

An imaging technique called elastography exploits the difference in the mechanical properties between healthy tissue and tumor tissue as a contrast mechanism [138,139]. Elastography uses an either static or dynamic mechanical stimulation to induce motion in the tissue, which differs between healthy and malignant tissue. The resulting tissue deformation is measured by conventional medical imaging and used to reconstruct the spatial distribution of the elastic properties [140]. Microwave imaging has been suggested for imaging the strain distribution induced in tissue by an external pressure (compression microwave imaging) [141,142] and for imaging the motion induced by ultrasonic mechanical stimulation (harmonic motion microwave Doppler imaging) [143]. A schematic representation of a harmonic motion microwave Doppler imaging system is show in Fig. 3.9.

Furthermore, hybrid systems have been suggested for which information from an imaging method other than microwave imaging, e.g., ultrasonic imaging, is used to provide *a priori* information for image reconstruction in microwave tomography or UWB radar imaging [144].

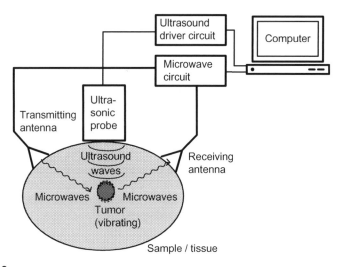

FIGURE 3.9

Schematic representation of a harmonic motion microwave Doppler imaging system.

3.7 MICROWAVE IMAGING FOR BREAST CANCER SCREENING

Breast cancer is the second most frequently diagnosed cancer (after cancer of the skin) in women in the United States and the second most common cause of cancer-related deaths overall (after lung cancer) [145]. Early detection through screening enables more treatment options and ultimately may save lives. Nowadays X-ray mammography is widely used for breast cancer screening and can detect early stage breast tumors. However, mammography misses up to 15% of tumors, especially in women with dense breast tissue. Furthermore, it exposes the women to ionizing radiation which can cause cancer by itself, which is statistically relevant for a screening method for the whole population. X-ray mammography also causes psychological stress because of its low specificity resulting in overdiagnoses; i.e., in the case of X-ray mammography up to 95% of abnormal mammograms turn out to be false-positives after additional imaging or biopsy [146,145]. Magnetic resonance imaging can also be used for breast cancer imaging, in particular as supplement to X-ray mammography, but due to its high cost it is only used for screening women at high risk.

After more than 20 years of development and research [147,148], hopes are high that microwave imaging systems can offer a valid alternative or at least a complement for breast cancer screening [9] as it uses nonionizing radiation and relatively inexpensive instrumentation.

The breast has some features that make it a good candidate for imaging using microwaves. Generally the absorption of microwaves is lower in adipose breast tissue than in most other tissues, and the breast is accessible from many sides. This allows for penetration of several centimeters and makes transmission measurements feasible.

However, penetration depends on wavelength and decreases at higher frequencies, thus limiting the imaging resolution. As a compromise, microwave breast cancer imaging systems often work at frequencies between 0.5 and 8 GHz [8,9,36]. To measure the weak signals which result from the high attenuation at these frequencies, and in order to detect small tumors or objects with small contrast, highly sensitive detectors are required, and the dynamic range of mature microwave breast cancer imaging systems is typically better than 100 dB [43,86].

The high contrast between healthy tissue and malignant tissue reported by early studies [18] was probably one of the main reasons for researchers to dedicate time and effort to the development of microwave breast cancer imaging systems [8]. More recent measurements [16], however, revealed a much lower contrast between malignant tissue and healthy fibroglandular tissue (<10%) while it confirmed a high contrast between malignant tissue and adipose breast tissue (for a detailed discussion of the contrast between different breast tissue types see Section 3.4.1). The high variation in contrast between different types of breast tissue and the fact that breast tissue often is highly heterogeneous imposes more stringent demands on the sensitivity and resolution of microwave imaging systems than initially thought. Researchers meet these challenges with, e.g., custom-build microwave electronics [43,86,149] and improved image reconstruction algorithms [112,114,116,150].

An approach to circumvent the challenges that arise from the small contrast between tumor tissue and fibroglandular tissue is the use of contrast agents such as magnetic nanoparticles [151,152] and single-walled carbon nanotubes [153].

Image reconstruction in breast cancer imaging is challenging and requires high computational resources due to several reasons. Breast tissue is highly heterogeneous and full-wave electromagnetic models are needed to accurately represent the complex field distributions. Interferences might be caused by the skin, nipple, or chest wall. Size, shape, and anatomical features differ widely between patients and also vary over time within a single person; thus, prior knowledge for image reconstruction is difficult to obtain.

A practical challenge for the imaging of the breast is small movements of the patient during the data acquisition, for instance from breathing, which result in image blurring and impact image resolution and diagnostic sensitivity (defined as the ratio of patients correctly diagnosed with a tumor to the number of all the patients with a tumor). There are several approaches to minimizing artefacts from motion. The breast can be physically restrained between two plates or in a constraining cup, though this might be uncomfortable for the screened person. Another possibility is that the person lies in a prone (face down) position and the breast is hanging freely into the imaging system. A clinical prototype of an imaging system for the patient in prone position is shown in Fig. 3.10. The different possibilities typically result in different surfaces over which data is collected. The most common data acquisition surfaces for the above cases are planar,

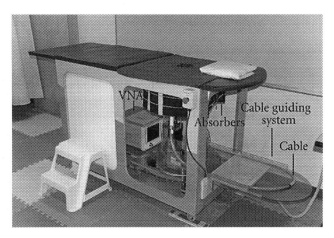

FIGURE 3.10

Example of a clinical prototype of a microwave breast cancer imaging system [38].

hemispherical, and cylindrical, respectively. If the antennas are placed at some distance to the breast surface, a coupling liquid with dielectric properties close to the properties of tissue is used to avoid strong reflections from the skin surface [9,40,154]. A clinical setup should allow for easy exchange of the liquid before imaging the next patient, and care must be taken to avoid air bubbles.

Microwave techniques that have been proposed for breast cancer detection include radiometry [129,130], tomography [84,86,103,155], UWB radar [37,40,43,115,117,156], a method that combines tomography and radar [157], as well as near-field microwave holography [158−161]. Furthermore hybrid methods such as thermoacoustic imaging [137] and microwave Doppler imaging [143] are tested. The basics of most of these techniques have been discussed in Section 3.6; in the following, some active microwave imaging systems of which clinical or advanced prototypes exist are presented.

A microwave tomography system has been developed at Dartmouth College (Hanover, NH, United States), [162,163] and in 2014 an improved clinical prototype has been presented [86]. The imaging system operates in a frequency range from 250 kHz to 3 GHz, and consists of 16 independent channels, each with an antenna and a transceiver unit. The 16 vertically oriented monopole-antennas are arranged in a cylindrical array, which can be adapted to different breast sizes. The antenna array is divided into two sub-arrays, which can be moved independently to different heights. The microwave electronics is custom-built in order to achieve the needed signal-isolation and a dynamic range of better than 110 dB. A commercial RF signal generator is used as signal source. The sensitivity of the system is −130 dBm up to 2300 MHz, and −110 dBm at higher frequencies, with these lowest detectable signals having a phase root-mean-square error of less than 1 degree. The measurements are performed with the patient in a prone position and the breast hanging into the tank with the antenna array. Hydraulic valves and an optical sensor are used for automated filling and draining of the tank with the coupling medium. The coupling medium used is lossy. This reduces multipath signals and mitigates mutual coupling. Two imaging strategies have been implemented: the first, 2D imaging, is performed with both sub-arrays in the same plane for data-acquisition. Subsequently, each antenna element acts as a transmitter and the remaining ones as receivers. Two-hundred forty amplitude and phase measurements at a single frequency are recorded, and measurements at several frequencies are done. The time for a single frequency scan in one plane is 5.8 seconds. For full 3D imaging, measurements are also performed with the two sub-arrays in different vertical planes. The system has seven imaging planes, therefore there are 7^2 combinations of sub-array positions. For the offline image reconstruction a finite-difference time-domain algorithm is used to obtain the forward solution, and an iterative Gauss−Newton algorithm is used to find the inverse solution. The use of a log transformation makes inverse problems more linear and avoids convergence to local minima [98,150].

A clinical prototype of a multistatic UWB radar using a hemispherical 60-element antenna array has been developed by a group at the University of Bristol

(United Kingdom) [37,112,164]. The system operates in a frequency range from 4 to 8 GHz; the measurements are conducted in the FD and the data is transformed into the TD before further processing. UWB aperture-coupled stacked-patch antennas with a cavity at the back to shield from the surroundings are used, and the antenna array is connected to a VNA through a custom-built network of electromechanical switches. The use of an 8-port VNA enables the conducting of 15 transmission S-parameter measurements simultaneously. Thanks to the highly parallelized data acquisition a complete scan of 1770 measurements takes no more than 10 seconds. No monostatic measurements are performed, i.e., only S_{21} is measured for all possible antenna combinations. The S_{11} signals are omitted because they require the use of directional couplers and, in any case, comprise the less important part of the measurement data. For the measurement the patient is lying in a prone position with the breast immerged into the hemispherical antenna array, which is filled with a matching liquid. The whole array can be rotated around the breast. Two sets of data are taken at two different rotation angles in order to be able to extract the tumor response. The unwanted signals from antenna coupling and skin reflections are assumed to be identical for both measurement sets and can be removed from the raw data by subtracting the signals from each other. The data is preprocessed in order to equalize the scattered tumor responses from different signals. For postprocessing, two different focusing algorithms are considered: DAS beamforming and the data-adaptive algorithm known as MAMI. In simulations, the performances of the two algorithms were compared, and the degradation of the results arising from variations in the signal velocity with different breasts was analyzed. Which algorithm performs better depends on the imaging scenario. MAMI showed improved focusing quality over DAS; however, DAS provided better results for real-life situations where experimental errors and variations between patients need to be taken into account. In experimental measurements on a breast phantom with two spherical tumors, a better signal-to-clutter ratio was achieved with MAMI as compared to DAS; for instance, for a 6 mm tumor the signal-to-clutter ratio was 8.2 and 3.8 respectively (numbers depending on the imaging plane). However, the breast phantom mimicked tumors in an otherwise homogenous adipose-dominated breast. The authors themselves point out that the response from glandular tissue, with only slightly different dielectric properties than tumor tissue, is a major challenge for microwave breast cancer imaging.

A UWB radar imaging system developed at the McGill University (Montreal, QC, Canada) [40,115,165], unlike most other systems, performs the measurements in the TD. It employs pulses of 70 ps duration, which covers the frequency range up to 14 GHz. Its main components are a 16-element broadband-antenna array, an impulse generator, a clock, and a picoscope. The antennas are arranged in a hemispherical bowl-shaped radome. Traveling wave tapered and loaded transmission line antennas [166] are used. For imaging, the patient lies in the prone position and the breast is immerged into the radome with an ultrasound gel used as the coupling medium. With each of the 16 antennas subsequently acting as a transmitter, 240 bistatic measurements are performed, and a complete scan takes up to

18 minutes. In a preprocessing step the measurement data is cropped, filtered, and time-aligned. An extensive system evaluation was conducted using homogeneous and heterogeneous breast phantoms [40]. Using phantoms mimicking adipose breast tissue with a tumor, the localization error was smaller than 9.2 mm, and a signal-to-clutter ratio of 6.8 dB could be achieved. In heterogeneous phantoms mimicking a breast containing glandular tissue, tumors down to a size of 5 mm were detectable. However, the data processing mainly consisted of comparing the measured response in the absence of the tumor to the response when the tumor-mimicking object was placed inside the phantom. In reality, this would require comparison of the measured response before and after tumor growth. In a later clinical trial on healthy volunteers a DMAS imaging algorithm [116] was employed for image reconstruction with data from single scans. In that study differences between reconstructed images of subsequent scans on the same patient after repositioning have been found to be small enough to have no influence on the interpretation of the images [115].

A monostatic UWB radar imaging system using a single antenna which is mechanically scanned (Fig. 3.11) has been introduced as a tissue sensing adaptive

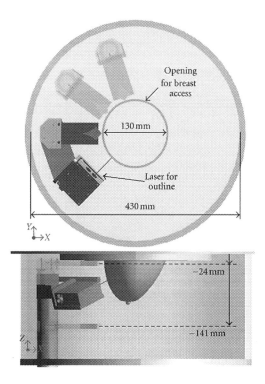

FIGURE 3.11

Schematic representation of the tissue sensing adaptive radar system of the University of Calgary, top and side view [38].

radar prototype system by a group at the University of Calgary (Canada) [38,167]. The size of the sensor is not critical in this monostatic system and a balanced antipodal Vivaldi antenna with a director (BAVA-D) [168] is used. The measurement is done in the FD. A VNA is used to measure S_{11} between 50 MHz and 15 GHz. The measured FD data is weighted with the spectrum of the differentiated Gaussian pulse and then transformed into the TD. For imaging, the patient is lying in the prone position with the breast extending into the cylindrical tank. The tank is filled with canola oil as a coupling medium. The sensor is mounted to a positioning arm and can be moved vertically inside the tank. The whole tank can be rotated in order to place the sensor at different positions. A laser is used to measure the distance to the breast surface. The system is calibrated by measuring the reflections of metal plates at known positions. A complete scan covers 200 antenna locations and takes up to 30 minutes. Signals from two scans are collected, with and without the breast in place. The signals recorded from the scan with coupling medium alone are then subtracted from the signals measured with the breast in place. Reflections from the skin surface are removed from the signals before the data is postprocessed using a DAS beamforming algorithm for image reconstruction. Measurements on breast phantoms as well as on volunteers have been performed, and the correlation between measured and simulated signal on volunteers was better that 0.9 for most signals.

A radar module which uses custom-built hardware and is tailored for breast cancer imaging has been presented by a group at the University of Padova (Italy) [43,149]. The module consists of a transceiver and two antennas and is designed for monostatic or bistatic FD measurements (stepped frequency continuous wave radar concept). For monopole antennas, specially shaped radiating patches with wideband performance from 2 to 16 GHz are used. The sending and the receiving antennas are placed very close to each other and decoupled by a T-shaped structure. The integrated transceivers were designed based on a thorough analysis of the circuit-level requirements [149] and built in 65 nm complementary metal-oxide-semiconductor technology. The dynamic range of the system is better than 107 dB, while the bandwidth is 14 GHz. As a proof-of-concept, a planar antenna array was built and measurements on phantoms conducted in which tumor-mimicking structures with sizes of 6 mm and 9 mm could be clearly located. This small and cost-efficient radar system could replace the VNA which is used in many microwave imaging systems.

The fact that a number of systems for microwave breast cancer imaging have already reached clinical trials allows hope for the implementation of microwave-based systems as widespread screening tools in the near future.

3.8 SKIN CANCER DIAGNOSIS USING MICROWAVE TECHNOLOGY

Microwave measurements on skin have been undertaken by numerous research groups and for different applications, including the plain determination of the skin's permittivity, the evaluation of skin hydration for cosmetic research, and the monitoring of wound healing. Microwave reflectometry with open-ended coaxial probes or rectangular microwave probes has so far been the most commonly used technique [169]. Data for the permittivity of skin and skin tumors was presented in Section 3.4.2, and revealed a need for more data especially for the different types of skin cancer. However, the existing data indicates that there is a sufficient contrast between cancer and healthy skin tissue in the microwave band, and several research groups have presented microwave probes and systems for the detection or imaging of skin cancer [30,33,34,39,62,76]. This section first gives some background information on skin and skin cancer, as well as aspects and requirements specific to skin cancer diagnostic systems. General considerations for the design of microwave sensors and imaging systems have been presented in detail in Section 3.5 and are not discussed here.

The skin is the outermost layer of the body, and as such it is easily accessible for measurements of all kinds. The skin is not a homogenous structure but consists of a number of different layers whose boundaries are not necessarily planar. The three main layers—epidermis, dermis, and hypodermis—themselves are divided into several layers. The water content is lowest in the stratum corneum, which is the outermost layer of the skin and part of the epidermis, and hydration increases toward the inner layers. Fig. 3.12 depicts the general skin anatomy and the hydration profile as it was measured by Bennett et al. [27].

The three most common skin cancer types are BCC, SCC, and malignant melanoma [145,171]. BCC and SCC, the nonmelanoma skin cancers, affect a large number of people; e.g., in the United States 3.5 million cases per year are estimated. BCC and SCC rarely spread to other body parts and are highly curable. They do, however, require treatment, i.e., the removal of the tumor, by, e.g.,

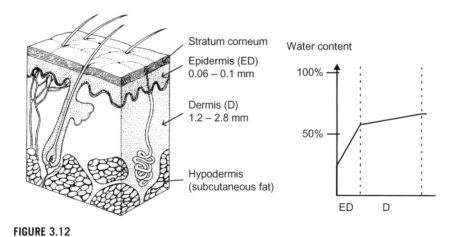

FIGURE 3.12

(A) Skin anatomy, with the three main layers and the stratum corneum [170]. (B) Skin hydration profile as measured by [27].

surgical excision or cryosurgery. Malignant melanomas make up only less than 5% of all skin cancer cases in the United States; total numbers are nevertheless high, and the probability to be diagnosed with melanoma at some point during one's lifetime is 1.9% for women and 3.0% for men in the United States. The risk that malignant melanoma metastasizes is much higher than it is for BCC and SCC, and therefore malignant melanoma accounts for most skin cancer−related deaths. The risk for metastases increases with the depth of tumor invasion [28,172], and early detection and treatment is critical for the survival of the patient. For early-stage localized melanoma the 5-year survival rate is 98%, but for advanced stages, i.e., if the malignant melanoma has already spread, the 5-year survival rate is only 16%. For the treatment of malignant melanoma the tumor, together with a margin of surrounding normal tissue, needs to be removed. If the tumor has spread, immunotherapy, chemotherapy, and/or radiation therapy might be additionally used for treatment [145].

BCC, SCC, and malignant melanoma all develop from cells within the epidermis [28], of which the thickness depends on the body site but is generally less than 100 μm [173]. Even within one type of skin cancer, the tumors can vary considerably in appearance and histological type. It can be difficult to distinguish a skin tumor from benign lesions such as intradermal navi. Diagnosis of skin tumors is nowadays most commonly done by visual inspection. This requires training and experience, and the diagnostic accuracy for malignant melanoma can be as low as 80% even for experienced dermatologists [174]. Diagnosis is especially difficult for shallow, early-stage tumors. An easy and reliable screening technique for skin tumors could help to diagnose skin cancer before it has grown too deeply and thus before it has started to metastasize. Furthermore, in order to decide which removal procedure is best suitable for a particular tumor, information about the size, shape, and the histological type, e.g., if the tumor is well- or ill-defined, is beneficial. A high-resolution skin cancer imaging system could provide this information.

For in vivo evaluation of the skin, reflection measurements are easy to perform and are generally the method of choice. Here, the high attenuation of microwaves in tissue poses a relatively small challenge as tumor growth usually starts at a depth of less than 100 μm. In fact, a limited sensing depth can even prove advantageous for the detection of small tumors since the contribution of the malignant tissue to the overall signal is not hidden as it would if the sensor interaction volume was dominated by the subcutaneous (healthy) fat layer.

For the design of a diagnostic system, the needed spatial resolution and sensing depth play an important role and influence, e.g., the choice of the frequency range. For sensors that target the diagnosis/detection of skin tumors, a sensing depth in the range of several hundred micrometers up to one millimeter and submillimeter spatial resolution is sufficient, and the most commonly suggested technique is near-field microwave reflectometry with the millimeter-wave spectrum being especially suitable for early-stage tumors [39]. Permittivity determination based on microwave-induced heating kinetics [132] is another potential method for skin cancer screening. For the imaging of skin tumors for treatment planning, higher resolution and sensing depth is needed; here terahertz pulsed imaging has been tested [30]. In order to optimize the sensor design with respect to measurement contrast, sensing depth, and resolution, a good understanding of the microwaves' interaction with skin is needed. In measurements and thus also in simulations the inhomogeneous nature of the skin might not be negligible; e.g., when modeling the reflection of a radiated broadband pulse, the varying water content and boundaries between layers need to be taken into account. However, for near-field probes the measured volume is often considered homogenous. This simplification often is sufficient for probes with a small sensing volume [59].

Several aspects are important for use in clinical practice. For a probe in direct contact to the skin, the measurement must be insensitive to the contact force between probe and skin (or alternatively, this force has to be controlled by some means) and air pockets between skin and probe must be prevented, preferably without using any coupling medium. In a free-space measurement, reflection at the air-skin interface is a challenge and good coupling to the skin and a controlled illumination angle can be achieved, e.g., by the use of a quartz window that is pressed onto the skin [175]. As the dielectric properties of skin vary between patients as well as between different body sites, a differential measurement that compares the permittivity of a suspicious lesion to its surrounding healthy tissue is advisable. Therefore in order to be able to examine a certain skin area, sensor arrays or scanning the probe over the suspicious area should be employed. Overall, a short total screening time must be the goal for relevance to clinical applications.

The following selection of systems for the microwave diagnosis and imaging of skin cancer might not be complete, but it presents several different approaches for suitable methods and sensors found in the literature.

Mehta et al. [33] used a Teflon-filled open-ended coaxial probe with an outer conductor diameter of 3.62 mm connected to a VNA for reflection measurements on skin in the frequency range from 300 MHz to 6 GHz. They conducted in vivo measurements on five malignant melanomas and four benign lesions and compared the magnitude and phase of the reflection coefficient with measurements taken both adjacent to the lesion and on healthy skin away from to the lesions. In their measurements the magnitude of the reflection was higher for malignant melanoma compared to normal skin, and lower for benign lesions compared to normal skin, in the measurement frequency range from 0.3 to 3 GHz. This suggests that these kinds of reflection measurements can be used to distinguish malignant lesions from benign ones. However, the number of samples was relatively small and further measurements are needed to support the findings. Furthermore, the probe has a diameter of 3.62 mm and therefore is presumably not suitable for the diagnosis of small, early-stage tumors.

Low cost, compact, scalar reflectometers—one with an open-ended WR-22 waveguide probe operating at 42 GHz and one with a WR-15 waveguide probe operating at 70 GHz—were presented by Taeb et al. [34]. The main system components are a gun diode source, a Y-circulator, a Schottky diode detector, and an open-ended waveguide which is slightly pressed onto the skin for measurements. A resolution of 1.9 mm can be achieved with the 70 GHz probe. The output is a DC voltage which is proportional to the amplitude of the reflected signal; phase information is not recorded, and the whole system is very compact due to its simplicity. Measurements on 15 patients with either BCC or benign lesions using the 42 GHz reflectometer were performed. The average recorded output voltage for measurements on BBC, benign lesions, and healthy skin close to the lesion were $6.8 - 7.7$, $10.5 - 11$, and $12 - 12.8$ V, respectively. Taeb et al. hypothesize that the measurement contrast between BCC and healthy skin is caused by their different tissue permittivity as well as an increased stratum corneum thickness for BCC.

Smulders [62] presented a similar approach using a WR-19 waveguide probe and a VNA for the frequency range of $40 - 60$ GHz, but has not presented results from measurement on skin tumors yet. His measurements on healthy skin of different locations confirm that the magnitude information is sufficient for differentiation of skin tissue.

The above presented systems all use a commercially available open-ended waveguide as probe, and their resolution might not be suitable for the diagnosis of early-stage tumors.

Caratelli et al. [76] presented a resonant near-field sensor, which resembles probes for microwave microscopy. A reconfigurable piezoelectric-needle antenna connected to the inner conductor of a coaxial line with a ground plane, which is connected to the outer conductor, is placed at an elevation of 2.6 mm above the skin. The curvature radius of the antenna can be changed in order to steer the near-field of the probe. The resonance frequency and the bandwidth are expected to vary depending on the permittivity of the skin below. Extensive simulations but no measurements have been published so far. Since the sensor operates at a fixed distance to the skin, control of the distance would need to be part of the measurement system.

The authors of this chapter presented a micromachined near-field probe (Fig. 3.13) with a sensing depth and resolution optimized for early-stage malignant melanoma diagnosis [39]. In contrast to most previous work, the probe is well-matched in impedance to the skin, both by material and by frequency choice, in order to enhance its sensitivity to permittivity variation in the skin. The broadband probe works at frequencies from 90 to 104 GHz. It consists of a metallized dielectric-rod waveguide which is tapered toward the tip, and which is fed from a WR-10 rectangular waveguide probe connected to a VNA. By using high resistivity silicon with a relative permittivity of 11.6 as core material and by tapering the waveguide, a reduced tip size of only 6% of the area of a conventional rectangular waveguide probe of the same frequency range can be achieved, which provides high lateral resolution. The sensing depth in tissue was estimated by simulation and measurements to be $0.3 - 0.4$ mm. The probe was tested on phantom material which mimicked healthy and cancerous skin, and tumor structures

FIGURE 3.13

Micromachined near-field microwave probe for high-resolution skin cancer diagnosis, tested on a murine cancer model.

down to a width of 0.2 μm could be identified in scanning measurements by the probe. From the measurements, a difference of 2 dB in the return loss measured on normal tissue compared to that measured on tumor tissue is expected.

Wallace et al. used a quasi-optical terahertz pulsed imaging system (TeraView Ltd, Cambridge, United Kingdom) with a usable frequency range of 100 GHz to 3 THz to image ex vivo and in vivo samples of BCC [176]. A detailed description of the imaging system can be found in [177]. The THz pulses are generated by an optically excited gallium arsenide emitter and focused onto a quartz imaging plate onto which the sample is placed. The reflected pulse is focused onto a bow-tie photoconductive receiver. Off-axis parabolic mirrors are used to collimate and focus the beams. Light from an ultrafast fiber laser is separated into two beams, of which one is used for the THz pulse excitation and the other is delayed and focused on to the receiver. The system obtains the TD THz waveforms, and images are taken by scanning over the sample area. The lateral and axial resolutions are 150 μm and 20 μm at 3 THz, and the probing depth is in the order of 1 mm. Scans of 18 BCCs ex vivo and five BCCs in vivo were taken, and the waveform of the reflected pulse was recorded for each scanned pixel and analyzed in two different manners: False color images were generated (1) from the maximum amplitude of the reflected pulse and (2) from the amplitude at a certain time, normalized by the amplitude of the minimum peak. In this way a sufficient contrast to identify the tumor could be generated. Alternatively, a Fourier transform can be performed on the TD data to obtain the spectra containing amplitude and phase information and calculating the complex permittivity of the tissue [29]. A similar, slightly improved system was used later to obtain the refractive index and absorption for 13 excised BCC and 10 excised normal tissue samples; a statistically significant difference for the absorption coefficient and the refractive index in the frequency range from 0.2 to 2 THz was found [30].

Studies with any of the above presented microwave skin cancer probes have had at most exploratory characters so far, and their significance is limited by the small number of samples used. Furthermore, many probes were only tested for one type of skin cancer. Large scale clinical trials are necessary in order to study the probes' ability to reliably differentiate between the different skin cancer types, other skin irregularities, and healthy skin.

3.9 FURTHER POSSIBLE APPLICATION AREAS OF MICROWAVE CANCER DIAGNOSIS

Even if breast cancer imaging and skin cancer diagnosis have been the main focus for the development of microwave systems for cancer diagnosis, other organs and body regions have also been targeted.

Like the breast, the head has a relatively small size and it is accessible from many sides. Consequently, microwave imaging systems for the brain have also been suggested. These generally work in the same frequency range as the systems for the breast, i.e., the lower GHz range (1−4 GHz) [178]. The characteristics of the head pose some challenges beyond the ones relevant for breast cancer imaging. Signal absorption is high in brain tissue [15,179], and only a small amount of the incident radiation is backscattered. The head is composed of several layers (skin, fat, skull, gray matter, and white matter) [180], and strong signal reflection occurs not only at the skin-air interface but also at the other tissue interfaces. Furthermore, for practical reasons, the head cannot be simply immersed in a coupling liquid. Therefore, a microwave imaging system for the brain is subjected to even more stringent requirements of sensitivity and dynamic range when compared to breast cancer imaging systems.

Most advanced microwave imaging systems for the brain aim at the detection and classification of strokes [105−108,178,181,182] and traumatic brain injuries [183], and exploit the dielectric contrast between normal brain tissue (white and gray brain matter) and blood [15,179].

A dielectric contrast exists also between normal brain tissue and tumor tissue at microwave frequencies. The permittivity of brain tumor tissue has been measured to be 10−30% higher than the permittivity of normal brain tissue for frequencies between 0.1 and 5 GHz [184]. However, efforts for developing microwave systems for brain tumor detection have been limited so far, especially in comparison to the advanced microwave imaging systems for breast cancer. Most of the suggested systems have only been studied using simplified models in computer simulations and/or phantom studies. The publications often concentrate on a certain aspect of the imaging system such as the antenna design [185−188] or the signal processing [122,189,190]. Further research and development is needed to make microwave imaging systems for brain cancer detection clinically applicable. Experience and results from systems for breast cancer imaging as well as stroke detection might contribute to this development.

Data for normal and malignant lung tissue [6,191] has shown that microwaves can theoretically be used for the discrimination of lung malignancies from healthy lung tissue, and microwave tomography has been suggested for lung cancer detection [103,192]. However, so far only theoretical considerations along with simulations [103] and initial tests on a phantom [192] have been published. Microwave imaging of tumors in the lung is especially challenging due to several reasons. The upper body has a large volume, with the above explained consequences for microwave absorption and transmitted signal strength. It is structurally complex, and consequently it is dielectrically heterogeneous with a high contrast between different organs and tissues, e.g., bones, skin, and the heart. This makes the detection of the relatively small permittivity differences between healthy and malignant tissue extremely challenging. Motion from breathing and the variation of lung tissue properties during the respiratory cycle pose a further challenge. Utilization of microwave imaging for lung cancer detection appears to be much more challenging than microwave imaging of the breast, and systems are not expected to be applied clinically any time soon.

Studies on excised liver tissue indicated a dielectric contrast between healthy liver tissue and liver tumors [6,193]. A more recent study performing ex vivo as well as in vivo measurement at frequencies from 0.5 to 20 GHz confirmed the contrast for excised tissue. For data determined in vivo, however, almost no statistically significant differences were found [17]. Therefore, currently microwave imaging and in vivo measurements appear not to be applicable to liver cancer diagnosis.

REFERENCES

[1] Kharkovsky S, Zoughi R. Microwave and millimeter wave nondestructive testing and evaluation—overview and recent advances. IEEE Instrum Meas Mag 2007;10(2): 26—38.

[2] Bakhtiari S, Qaddoumi N, Ganchev S, Zoughi R. Microwave noncontact examination of disbond and thickness variation in stratified composite media. IEEE Trans Microw Theory Tech 1994;42(3):389—95.

[3] Kaatze U. Measuring the dielectric properties of materials. ninety-year development from low-frequency techniques to broadband spectroscopy and high-frequency imaging. Meas Sci Technol 2013;24(1):012005.

[4] Joines WT, Jirtle RL, Rafal MD, Schaefer DJ. Microwave power absorption differences between normal and malignant tissue. Int J Radiat Oncol Biol Phys 1980;6 (6):681—7.

[5] Schepps JL, Foster KR. The UHF and microwave dielectric-properties of normal and tumor-tissue—variation in dielectric-properties with tissue water-content. Phys Med Biol 1980;25(6):1149—59.

[6] Joines WT, Zhang Y, Li C, Jirtle RL. The measured electrical properties of normal and malignant human tissues from 50 to 900 mhz. Med Phys 1994;21(4):547−50.

[7] Rosen A, Stuchly M, Vander Vorst A. Applications of RF/microwaves in medicine. IEEE Trans Microw Theory Tech 2002;50(3):963−74.

[8] Fear E, Hagness S, Meaney P, Okoniewski M, Stuchly M. Enhancing breast tumor detection with near-field imaging. IEEE Microw Mag 2002;3(1):48−56.

[9] Nikolova N. Microwave imaging for breast cancer. IEEE Microw Mag 2011;12 (7):78−94.

[10] Foster KR, Schwan HP. Dielectric-properties of tissue and biological-materials—a critical review. Crit Rev Biomed Eng 1989;17(1):25−104.

[11] Hurt W. Multiterm debye dispersion relations for permittivity of muscle. IEEE Trans Biomed Eng 1985;BME-32(1):60−4.

[12] Cole KS, Cole RH. Dispersion and absorption in dielectrics I. Alternating current characteristics. J Chem Phys 1941;9(4):341−51.

[13] Davidson DW, Cole RH. Dielectric relaxation in glycerol, propylene glycol, and n-propanol. J Chem Phys 1951;19(12):1484−90.

[14] Havriliak S, Negami S. A complex plane representation of dielectric and mechanical relaxation processes in some polymers. Polymer (Guildf) 1967;8:161−210.

[15] Gabriel S, Lau RW, Gabriel C. The dielectric properties of biological tissues: III. Parametric models for the dielectric spectrum of tissues. Phys Med Biol 1996;41 (11):2271.

[16] Lazebnik M, Popovic D, McCartney L, Watkins CB, Lindstrom MJ, Harter J, et al. A large-scale study of the ultrawideband microwave dielectric properties of normal, benign and malignant breast tissues obtained from cancer surgeries. Phys Med Biol 2007;52(20):6093.

[17] O'Rourke AP, Lazebnik M, Bertram JM, Converse MC, Hagness SC, Webster JG, et al. Dielectric properties of human normal, malignant and cirrhotic liver tissue: in vivo and ex vivo measurements from 0.5 to 20 GHz using a precision open-ended coaxial probe. Phys Med Biol 2007;52(15):4707.

[18] Sha L, Ward E, Stroy B. A review of dielectric properties of normal and malignant breast tissue. IEEE SoutheastCon 2002457−62.

[19] Lazebnik M, McCartney L, Popovic D, Watkins CB, Lindstrom MJ, Harter J, et al. A large-scale study of the ultrawideband microwave dielectric properties of normal breast tissue obtained from reduction surgeries. Phys Med Biol 2007;52(10):2637.

[20] Lazebnik M, Okoniewski M, Booske J, Hagness S. Highly accurate debye models for normal and malignant breast tissue dielectric properties at microwave frequencies. IEEE Microw Wireless Compon Lett 2007;17(12):822−4.

[21] Sugitani T, Kubota S-i, Kuroki S-i, Sogo K, Arihiro K, Okada M, et al. Complex per-mittivities of breast tumor tissues obtained from cancer surgeries. Appl Phys Lett 2014;104(25). p. 253702.

[22] Khan U, Al-Moayed N, Nguyen N, Korolev K, Afsar M, Naber S. Broadband dielec-tric characterization of tumorous and nontumorous breast tissues. IEEE Trans Microw Theory Tech 2007;55(12):2887−93.

[23] Martellosio A, Pasian M, Bozzi M, Perregrini L, Mazzanti A, Svelto F, et al. 0.5−50 GHz dielectric characterisation of breast cancer tissues. Electron Lett 2015;51(13):974−5.

[24] Ghodgaonkar D, Gandhi OP, Iskander M. Complex permittivity of human skin in vivo in the frequency band 26.5−60 GHz. IEEE Antennas and Propagation Society International Symposium 2000;2:1100−3.

[25] Alabaster C. Permittivity of human skin in millimetre wave band. Electron Lett 2003;39(21):1521−2.

[26] Chahat N, Zhadobov M, Augustine R, Sauleau R. Human skin permittivity models for millimetre-wave range. Electron Lett 2011;47(7):427−8.

[27] Bennett D, Li W, Taylor Z, Grundfest W, Brown E. Stratified media model for terahertz reflectometry of the skin. IEEE Sens J 2011;11(5):1253−62.

[28] Nouri K. In: Nouri K, editor. Skin Cancer. McGraw-Hill Professional Publishing; 2007.

[29] Pickwell E, Fitzgerald AJ, Cole BE, Taday PF, Pye RJ, Ha T, et al. Simulating the response of terahertz radiation to basal cell carcinoma using ex vivo spectroscopy measurements. J Biomed Opt 2005;10(6) 064021−064021−7.

[30] Wallace VP, Fitzgerald AJ, Pickwell E, Pye RJ, Taday PF, Flanagan N, et al. Terahertz pulsed spectroscopy of human basal cell carcinoma. Appl Spectrosc 2006;60(10):1127−33.

[31] Pickwell E, Cole BE, Fitzgerald AJ, Pepper M, Wallace VP. In vivo study of human skin using pulsed terahertz radiation. Phys Med Biol 2004;49(9):1595.

[32] Truong B, Tuan H, Fitzgerald A, Wallace V, Nguyen H. High correlation of double debye model parameters in skin cancer detection. 36th Annual International Conference of the IEEE Engineering in Medicine and Biology Society 2014;718−21.

[33] Mehta P, Chand K, Narayanswamy D, Beetner D, Zoughi R, Stoecker W. Microwave reflectometry as a novel diagnostic tool for detection of skin cancers. IEEE Trans Instrum Meas 2006;55(4):1309−16.

[34] Taeb A, Gigoyan S, Safavi-Naeini S. Millimetre-wave waveguide reflectometers for early detection of skin cancer. IET Microw, Antennas Propagation 2013;7(14): 1182−6.

[35] Lin J. Frequency optimization for microwave imaging of biological tissues. Proc IEEE 1985;73(2):374−5.

[36] Semenov S, Svenson R, Boulyshev A, Souvorov A, Borisov V, Sizov Y, et al. Microwave tomography: two-dimensional system for biological imaging. IEEE Trans Biomed Eng 1996;43(9):869−77.

[37] Klemm M, Gibbins D, Leendertz J, Horseman T, Preece A, Benjamin R, et al. Development and testing of a 60-element UWB conformal array for breast cancer imaging. In: 5th European Conference on Antennas and Propagation (EUCAP); April 2011, pp. 3077−79.

[38] Bourqui J, Sill JM, Fear EC. A prototype system for measuring microwave frequency reflections from the breast. Int J Biomed Imaging 2012;2012.

[39] Töpfer F, Dudorov S, Oberhammer J. Millimeter-wave near-field probe designed for high-resolution skin cancer diagnosis. IEEE Trans Microw Theory Tech 2015;63 (6):2050−9.

[40] Porter E, Santorelli A, Coates M, Popovi M. Time-domain microwave breast cancer detection: Extensive system testing with phantoms. Technol Cancer Res Treat 2013;12(2):131−43.

[41] Hines ME, Stinehelfer HE. Time-domain oscillographic microwave network analysis using frequency-domain data. IEEE Trans Microw Theory Tech 1974;22(3):276−82.

[42] Zhurbenko V. Challenges in the design of microwave imaging systems for breast cancer detection. Adv Electr Comput Eng 2011;11(1).

[43] Bassi M, Caruso M, Khan M, Bevilacqua A, Capobianco A, Neviani A. An integrated microwave imaging radar with planar antennas for breast cancer detection. IEEE Trans Microw Theory Tech 2013;61(5):2108–18.

[44] Stuchly M, Stuchly S. Coaxial line reflection methods for measuring dielectric properties of biological substances at radio and microwave frequencies—a review. IEEE Trans Instrum Meas 1980;29(3):176–83.

[45] Popovic D, McCartney L, Beasley C, Lazebnik M, Okoniewski M, Hagness S, et al. Precision open-ended coaxial probes for in vivo and ex vivo dielectric spectroscopy of biological tissues at microwave frequencies. IEEE Trans Microw Theory Tech 2005;53(5):1713–22.

[46] Krupka J. Frequency domain complex permittivity measurements at microwave frequencies. Meas Sci Technol 2006;17(6):R55.

[47] Kaatze U. Techniques for measuring the microwave dielectric properties of materials. Metrologia 2010;47(2):S91.

[48] Brown ER. Fundamentals of terrestrial millimeter-wave and THz remote sensing. Int J High Speed Electron Syst 2003;13(04):995–1097.

[49] Bennett DB, Taylor ZD, Tewari P, Singh RS, Culjat MO, Grundfest WS, et al. Terahertz sensing in corneal tissues. J Biomed Opt 2011;16(5). pp. 057003–057003–8.

[50] Taylor Z, Singh R, Bennett D, Tewari P, Kealey C, Bajwa N, et al. THz medical imaging: in vivo hydration sensing. IEEE Trans Terahertz Sci Technol 2011;1(1):201–19.

[51] Sung S, Bennett D, Taylor Z, Bajwa N, Tewari P, Maccabi A, et al. Reflective measurement of water concentration using millimeter wave illumination,". Proceedings of SPIE 7984, Health Monitoring of Structural and Biological Systems 2011 2011;7984. pp. 798434–798434–6.

[52] Burdette E, Cain F, Seals J. In vivo probe measurement technique for determining dielectric properties at vhf through microwave frequencies. IEEE Trans Microw Theory Tech 1980;28(4):414–27.

[53] Alanen E, Lahtinen T, Nuutinen J. Variational formulation of open-ended coaxial line in contact with layered biological medium. IEEE Trans Biomed Eng 1998;45(10):1241–8.

[54] Zajícek R, Oppl L, Vrba J. Broadband measurement of complex permittivity using reflection method and coaxial probes. Radioengineering 2008;17(1):14–19.

[55] Alanen E, Nuutinen J, Nicklén K, Lahtinen T, Mönkkönen J. Measurement of hydration in the stratum corneum with the moisturemeter and comparison with the corneometer. Skin Res Technol 2004;10(1):32–7.

[56] Hayashi Y, Miura N, Shinyashiki N, Yagihara S. Free water content and monitoring of healing processes of skin burns studied by microwave dielectric spectroscopy in vivo. Phys Med Biol 2005;50(4):599.

[57] Hwang H, Yim J, Cho J-W, Changyul-Cheon, Kwon Y. 110 GHz broadband measurement of permittivity on human epidermis using 1 mm coaxial probe. IEEE MTT-S Int Microw Symp 2003;1:399–402.

[58] Kharkovsky S, Ghasr M, Abou-Khousa M, Zoughi R. Near-field microwave and mm-wave noninvasive diagnosis of human skin. IEEE Int Workshop Med Meas Appl 20095–7.

[59] Alekseev S, Ziskin M. Human skin permittivity determined by millimeterwave reflection measurements. Bioelectromagnetics 2007;28(5):331−9.

[60] Alekseev S, Szabo I, Ziskin M. Millimeter wave reflectivity used for measurement of skin hydration with different moisturizers. Skin Res Technol 2008;14(4):390−6.

[61] Janssen N, Smulders P. Design of millimeter-wave probe for diagnosis of human skin. In: 42nd European Microwave Conference (EuMC); 2012, 440−43.

[62] Smulders P. Analysis of human skin tissue by millimeter-wave reflectometry. Skin Res Technol 2013;19(1):e209−16.

[63] Takeyama T, Nikawa Y. Change of electromagnetic field distribution in millimeter waves by dental caries appearance. Asia-Pac Microwave Conf Proc (APMC) 20101372−5.

[64] Qaddoumi N, Abou-Khousa M, Saleh W. Near-field microwave imaging utilizing tapered rectangular waveguides. IEEE Trans Instrum Meas 2006;55(5):1752−6.

[65] Jeong E, Jeong G, Kim J-M, Park J-H, Cho J-W, Changyul-Cheon, et al. Multi-layer processed probes for permittivity measurement. 2004 IEEE MTT-S Int Microw Symp 2004;3:1813−16.

[66] Kim J-M, Oh D, Yoon J, Cho S, Kim N, Cho J, et al. In vitro and in vivo measurement for biological applications using micromachined probe. IEEE Trans Microw Theory Tech 2005;53(11):3415−21.

[67] Kim J-M, Oh DH, Baek C-W, Cho J-W, Kwon Y, Cheon C, et al. In vitro measurement using a MEMS probe array with five-strip lines for permittivity measurement. J Micromech Microeng 2006;16(1):173.

[68] Kang B, Park J-H, Cho J, Kwon K, Lim S, Yoon J, et al. Novel low-cost planar probes with broadside apertures for nondestructive dielectric measurement of biological materials at microwave frequencies. IEEE Trans Microw Theory Tech 2005;53(1):134−43.

[69] Kim J-M, Cho S, Kim N, Yoon J, Cho J, Cheon C, et al. Planar type micromachined probe with low uncertainty at low frequencies. Sens Actuators, A 2007;139(1−2):111−17 Selected Papers From the Asia-Pacific Conference of Transducers and Micro-Nano Technology (APCOT 2006) Asia-Pacific Conference of Transducers and Micro-Nano technology.

[70] Kim K, Kim N, Hwang S-H, Kim Y-K, Kwon Y. A miniaturized broadband multi-state reflectometer integrated on a silicon MEMS probe for complex permittivity measurement of biological material. IEEE Trans Microw Theory Tech 2013;61(5):2205−14.

[71] Kim J-M, Cheon C, Kwon Y, Kim Y-K. A hybrid RF MEMS probe array system with a SP3T RF MEMS silicon switch for permittivity measurement. J Micromech Microeng 2008;18(8).

[72] Klein N, Lahl P, Poppe U, Kadlec F, Kuzel P. A metal-dielectric antenna for tera-hertz near-field imaging. J Appl Phys 2005;98(1). pp. 014910−014910−5.

[73] Danylyuk S, Poppe U, Kadlec F, Kuzel P, Berta M, Klein N, et al. Broadband microwave-to-terahertz near-field imaging. IEEE/MTT-S Int Microw Symp 20071383−6.

[74] Caratelli D, Yarovoyo AG, Massaro A, Lay-Ekuakille A. Design and full-wave analysis of piezoelectric micro-needle antenna sensors for enhanced near-field detection of skin cancer. Prog Electromagn Res 2012;125:391−413.

[75] Caratelli D, Massaro A, Cingolani R, Yarovoy A. Accurate time-domain modeling of reconfigurable antenna sensors for non-invasive melanoma skin cancer detection. IEEE Sens J 2012;12(3):635−43.

[76] Caratelli D, Lay-Ekuakille A, Vergallo P. Non-invasive reflectometry-based detection of melanoma by piezoelectric micro-needle antenna sensors. Prog Electromagn Res 2013;135:91−103.

[77] Wang CR, Yan LP, Deng YD, Liu CJ. A double-armed planar equiangular spiral patch probe for biomedical measurements. J Electromagn Waves Appl 2008;22 (8−9):1258−66.

[78] Forgues P, Goldberg M, Smith A, Stuchly S. Medical computed tomography using microwaves. Frontiers of Engineering in Health Care. IEEE Engineering in Medicine and Biology Society; 1980. p. 240−74.

[79] Rao PS, Santosh K, Gregg EC. Computed tomography with microwaves.. Radiology 1980;135(3):769−70.

[80] Ermert H, Fulle G, Hiller D. Microwave computerized tomography. In: 11th European Microwave Conference; 1981, 421−26.

[81] Maini R, Iskander M, Durney C. On the electromagnetic imaging using linear reconstruction techniques. Proc IEEE 1980;68(12):1550−2.

[82] Chandra R, Zhou H, Balasingham I, Narayanan R. On the opportunities and challenges in microwave medical sensing and imaging. IEEE Trans Biomed Eng 2015;62 (7):1667−82.

[83] Bulyshev AE, Souvorov AE, Semenov SY, Svenson RH, Nazarov AG, Sizov YE, et al. Three-dimensional microwave tomography. Theory and computer experiments in scalar approximation. Inverse Probl 2000;16(3):863.

[84] Yu C, Yuan M, Stang J, Bresslour E, George R, Ybarra G, et al. Active microwave imaging ii: 3-D system prototype and image reconstruction from experimental data. IEEE Trans Microw Theory Tech 2008;56(4):991−1000.

[85] Semenov S, Svenson R, Bulyshev A, Souvorov A, Nazarov A, Sizov Y, et al. Three-dimensional microwave tomography: experimental prototype of the system and vector born reconstruction method. IEEE Trans Biomed Eng 1999;46(8):937−46.

[86] Epstein NR, Meaney PM, Paulsen KD. 3d parallel-detection microwave tomography for clinical breast imaging. Rev Sci Instrum 2014;85(12).

[87] Meaney P, Paulsen K, Chang J, Fanning M, Hartov A. Nonactive antenna compensation for fixed-array microwave imaging. ii. Imaging results. IEEE Trans Med Imaging 1999;18(6):508−18.

[88] Shrestha S, Agarwal M, Ghane P, Varahramyan K. Flexible microstrip antenna for skin contact application. Int J Antennas Propagation 2012;5:745426.

[89] Bolomey J, Izadnegahdar A, Jofre L, Pichot C, Peronnet G, Solaimani M. Microwave diffraction tomography for biomedical applications. IEEE Trans Microw Theory Tech 1982;30(11):1998−2000.

[90] Peronnet G, Pichot C, Bolomey J, Jofre L, Izadnegahdar A, Szeles C, et al. A microwave diffraction tomography system for biomedical applications. In: 13th European Microwave Conference; 1983, 529−33.

[91] Tabbara W, Duchene B, Pichot C, Lesselier D, Chommeloux L, Joachimowicz N. Diffraction tomography: contribution to the analysis of some applications in microwaves and ultrasonics. Inverse Probl 1988;4(2):305.

[92] Slaney M, Kak A, Larsen L. Limitations of imaging with first-order diffraction tomography. IEEE Trans Microw Theory Tech 1984;32(8):860−74.

[93] Pastorino M. In: Hoboken N, editor. Microwave imaging. Wiley; 2010.

[94] Joachimowicz N, Pichot C, Hugonin J. Inverse scattering: an iterative numerical method for electromagnetic imaging. IEEE Trans Antennas Propagation 1991;39 (12):1742−53.

[95] De Zaeytijd J, Franchois A, Eyraud C, Geffrin JM. Full-wave three-dimensional microwave imaging with a regularized gauss−newton method—theory and experiment. IEEE Trans Antennas Propagation 2007;55(11):3279−92.

[96] Rubk T, Meaney P, Meincke P, Paulsen K. Nonlinear microwave imaging for breast-cancer screening using gauss: Newton's method and the CGLS inversion algorithm. IEEE Trans Antennas Propagation 2007;55(8):2320−31.

[97] Franchois A, Pichot C. Microwave imaging-complex permittivity reconstruction with a Levenberg−Marquardt method. IEEE Trans Antennas Propagation 1997;45 (2):203−15.

[98] Meaney P, Paulsen K, Pogue B, Miga M. Microwave image reconstruction utilizing log-magnitude and unwrapped phase to improve high-contrast object recovery. IEEE Trans Med Imaging 2001;20(2):104−16.

[99] Chew W, Wang Y. Reconstruction of two-dimensional permittivity distribution using the distorted born iterative method. IEEE Trans Med Imaging 1990;9 (2):218−25.

[100] Kleinman R, den Berg P. A modified gradient method for two- dimensional problems in tomography. J Comput Appl Math 1992;42(1):17−35.

[101] Abubakar A, van den Berg P, Mallorqui J. Imaging of biomedical data using a multiplicative regularized contrast source inversion method. IEEE Trans Microw Theory Tech 2002;50(7):1761−71.

[102] Zhang ZQ, Liu QH. Three-dimensional nonlinear image reconstruction for microwave biomedical imaging. IEEE Trans Biomed Eng 2004;51(3):544−8.

[103] Semenov S. Microwave tomography: review of the progress towards clinical applications. Philos Trans A Math Phys Eng Sci 2009;367(1900):3021−42.

[104] Meaney P, Fanning M, di Florio-Alexander R, Kaufman P, Geimer S, Zhou T, et al. Microwave tomography in the context of complex breast cancer imaging. In: 2010 Annual International Conference of the IEEE Engineering in Medicine and Biology Society (EMBC); 2010, 3398−401.

[105] Ireland D, Bialkowski ME. Microwave head imaging for stroke detection. Prog Electromagn Res M 2011;21:163−75.

[106] Scapaticci R, Donato LD, Catapano I, Crocco L. A feasibility study on microwave imaging for brain stroke monitoring. Prog Electromagn Res B 2012;40:305−24.

[107] Mohammed BJ, Abbosh AM, Ireland D, Bialkowski ME. Compact wideband antenna immersed in optimum coupling liquid for microwave imaging of brain stroke. Prog Electromagn Res 2012;27:27−39.

[108] Persson M, Fhager A, Trefna H, Yu Y, McKelvey T, Pegenius G, et al. Microwave-based stroke diagnosis making global prehospital thrombolytic treatment possible. IEEE Trans Biomed Eng 2014;61(11):2806−17.

[109] Fear E, Li X, Hagness S, Stuchly M. Confocal microwave imaging for breast cancer detection: localization of tumors in three dimensions. IEEE Trans Biomed Eng 2002;49(8):812−22.

[110] Bond EJ, Li X, Hagness S, Van Veen B. Microwave imaging via space-time beamforming for early detection of breast cancer. IEEE Trans Antennas Propagation 2003;51(8):1690−705.

[111] Li D, Meaney P, Paulsen K. Conformal microwave imaging for breast cancer detection. IEEE Trans Microw Theory Tech 2003;51(4):1179−86.

[112] Klemm M, Craddock I, Leendertz J, Preece A, Benjamin R. Radar-based breast cancer detection using a hemispherical antenna array: experimental results. IEEE Trans Antennas Propagation 2009;57(6):1692−704.

[113] Chew KM, Sudirman R, Mahmood NH, Seman N, Yong CY. Human brain microwave imaging signal processing: frequency domain (S-parameters) to time domain conversion. Engineering 2013;5(5B):31−6.

[114] Byrne D, O'Halloran M, Glavin M, Jones E. Data independent radar beamforming algorithms for breast cancer detection. Prog Electromagn Res 2010;107(331−348).

[115] Porter E, Santorelli A, Popovic M. Time-domain microwave radar for breast screening: initial testing with volunteers. In: 8th European Conference on Antennas and Propagation (EuCAP); 2014, 104−7.

[116] Lim HB, Nhung NTT, Li E-P, Thang ND. Confocal microwave imaging for breast cancer detection: delay-multiply-and-sum image reconstruction algorithm. IEEE Trans Biomed Eng 2008;55(6):1697−704.

[117] Li X, Bond EJ, Van Veen B, Hagness S. An overview of ultra-wideband microwave imaging via space-time beamforming for early-stage breast-cancer detection. IEEE Antennas Propagation Mag 2005;47(1):19−34.

[118] Moll J, Kexel C, Krozer V. A comparison of beamforming methods for microwave breast cancer detection in homogeneous and heterogeneous tissue. In: European Microwave Conference (EuMC); 2013, 1839−42.

[119] Kosmas P, Rappaport C. Time reversal with the FDTD method for microwave breast cancer detection. IEEE Trans Microw Theory Tech 2005;53(7):2317−23.

[120] Kosmas P, Rappaport C. FDTD-based time reversal for microwave breast cancer detection-localization in three dimensions. IEEE Trans Microw Theory Tech 2006;54(4):1921−7.

[121] Chen Y, Gunawan E, Low KS, Wang SC, Kim Y, Soh CB. Pulse design for time reversal method as applied to ultrawideband microwave breast cancer detection: A two-dimensional analysis. IEEE Trans Antennas Propagation 2007;55(1):194−204.

[122] Zhang H, Arslan T, Flynn B. Wavelet de-noising based microwave imaging for brain cancer detection. In: Antennas and Propagation Conference (LAPC); 2013, 482−85.

[123] Guo L, Abbosh A. Optimization-based confocal microwave imaging in medical applications. IEEE Trans Antennas Propagation 2015;63(8):3531−9.

[124] Carr K. Microwave radiometry: its importance to the detection of cancer. IEEE Trans Microw Theory Tech 1989;37(12):1862−9.

[125] Cheever E, Foster K. Microwave radiometry in living tissue: what does it measure? IEEE Trans Biomed Eng 1992;39(6):563−8.

[126] Mouty S, Bocquet B, Ringot R, Rocourt N, Devos P. Microwave radiometric imaging (MWI) for the characterisation of breast tumours. Eur Phys J Appl Phys 2000;10:73−8, 4.

[127] Stec B, Dobrowolski A, Susek W. Multifrequency microwave thermograph for biomedical applications. IEEE Trans Biomed Eng 2004;51(3):548−50.

[128] Klemetsen O, Birkelund Y, Jacobsen S. Low-cost and small-sized medical microwave radiometer design. In: IEEE Antennas and Propagation Society International Symposium (APSURSI); 2010, 1—4.

[129] Caferova S, Uysal F, Balc P, Saydam S, Canda T. Efficacy and safety of breast radiothermometry in the differential diagnosis of breast lesions. Contemp Oncol (Pozn) 2014;18(3):197—203.

[130] "Medical radiometer - RTM - 01 - RES and its use in detecting hotspots in female breast." [Online]. Available from: https://clinicaltrials.gov/ct2/show/NCT02286583.

[131] Carr K, El-Mahdi A, Shaeffer J. Dual-mode microwave system to enhance early detection of cancer. IEEE Trans Microw Theory Tech 1981;29(3):256—60.

[132] Chahat N, Zhadobov M, Sauleau R, Alekseev S. New method for determining dielectric properties of skin and phantoms at millimeter waves based on heating kinetics. IEEE Trans Microw Theory Tech 2012;60(3):827—32.

[133] Ku G, Wang LV. Scanning microwave-induced thermoacoustic tomography: signal, resolution, and contrast. Med Phys 2001;28(1):4—10.

[134] Kruger RA, Reinecke DR, Kruger GA. Thermoacoustic computed tomography—technical considerations. Med Phys 1999;26(9):1832—7.

[135] Kruger RA, Miller KD, Reynolds HE, William J, Kiser L, Reinecke DR, et al. Breast cancer in vivo: contrast enhancement with thermoacoustic CT at 434 mhz—feasibility study. Radiology 2000;216(1):279—83.

[136] Lü-Ming Z, Da X, Huai-Min G, Di-Wu Y, Si-Hua Y, Liang-Zhong X. Fast microwave-induced thermoacoustic tomography based on multi-element phase-controlled focus technique. Chin Phys Lett 2006;23(5):1215.

[137] Nie L, Xing D, Zhou Q, Di-Wu Z, Guo H. Microwave-induced thermoacoustic scanning CT for high-contrast and noninvasive breast cancer imaging. Med Phys 2008;35(9):4026—32.

[138] Kirkpatrick SJ, Wang RK, Duncan DD, Kulesz-Martin M, Lee K. Imaging the mechanical stiffness of skin lesionsby in vivo acousto-optical elastography. Opt Express 2006;14(21):9770—9.

[139] Samani A, Zubovits J, Plewes D. Elastic moduli of normal and pathological human breast tissues: an inversion-technique-based investigation of 169 samples. Phys Med Biol 2007;52(6):1565.

[140] Wang Z. Mechanical and optical methods for breast cancer imaging. PhD dissertation, University of Iowa, 2010. [Online]. Available from: http://ir.uiowa.edu/etd/618.

[141] Abbosh A, Crozier S. Strain imaging of the breast by compression microwave imaging. IEEE Antennas Wireless Propagation Lett 2010;9:1229—32.

[142] Henin B, Abbosh A. Microwave imaging of breast using combination of biomechanical and electrical properties. In: International Biomedical Engineering Conference (CIBEC); 2012, 183—6.

[143] Top C, Tafreshi A, Gencer N. Harmonic motion microwave doppler imaging method for breast tumor detection. In: 36th Annual International Conference of the IEEE Engineering in Medicine and Biology Society (EMBC); 2014, 6072—75.

[144] Jiang H, Li C, Pearlstone D, Fajardo LL. Ultrasound-guided microwave imaging of breast cancer: tissue phantom and pilot clinical experiments. Med Phys 2005;32 (8):2528—35.

[145] Society AC, editor. Cancer facts & figures 2015. Atlanta: American Cancer Society; 2015.

[146] I. o. M. D. o. E. Committee on Technologies for the Early Detection of Breast Cancer, National Cancer Policy Board and N. R. C. Life Studies, Mammography and Beyond: Developing Technologies for the Early Detection of Breast Cancer, I. C. H. Sharyl J. Nass and J. C. Lashof, Eds. Washington, D.C.: The National Academies Press, 2001.

[147] Sepponen R. Medical diagnostic microwave scanning apparatus. US Patent 4641659, 1987.

[148] Bocquet B, van de Velde J, Mamouni A, Leroy Y, Giaux G, Delannoy J, et al. Microwave radiometric imaging at 3 ghz for the exploration of breast tumors. IEEE Trans Microw Theory Tech 1990;38(6):791−3.

[149] Bassi M, Bevilacqua A, Gerosa A, Neviani A. Integrated SFCW transceivers for UWB breast cancer imaging: architectures and circuit constraints. IEEE Trans Circuits Syst I: Regul Pap 2012;59(6):1228−41.

[150] Grzegorczyk TM, Meaney PM, Jeon SI, Geimer SD, Paulsen KD. Importance of phase unwrapping for the reconstruction of microwave tomographic images. Biomed Opt Express 2011;2(2):315−30.

[151] Bucci OM, Bellizzi G, Catapano I, Crocco L, Scapaticci R. MNP enhanced microwave breast cancer imaging: Measurement constraints and achievable performances. IEEE Antennas Wireless Propagation Lett 2012;11:1630−3.

[152] Scapaticci R, Bellizzi G, Catapano I, Crocco L, Bucci O. An effective procedure for MNP-enhanced breast cancer microwave imaging. IEEE Trans Biomed Eng 2014;61(4):1071−9.

[153] Gao F, Van Veen B, Hagness S. Microwave imaging of breast tumors with contrast agents: theoretical study of dielectric contrast enhancement requirements. In: IEEE Antennas and Propagation Society International Symposium (APSURSI); 2013, 2030−31.

[154] Sill J, Fear E. Tissue sensing adaptive radar for breast cancer detection: study of immersion liquids. Electron Lett 2005;41(3):113−15.

[155] Chao L, Afsar M. A millimeter wave breast cancer imaging methodology.In: Conference on Precision Electromagnetic Measurements (CPEM); 2012, 74−5.

[156] Li X, Hagness S, Van Veen B, van der Weide D. Experimental investigation of microwave imaging via space-time beamforming for breast cancer detection. In: IEEE MTT-S International Microwave Symposium; 2003, 379−82, 1.

[157] Sabouni A, Flores-Tapia D, Noghanian S, Thomas G, Pistorius S. Hybrid microwave tomography technique for breast cancer imaging. In: 28th Annual International Conference of the IEEE Engineering in Medicine and Biology Society (EMBS '06); 2006, 4273−76.

[158] Elsdon M, Smith D, Leach M, Foti SJ. Experimental investigation of breast tumor imaging using indirect microwave holography.. Microw Opt Technol Lett 2006;48 (3):480−2.

[159] Elsdon M, Leach M, Skobelev S, Smith D. Microwave holographic imaging of breast cancer. In: International Symposium on Microwave, Antenna, Propagation and EMC Technologies for Wireless Communications; 2007, 966−9.

[160] Amineh RK, Ravan M, Khalatpour A, Nikolova N. Three-dimensional near-field microwave holography using reflected and transmitted signals. IEEE Trans Antennas Propagation 2011;59(12):4777−89.

[161] Amineh RK, Ravan M, McCombe J, Nikolova NK. Three-dimensional microwave holographic imaging employing forward-scattered waves only. Int J Antennas Propagation, 2013.

[162] Meaney P, Paulsen K, Hartov A, Crane R. An active microwave imaging system for reconstruction of 2-d electrical property distributions. IEEE Trans Biomed Eng 1995;42(10):1017–26.

[163] Meaney P, Fanning M, Li D, Poplack SP, Paulsen K. A clinical prototype for active microwave imaging of the breast. IEEE Trans Microw Theory Tech 2000;48 (11):1841–53.

[164] Klemm M, Leendertz J, Gibbins D, Craddock I, Preece A, Benjamin R. Microwave radar-based differential breast cancer imaging: imaging in homogeneous breast phantoms and low contrast scenarios. IEEE Trans Antennas Propagation 2010;58 (7):2337–44.

[165] Porter E, Kirshin E, Santorelli A, Popovic M. A clinical prototype for microwave breast imaging using time-domain measurements. In: 7th European Conference on Antennas and Propagation (EuCAP); 2013, 830–2.

[166] Kanj H, Popovic M. A novel ultra-compact broadband antenna for microwave breast tumor detection. Prog Electromagn Res 2008;86:169–98.

[167] Fear E, Sill J. Preliminary investigations of tissue sensing adaptive radar for breast tumor detection. In: Proc 25th Annual International Conference of the IEEE Engineering in Medicine and Biology Society; 2003, 3787–90 Vol. 4.

[168] Bourqui J, Okoniewski M, Fear E. Balanced antipodal Vivaldi antenna with dielectric director for near-field microwave imaging. IEEE Trans Antennas Propagation 2010;58(7):2318–26.

[169] Töpfer F, Oberhammer J. Millimeter-wave tissue diagnosis: the most promising fields for medical applications. IEEE Microw Mag 2015;16(4):97–113.

[170] Skin (layers, glands, vessels). The Web site of the National Cancer Institute. Available from: http://www.cancer.gov.

[171] Howlader N, Noone A, Krapcho M, Garshell J, Miller D, Altekruse S, Kosary C, Yu M, Ruhl J, Tatalovich Z, Mariotto A, Lewis D, Chen H, Feuer E, Cronin K, editors. SEER cancer statistics review 1975–2011. Bethesda: National Cancer Institute; 2014based on November 2013 SEER data submission, posted to the SEER web site. [Online]. Available from: http://seer.cancer.gov/csr/1975_2011/.

[172] Clark WH, From L, Bernardino EA, Mihm MC. The histogenesis and biologic behavior of primary human malignant melanomas of the skin. Cancer Res 1969;29 (3):705–27.

[173] Sandby-Moller J, Poulsen T, Wulf H. Epidermal thickness at different body sites: relationship to age, gender, pigmentation, blood content, skin type and smoking habits. Acta Derm Venereol 2003;83(6):410–13.

[174] Morton CA, MacKie RM. English clinical accuracy of the diagnosis of cutaneous malignant melanoma. Br J Dermatol 1998;138(2):283–7.

[175] Pickwell-MacPherson E. (Invited paper) practical considerations for in vivo THz imaging. Int J Terahertz Sci Technol 2010;3(4):163–71.

[176] Wallace V, Fitzgerald A, Shankar S, Flanagan N, Pye R, Cluff J, et al. Terahertz pulsed imaging of basal cell carcinoma ex vivo and in vivo. Br J Dermatol 2004;151(2):424–32.

[177] Wallace V, Taday P, Fitzgerald A, Woodward R, Cluff J, Pye R, et al. Terahertz pulsed imaging and spectroscopy for biomedical and pharmaceutical applications. Faraday Discuss 2004(126):255–63.

[178] Abbosh A. Microwave systems for head imaging: challenges and recent developments. In: 2013 IEEE MTT-S International Microwave Workshop Series on RF and Wireless Technologies for Biomedical and Healthcare Applications (IMWS-BIO); 2013, 1−3.

[179] Foster KR, Schepps JL, Stoy RD, Schwan HP. Dielectric properties of brain tissue between 0.01 and 10 ghz. Phys Med Biol 1979;24(6):1177.

[180] Zubal IG, Harrell CR, Smith EO, Rattner Z, Gindi G, Hoffer PB. Computerized three-dimensional segmented human anatomy. Med Phys 1994;21:299−302.

[181] Semenov SY, Corfield DR. Microwave tomography for brain imaging: feasibility assessment for stroke detection. Int J Antennas Propagation 2008;2008:8 article ID 254830.

[182] Fhager A, Yu Y, McKelvey T, Persson M. Stroke diagnostics with a microwave helmet.In: 7th European Conference on Antennas and Propagation (EuCAP); 2013, 845−6.

[183] Mobashsher A, Abbosh A, Wang Y. Microwave system to detect traumatic brain injuries using compact unidirectional antenna and wideband transceiver with verification on realistic head phantom. IEEE Trans Microw Theory Tech 2014;62 (9):1826−36.

[184] Yoo D-S. The dielectric properties of cancerous tissues in a nude mouse xenograft model. Bioelectromagnetics 2004;25(7):492−7.

[185] Zhang H, El-Rayis A, Haridas N, Noordin N, Erdogan A, Arslan T. A smart antenna array for brain cancer detection. In: Antennas and Propagation Conference (LAPC); 2011, 1−4.

[186] Zhang H, Flynn B, Erdogan A, Arslan T. Microwave imaging for brain tumour detection using an UWB vivaldi antenna array. In: Antennas and Propagation Conference (LAPC); 2012, 1−4.

[187] Jamlos M, Jamlos M, Ismail A. High performance novel UWB array antenna for brain tumor detection via scattering parameters in microwave imaging simulation system. In: 9th European Conference on Antennas and Propagation (EuCAP); 2015, 1−5.

[188] Rezaeieh S, Zamani A, Abbosh A. 3-D wideband antenna for head-imaging system with performance verification in brain tumor detection. IEEE Antennas Wireless Propagation Lett 2015;14:910−14.

[189] Mustafa S, Mohammed B, Abbosh A. Novel preprocessing techniques for accurate microwave imaging of human brain. IEEE Antennas Wireless Propagation Lett 2013;12:460−3.

[190] Chew KM, Sudirman R, Seman N, Yong CY. Signal processing of microwave imaging brain tumor detection using superposition windowing. Appl Mech Mater 2014;654:321−6.

[191] Semenov SY, Buleyshev AE, Posukh VG, Williams T, Sizov YE, Souvorov AE, et al. Microwave tomography for detection of tissue malignancies: lung, liver and kidney sites. Prog Electromagn Res Symp 2003.

[192] Zamani A, Rezaeieh S, Abbosh A. Lung cancer detection using frequency-domain microwave imaging. Electron Lett 2015;51(10):740−1.

[193] Rossetto F, Stauffer PR, Prakash M, Neuman DG, Lee T. Phantom and animal tissues for modelling the electrical properties of human liver. Int J Hyperthermia 2003;19(1):89−101.

Wireless closed-loop stimulation systems for symptom management

J.-C. Chiao

University of Texas at Arlington, Arlington, TX, United States

CHAPTER OUTLINE

4.1 NEEDS IN HEALTHCARE SYSTEMS

The issue of healthcare cost is a grand challenge globally, particularly for aging societies and underdeveloped areas. Even with developed countries, the costs are a major burden for personal and societal finance, directly and indirectly. Many efforts have been taken in different aspects of healthcare systems to improve management and delivery. Information technologies have been utilized greatly to increase personal care efficiency, improve quality in hospitals and homecare by electronic patient records, administer drug and equipment inventory, and organize scheduling and staffing. Wireless and cloud computing technologies have further

C. Li, M. Tofighi, D. Schreurs and T-Z. J. Horng (Eds): *Principles and Applications of RF/Microwave in Healthcare and Biosensing.*

made electronic information more accessible and ubiquitous. These efficient communication methods combined with electronic records and processing have obviously broken down the first layer of inefficiency in labor-intensive tasks.

However, major bottlenecks still exist. Caregivers are required to physically interact with patients for the acquisition of physiological parameters. Each caregiver can only handle a limited number of patients within a certain period. The patients have to go to a clinic or hospital frequently. The patients' commuting and waiting time and nurses/doctors' excessive administrative efforts decrease productivity for our society, not to mention the frustration and potential mistakes in crowded clinics leading to the waste of resources for urgent cases. The sampling of vital signs can only be carried out during the interaction period. This restricts physiological parameter data acquisition to a short period of time and small sampling numbers. The accuracy of diagnosis often is limited, especially for chronic diseases. Better care with higher diagnosis accuracy can be provided if more and time-lapsed data can be obtained without causing patients discomfort or limiting their mobility. It will be the best case if the patients' biological signs can be continuously monitored in their regular daily activities at home or work. Portability and timely accessibility of the physiological information can further aid patients and caregivers in real-time management of symptoms. This potentially provides a solution of having an adaptive closed-loop health management system in or around the patient's body. The closed loop can be autonomous with portable electronics in implants or wearables to deliver therapy [1,2] or drugs such as insulin [3]. The closed loop can also be behavior modification–based, in which a portable electronic device such as smart phone delivers audio or text messages to remind patients about their physical or emotional states. One can use the reminder to change his or her gestures to reduce possibility for back pain [4], or to take a few deep breaths because one is under stress [5].

4.2 WIRELESS IMPLANT SYSTEMS

Wireless technologies bring promising solutions to the issues mentioned before. Low-cost portable electronics and wireless devices have been widely used around the world and have made a significant impact on our societies. Smart phones are commonly equipped with 3G/4G, Wi-Fi, and Bluetooth communication functions, with ports that can be connected to other electronic devices. The mobile devices become a powerful tool as an interface to communicate with sensor or stimulator implants when they are equipped with wireless communication capabilities. Of course, smart phones may not have the desired frequencies or protocols to communicate with an implant, but they can serve as an interface to relay the internal biological signals transduced by an application-specific module outside the body onto the smart phone and then into the wireless network. Such a wearable system and chronic implants embedded with wireless signal transduction functions can be implemented to acquire patients' physiological data over a long period while they resume normal activities without physical constraints. Once signals are transduced, there are many data processing techniques to handle the information. Via networks, it can be transmitted to primary caregivers in real time, or the massive data can be stored in personal files for trend tracking. Patients can also gain access to interactions with doctors based on the quantitative data or manage their own treatment methods, such as adjusting stimulation dosages according to their own subjective feelings or comfort levels. The results will reduce workloads for health providers and thus the costs since most of the patients can be monitored without physically being in a clinic. The care outcomes will be better as the system empowers patients to manage their own chronic illnesses and enables quantitative documentation of symptoms for more precise diagnosis.

It is quite obvious that wireless communication to receive vital signals and remotely change an implant's settings is needed and will benefit patients. Combining wireless communication functionality in commercially available implants, however, requires redesigns and approvals for safety regulations. Therefore, the progress has been gradual instead of disruptive.

Among implants, cardiac pacemakers and implantable cardioverter defibrillators are now widely used to treat arrhythmias, which are heartbeat rhythm or rate problems, or to perform cardioversion and defibrillation for patients at risk for sudden cardiac death [6]. These modern implants can be programmed to detect abnormal rhythms of the heart and deliver electrical pacing or shocks. Simply put, when the heartbeat is too slow, they send weak electrical signals to regulate the tissues. When the heartbeat is too fast or chaotic, they give defibrillation shocks to stop the abnormal rhythm. To work in time to prevent cardiac arrest, heart rates are monitored continuously and detected arrhythmic events are stored for diagnostic purposes. The stimulation parameters are programmable so that the doctors can adjust for individual patients. Recently, real-time telemetry is embedded in the device which makes it capable of providing continuous or periodic information about the patient's condition and the device's status. The patient's information can be transmitted to a process center for emergency cases. Several studies have shown the feasibility of integrating healthcare resources with the wireless functions in the devices to maximize the potential of obtaining the patient's real-time heart conditions for the prevention of severe adverse events and optimization of therapy [7−9]. There have been relentless research and development efforts on system-on-chip designs toward efficient and robust signal processing and power management with wireless communication on single integrated circuit devices [10]. The goals focus on lower power consumption in order to reduce battery capacity and device size, integrated power management for wireless charging, sophisticated signal processing and adaptive programming to empower the devices with more intelligence, and robust low-power wireless communication.

The medical implant communication service band at 402−405 MHz has been commonly used for the pacemaker's bidirectional telemetry [11]. The maximum transmit power is limited at an equivalent isotropically radiated power of 25 μW within 300-kHz channels. The communication range is typically more than 2 m, which is much longer than the inductive coupling distance via an external coil in contact with skin. Currently, the Medical Device Radiocommunications Service (MedRadio) bands are available in the 401−406, 413−419, 426−432, 438−444, and 451−457 MHz ranges. The medical body area networks band in the 2360−2400 MHz range can be utilized for low power sensor networks around the body. These frequency bands allow implants to be wirelessly connected with a hub device on the body or with home monitoring electronics. This is particularly important for high-risk patients in at-home care.

Besides pacemakers for hearts, stimulators have been applied for pain inhibition in the spinal cord [12,13] and restoration of vision [14,15] or hearing [16,17] clinically. The safety and efficacy of spinal cord stimulation has been extensively studied [18]. Several commercial neurostimulators have already provided wireless programming and inductive coupling charging functionalities. Typically, spinal cord stimulators are implanted in the lower abdominal areas, with lead wires connected to the spinal cord. Cochlear Implants are implanted behind the ear with a multielectrode lead connected to the cochlea. Both implants with coil antennas are under the skin, facing directly toward the external antenna coils in the wearable devices so that the electromagnetic energy only needs to penetrate a thin layer of tissue. Spinal cord stimulators typically are bulkier because they contain large batteries. For example, one commercial product has a size of 54 mm × 54 mm × 9 mm and weighs 45 g. The data communication is through a carrier frequency of 175 kHz in the industrial, scientific, and medical (ISM) band with a data rate of 195 kbps. Another commercial product features a 55 mm × 46 mm × 10.8 mm size. It utilizes the 119 −131 kHz carrier for communication over a maximum distance of 91cm and charges the implant at the 77−90 kHz frequency for a 2-cm distance between the implant and the external charger coils.

A cochlear implant [19] requires a smaller size as the external module is attached behind the ear. One commercial product features 19 mm × 49 mm × 9 mm dimensions with a coil diameter of 20 mm. The communication of data and control are at 49 and 10.7 MHz with a bandwidth of 1 Mbps. Another product has a size of 36.5 mm × 48 mm × 8.7 mm and communicates at 12 MHz with a bandwidth of 0.6 Mbps. Retinal stimulator implants are required to be thin in order to deliver electrical signals onto the remaining retinal ganglion cells in the degenerated retina so as to generate visual percepts. Parylene-C has been a candidate substrate for such flexible implants with coil, application-specific integrated circuit (ASIC) chip, and circuitry embedded. Two examples demonstrate different device architectures at two distinct operating frequencies. At 15 MHz for the carrier frequency, a maximum transmission distance of 22 mm is achieved with an average power of 100 mW from the transmitter coil in air. With implanted devices, a reliable transmission over 10 mm between coils is achieved [20,21]. For another system, at the 500 kHz carrier frequency, a transfer of power of 10 mW is received at 4-mm coil spacing. The implant coil has an outer diameter of 9 mm and an inner diameter of 5 mm, with a quality factor of 0.45 [22,23].

Because of the short distances, the device architecture extends the coil antenna through flexible connection lines to couple energy with an external coil near the temple located on the glasses frame.

For deep brain stimulation, the stimulation site is far from the surface of the head, and thus it is difficult to directly transfer energy into the electrodes. A transcutaneous power and data transmission is proposed for inductive coupling to transfer energy to a stimulator, which is connected to the electrode lead placed in the deep brain [24]. The external and internal coils, having diameters of 4 cm and 1 cm respectively, align with a 1.5-cm spacing across a skull phantom. An ASIC chip with a high AC−DC power conversion efficiency of 87% produces adjustable DC outputs between 2.5 and 4.6 V at 2.8 m A loading from the 5-Vpp signals at 2 MHz in the external transmitter. Moving the operation to higher frequencies can reduce the antenna dimensions in order to decrease the implant size. A micro-electro-mechanical systems (MEMS)-based batteryless, wireless stimulator targeting pain inhibition with a size of 3.1 × 1.5 × 0.3 mm^3 is designed to operate at 394 MHz with inductive coupling over the skin as the transmitter coil is aligned directly with the implant [25]. The device contains an ASIC chip (1 × 1 × 0.15 mm^3) and a circular spiral inductor of 1 mm in diameter, 20 turns and a 20-μm pitch distance on Su-8 substrate and superstrate. The thickness of the encapsulated device is only 30 μm for subcutaneous implantation. Transmitting powers from 125 mW to 1 W generate clear evoked cortical responses in the nerves. The energy harvesting mechanism over skin, similar to the one for radio frequency identification (RFID) devices, also eliminates the need for a battery.

These wireless implant technologies not only provide promising solutions to the tough problems that could not be solved before, but also enable new applications toward closed-loop systems. They feature wireless signal transduction, wireless powering or charging mechanisms, low-power intelligent data processing, and miniaturization suitable for comfortable implementation. These features play a key role in autonomous health management systems.

4.3 BODY NETWORK APPLICATIONS

As mentioned, one of the solutions for autonomous management is the system architecture consisting of a passive transducer as implant, a wearable module to record and relay signals, and a receiver base station for data acquisition, as shown in Fig. 4.1. The passive transducer architecture allows a batteryless sensor to be implanted inside the body, enabling chronic sensing and recording. The device architecture allows radio frequency (RF) energy transfer across tissues for charging a capacitor or rechargeable battery in the implant. Therefore, it can also be used to provide sufficient energy for electrical stimulation of the target organ or tissues, which could be used for management of symptoms.

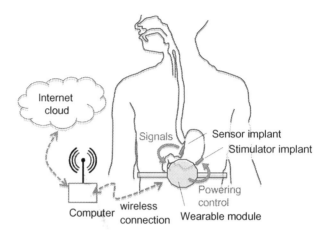

FIGURE 4.1

The concept of a wireless in vivo diagnosis and closed-loop treatment system.

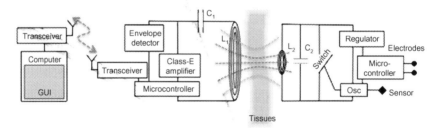

FIGURE 4.2

The batteryless wireless implant system blocks.

The wearable module serves as a control console for data storage, real-time signal processing, reprogramming, and changing dosages. With adaptive algorithms for machine learning and user's inputs through the interface, which mostly indicate subjective feelings, the wearable module becomes a powerful tool for autonomous closed-loop management of symptoms. Since low-power wireless communication protocols are available, it can serve as a relay for continuous acquisition and transmission of the essential signals to a receiving station, in which a large database can store the information for further off-line diagnosis. Such a system is useful for real-time patient or senior monitoring in a hospital or home, or in a laboratory to record signals from a large population of animals for data collection in drug trials.

The system blocks are illustrated in Fig. 4.2. The RF powering from the wearable unit to the implant utilizes a low RF frequency carrier. In our cases, it is 1.3 MHz or 6.78 MHz, which are in the ISM radio band. A higher operating frequency is possible but tissue losses and energy harvesting efficiency need to be taken into consideration. The efficiency depends on the coil quality factors, which are related to the skin effect and resistive losses, amplifier efficiency, tissue types, antenna cross sections, coil coupling efficiency, and current conversion efficiency. Some of these factors are also related to the signal frequency. Therefore, trade-offs in designs are necessary. A lower duty cycle, such as $10-30\%$, can be used to reduce the possible heating in tissues to further ensure safety.

The passive implant is similar to a conventional RFID tag with modifications. It consists of a coil antenna (L_2), a capacitor (C_2), a switch, an energy harvesting circuit, a relaxation oscillator, and a sensor for detection or a microcontroller (μC) for stimulation. The antenna and tuning capacitor are tuned to the resonant frequency for optimal coupling. A voltage multiplier with a series of diodes and capacitors is utilized to increase the potential for chip operation. The sensor signal transduction is conducted with load-modulation at the same operating frequency as for powering. However, simple amplitude modulation encounters issues with inductive coupling. The magnetic fields decay quickly with distance between coupling coils, which affects the induced current. Amplitude modulation becomes dependent on distance; thus the distance between the wearable module and the implant has to be fixed. Even so, the small variations of distance due to body motion generate noises in the transduced signals. A relaxation oscillator circuitry is used to convert the changes in potential, capacitance, or resistance of the sensor to frequency variations [26,27]. The switch modulates the carrier with a modulated frequency, much lower than the carrier frequency, generated by the relaxation oscillator. Multiple modulated signals can be carried by the same transceiving frequency to allow different sensor signals as long as the basebands are well separated. Since most biological signals are slow varying and individual neuronal action potentials (APs) are below 10 kHz, a minimum spectral spacing of 20 kHz is sufficient to separate the signals in these cases.

The wearable unit consists of a coil antenna (L_1) and capacitor (C_1), which are chosen for resonance at the carrier frequency, a class-E power amplifier, and an envelope detector to read the load-modulation signals. The modulated carrier passes through a low-pass filter to extract the modulated baseband signals. Then demodulated signals are fed into a microprocessor for digitalization. The communication between the wearable unit and the base station can be based on Wi-Fi, Bluetooth, or any communication protocol. In our cases, we choose eZ430RF2500 (*Texas Instrument*) and NRF24Le1 (*Nordic Semiconductor*) modules for their low power consumption features. MSP430 in the eZ430RF2500 module and the microprocessor in the NRF24Le1 module are programmed to process the modulated signals. The signals can be digitized by counting the number of pulses in a short duration. The text data are coded into the packet and transmitted wirelessly to the base station. Once the base station receives the data, signal processing and networking can be conducted easily.

The following two sections describe the demonstration of wireless closed-loop systems utilizing the proposed architecture for neurological and gastric applications.

4.4 NEUROLOGICAL APPLICATIONS

Neural networks in the body consist of neurons interconnecting to each other. Neuron axon terminals communicate via synapses by electrical or chemical means to dendrites on other neurons. Voltage-gated ion channels on the plasma membrane of a neuron open in response to transmembrane potential depolarization, allowing inward flows of sodium ions. This produces electrochemical gradients and thus electrical currents across the cell membrane, resulting in a sudden increase of potential. The sodium ion influx eventually reverses the polarity of the plasma membrane and the ion channels become inactive. As the channels close, sodium ions transport back out of the membrane. Potassium channels then become active resulting in outward electrical currents from the potassium ions. This action creates a negative potential change and makes the electrochemical gradients return back to the resting state. During the processes, the neuron sends the AP, with a positive swing and then a negative swing in voltage, at the axon hillock and transmits the electrical signal along the axon. Neurons communicate through the firing of APs in the network and the firing rates of certain neurons in specific signal pathways indicate neural activities in response to an internal or external stimulus. The symptoms of neurological disorders can often be recognized by the electrical activities in neurons.

4.4.1 NEUROSTIMULATORS

Neurostimulators have been used to treat and relieve symptoms for neurological disorders, including pain, Parkinson's disease, epilepsy, tremor, dystonia, tinnitus, stroke, incontinence, gastroparesis (GP), and obesity [28−45]. Neurostimulation has also been used for artificial retina, cochlea, and limbs to restore vision, hearing functions, or control prosthetic arms, respectively [46−50]. Trials and experiments now are conducted utilizing neurostimulation for depression, obsessive-compulsive disorder, amyotrophic lateral sclerosis, Huntington's disease, and irritable bowel syndrome, etc. [51−56]. Neurostimulation electrodes have been implanted in various nervous system locations for clinical therapies, including in the spinal cord, thalamic nuclei (e.g., ventroposterolateral or ventroposteromedial), periaqueductal gray (PAG), periventricular gray, anterior cingulate cortex, and other regions near the central gray [28,57], onto the nervous tissues, or onto organs such as the sacral nerve, vagus nerve, heart, or stomach. Typically the electrodes are placed at the target and connect to a sizable electrical pulse generator, called a stimulator, placed in a pocket formed beneath the skin. Peripheral nerve stimulation has also been utilized in the relief of pain in the extremities. These procedures are usually less dramatic than the ones for central nervous system.

Generally speaking, neurostimulation delivers electrical voltage or current into the excitable tissues, such as muscles and nerves, to reactivate tissue functions. With electrical signals delivered into nerves to intercept pain signals before they reach the brain, the feeling of pain can be inhibited. Different conditions of course require targeting specific location(s) for delivering the signals using particular parameters of electrical waveforms. Neurostimulation conventionally is delivered by conductive electrodes on a lead that is inserted into the tissues and placed at the target site. The lead connects the electrodes to the stimulator, consisting of circuits and battery to produce desired waveforms. Some stimulators can be reprogrammed to change the dosages wirelessly by placing a wireless programmer on the skin. Coil from this programmer couples radio frequency signals into the coil in the stimulator for communication. The distance between the coils through tissues is usually short so the near-field coupling is effective for low data rates at a low RF carrier frequency. The operation is not continuous and is mostly conducted when there is a need to change settings. Therefore, the consumption of battery energy by wireless communication is not a major concern.

There are a few limitations to the current neurostimulation methods. They are open-loop systems. When symptoms vary, manual adjustment is needed to change the dosage settings by the patient or doctor. This means interruption of daily activities or inconvenience. In some conditions, such as pain, the adjustment can be as frequent as several times an hour. In reality, the majority of patients cannot operate the programmer correctly. Furthermore, some neurological conditions require patients' comprehension of subjective feelings such as pain, depression, or addiction. Currently, the only method to understand the levels of symptoms is verbal communication between doctor and patient. It is not very objective, so caregivers can misinterpret patients' suffering if the patients cannot precisely describe the perception or sensation they recognize or feel. It is then difficult to standardize procedures for treatment or measure the outcomes for documentation purposes. There is also a chance of losing other sensations for the patient when he or she uses excessive dosages to eliminate pain as much as possible. The loss of other sensations can put the patient in danger because he or she will not be able to feel illness or injury. However there is no method for doctors to prevent this since only the patient himself or herself can recognize the subjective feeling. It is difficult to optimize the balance between comfort and risk without a measureable parameter to quantify the neurological condition.

4.4.2 CLOSED-LOOP SYSTEMS FOR PAIN MANAGEMENT

These mentioned shortcomings call for a closed-loop management system to maximize the benefits of neurostimulation. Like any engineering system, a closed-loop system reduces possible errors, improves stability, enhances robustness against noises or interferences, and creates reliable and repeatable performance. A body—computer loop consisting of sensors and stimulators connected through a wireless network can be such a system. The sensors collect physiological data, and the computer processes, recognizes, and responds with dosage commands to the stimulators according to the interpreted bodily condition. Patients can enter the desired level of comfort as an input to adjust the loop feedback factors.

This integrates patients' subjective feelings into the quantitatively operating loop. The sensor data documentation gives caregivers assessable and time-lapsed insights on physiological conditions for them to adjust the function parameters in the feedback loop. This can prevent misuse or abuse by the patient to minimize risk. Because the biological conditions modulated by neurostimulation do not change quickly, unlike in some engineering systems that require micro- or milli-seconds to provide feedback, a slow feedback loop in the time scale of seconds can be easily managed by currently-available electronics. Although the concept of a closed-loop system for neuro-modulation or neuro-feedback is simple, technical challenges in implants and system instrumentation as well as obstacles in human clinical trials need to be overcome.

Our work for a closed-loop neurostimulation system focuses on pain management. Chronic pain is an important public health issue [58–60]. More than 50 million Americans are suffering from pain. Economists estimate that the total direct and indirect costs of pain in the United States exceed 100 billion dollars annually [61]. The global costs can be easily orders of magnitude higher because the pain issue is common. Despite the costs, diagnosing chronic pain and deciding on appropriate management methods are still rudimentary. In clinical practice, physicians rely on patients' verbal description of pain or facial expressions [62,63]. Doctors do not have physiological markers to document pain signals in a quantitative way, which is particularly vital for locked-in syndrome patients [64], stroke patients [65], and children [66]. The open-loop stimulators are not adaptive to providing the neural feedback from the patient due to lack of sensory inputs. Therefore, they are not always effective. In some cases, overstimulation induces side effects [29]. In contrast, closed-loop, real-time neurostimulator systems have been proposed for epilepsy, in which the stimulator is synchronized to the onset of a seizure in order to prevent serious injury [67,68]. We believe a closed-loop feedback approach for nociception inhibition in the neural pathway can provide improved efficiency in pain management, based on a similar principle. This input-based approach offers improved comfort to patients and reduces negative side effects such as excitotoxicity [69]. It is also unrealizable in the current open-loop system to experiment with different stimulation waveforms for individual patients in real time, which may produce better outcomes [70].

In laboratory settings, researchers have used electrophysiological indices to measure pain and to differentiate painful feelings from other nonpainful sensory feelings, such as touch [71,72]. In animal models, single wide dynamic range (WDR) neuron recording in the spinal dorsal horn shows graded responses to a broad range of innocuous to noxious stimuli, including mechanical, thermal, and chemical stimuli [73,74]. Application of more intense mechanical stimuli, causing more noxious expression from the object, increases the firing rate of APs by the WDR neurons. Therefore, recording the WDR neurons in the brain or spinal cord provides a quantitative way to document nociception related to pain perception.

For pain inhibition, spinal cord stimulation has been demonstrated to be effective. In our demonstrations of a closed-loop system, however, we target brain stimulation and spinal cord recording. The reasons are twofold. First, due to the small animal models, electrical signals from the stimulator travel on the skin of the animals and couple into the recorder if they are close. This creates artifacts in the recorded signals and sometimes saturates the amplifiers. So we separate them with one in the brain and one in the spinal cord. Second, we are to demonstrate the very feature of a closed-loop system by which dynamic dosages can be implemented to achieve optimal effects. The target in the brain for pain inhibition is small in size and there is a chance that the electrode may not be in the precise location. The dependence of the target location gives the closed loop a chance to find the optimal setting of stimulation parameters. Separating the recording in the spine can reduce physical interference to the stimulation in the brain. Among the structures for neurostimulation to inhibit pain, the PAG plays a pivotal role in descending inhibitory pathways [75,76]. To inhibit spinal nociception, the PAG excites both the nucleus raphe magnus [77] and the locus coeruleus [76]. Several studies have demonstrated that electrical stimulation of the superior colliculus (SC) has some antinociceptive effects in animals [78—81]. However, the reported inhibitory effects by SC stimulation are less effective than those by PAG stimulation. Therefore, the sensor may detect less pain inhibition due to the possibility of the electrode being located in the vicinity of the PAG and SC. In such a case, the closed-loop algorithm should be able to adjust the setting in order to achieve better outcomes.

To demonstrate a closed-loop system for pain management, we combine extracellular single-unit neural recording with a wireless recorder in the spine and single-unit neural stimulation in the brain with a neurostimulator. The reason for single-electrode probe recording is to be comparable with the standard wired method established with the commercially available instrument, which collects the data simultaneously in parallel with the wireless system. Both sets of data are compared after experiments to validate the performance of the wireless closed-loop system. In practical applications, a multielectrode probe with the wireless recorder is desired to provide redundancy and better accuracy. The same reasons apply for stimulation. The recorder continuously transmits WDR neuron signals detected in the dorsal horn of the spinal cord to a computer. In the animal experiments utilizing rats, because the recording and stimulating targets are very close to the skin, the probes can reach them directly so passive signal transduction is not needed. For large animals or human uses, passive transducers are desired to eliminate battery need in implants. The portable computer recognizes and classifies nociceptive activities produced by mechanical stimuli mimicking pain. The algorithm initiates electrical stimulation dosages and the computer sends wireless commands to the stimulator, which delivers the electrical pulses into the targeted brain site to suppress nociception.

4.4.3 SYSTEM DESIGN

The wearable unit consists of an analog board and NRF24Le1, which has an analog to digital converter, a μC, and a 2.4-GHz transceiver. The analog board amplifies and filters the signals. The gains and filter passbands depend on respective applications. The amplified signals are sampled and digitized before being loaded into data packets and sent by the transceiver. Packets are received at the receiver and sent to a computer via serial communication, where a graphical user interface (GUI) is used to display data in real time and allow the user to change stimulation criteria. If the data packet is lost in air, the radio in the wearable unit re-transmits the data packet to the base station until attainment is verified. The re-transmitting procedure depends on the sampling rate of the microprocessor and the time for each packet to travel in air, which varies for different applications.

A GUI is developed in LabVIEW for convenient operation. The continuously acquired neuronal signals are displayed in real time while a histogram is calculated and displayed in a time-lapsed fashion. This helps to visualize nociception induced by different mechanical stimuli. The interface allows real-time changes of nociception recognition by changing detection thresholds and classification criteria. This sets the sensitivity and selectivity of the pain detection. In clinical applications, this method of setting can be done graphically by the patient's subjective comprehension of the pain. Another set of control in the interface determines the stimulation dosages. Different settings of predetermined stimulating pulses are programmed in the stimulator μC and listed in the graphical interface to be chosen with graded amplitudes and/or stimulation frequencies. The algorithm can first use a sequence of chosen dosages during the trial-and-error period, but later, dosages can be based on an adaptive procedure to figure out the optimal stimulation after the software learns about the patient's pattern. The interface is based on LabVIEW so the algorithm can be easily and graphically changed without writing new codes. Once the dosage is determined, the interface sends the command through the universal serial bus and transmits it via the wireless link.

Once the waveform is transmitted, AP spike detection is conducted. A sampling rate of 10,000 samples per second is chosen to acquire the WDR neuronal signals in order to have the pulse shapes preserved, given that the AP duration is in the millisecond range [82]. Different spike detection techniques are available. Simple threshold, energy-based, matched-filter based and template-matching are some examples [83,84]. The choice needs to be based on the computational capability. Since the algorithm should be implemented in a wearable module in order to accomplish a real-time closed loop, the simple threshold method is chosen owing to its acceptable accuracy and low computational cost. The algorithm applies a positive threshold to the raw recorded data. It recognizes peaks above the threshold as APs. The electrode may record from several neurons nearby at the same time. Each neuron fires at different times and amplitudes. The distances between individual neurons and the electrode also change the received waveforms. The threshold needs to be adjusted as part of the calibration. Generally speaking, the nearby neurons usually work in a similar manner, thus the stronger signals usually come from the closest one and a higher threshold can be set for it. A higher threshold improves the reliability because it distinguishes APs from noises.

To classify the neural activities, the interspike interval is used as the main feature for defining the rate of neuronal activities [85,86]. The interspike interval is the time difference between the detected AP peaks above the threshold. Interspike intervals are detected with a preset signal amplitude threshold and saved in a floating numerical array called a cluster. The cluster size, which is the number of spike intervals, is used to discriminate between nociceptive and nonnociceptive responses [87,88]. If the summation of the intervals inside the user-defined cluster exceeds a predetermined level called the critical threshold, the program recognizes this series of APs as nociceptive, or painful. This method is an alternative means to validate the simple histogram method. When the neuron fires more frequently in a short period of time, the histogram frequency in a small fixed bin window will be high. However, deciding on the bin size (for over how many milliseconds the μC should report the firing number) may not be straightforward. If the neuron generally fires frequently, the bin size can be small. But if it is not, the bin size needs to be large to collect accurate data in order to distinguish noises. The time required may be too long to close the loop. On the other hand, when the neuron fires often, the interspike intervals will be small. By counting the numbers of small interspike intervals, this can also indicate the frequency of neuron firing. Therefore, the algorithm does not need to wait. Users could change the size and critical threshold of the cluster to calibrate the accurate detection of different neural activities related to various stimuli. The cluster parameters can be calibrated with known graded mechanical stimuli from various nociceptive sources [89]. In the future applications, patients can set the parameters to calibrate the classification depending on their subjective feeling. In our studies, the parameters are set as the system recognizes both marginal noxious and noxious stimuli, defined as the pressure and pinch stimuli, as nociceptive. When noxious signals are identified, the program is triggered in the closed loop to initiate a neurostimulation command and transmit it wirelessly to the stimulator.

The algorithm can generate neurostimulation parameters that vary in frequency, amplitude, and pulse duration. Some examples are provided in [90–92]. If the firing rate returns below the threshold set for the nociceptive activity, the program ceases neurostimulation commands until the threshold is breached again. If firing rates remain constant, the program triggers further stimulation with the same parameters. To monitor the inhibitory effect of each delivered stimulation train, a time delay of 1 second is implemented in the loop between the pulse trains. In practical applications, this type of delay may be also needed in the calibration period for the patient to identify a feeling of pain. Once completed, the automatic feedback loop can work in the background and the delay will not be needed.

4.4.4 EXPERIMENTS AND RESULTS

The electrodes target WDR neurons in the left L5 region. Other L5 neurons nearby firing with smaller amplitudes are also recorded. As part of the calibration procedure, once a neuron is detected, three different graded mechanical stimuli—brush, pressure and pinch—from innocuous to noxious levels are applied to the rat's hind paw. If the neuron fires APs to all three stimuli, it is recognized as a WDR neuron. The threshold for AP peaks and interspike interval parameters for recognition of three different groups of stimuli then are set during the calibration procedure. As mentioned, raw signals are used for real-time processing both with the simpler algorithm in the wireless system and in the conventional wired neurorecording instrument with the sophisticated template-match algorithm by the Spike2 program (Cambridge Electronic Design Limited, United Kingdom) for comparison. The Spike2 signal processing task is conducted offline after experiments because of its significant computational time. The offline data are used to compare with the results from our real-time algorithm and identify the inhibitory effects and efficacy.

Innocuous stimuli are less intense and evoke lower rates of neuronal firing in contrast to noxious stimuli that indicate pain [73,93]. Graded mechanical stimuli of brush, pressure, and pinch are applied to a receptive field to evoke neuronal responses in the spinal cord. Each mechanical stimulus is only applied for a limited time of 10 seconds with a rest period of 20 seconds in order not to desensitize the neurons. The protocol is rigorously followed in order to have comparable data in statistical analysis. The protocol, however, also sets limits on experimental results.

The stimulation electrode is placed in the targeted PAG (7.04 mm caudal from the bregma, 0.5 mm lateral right from the midline, and 5 mm deep from the brain surface in the rat model) [94]. Because no MRI image is taken, the insertion depth of the electrode is rigorously followed, although each rat may be slightly different. Therefore, the exact electrode location may not be in the center of PAG. The electrode is connected to the wireless stimulator. After each experiment, the brain can be examined with histological slide verification to identify the location of the stimulating electrode. Fig. 4.3 shows the recorded locations of stimulating electrode tips in the PAG (black circles) and SC (gray squares), which is less than a millimeter above the PAG. Thus we expect that nociception inhibition effects are not uniform with the same dosage setting as they are less prominent in the SC area [76].

To illustrate the effects of the feedback loop, three segments of one experiment are shown in Fig. 4.4(A), (B), and (C). They include the control and

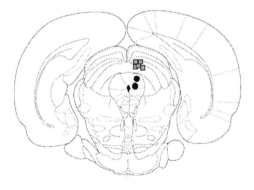

FIGURE 4.3

Location of stimulating electrode tips in the periaqueductal gray (PAG; black circles) and superior colliculus (SC; gray squares).

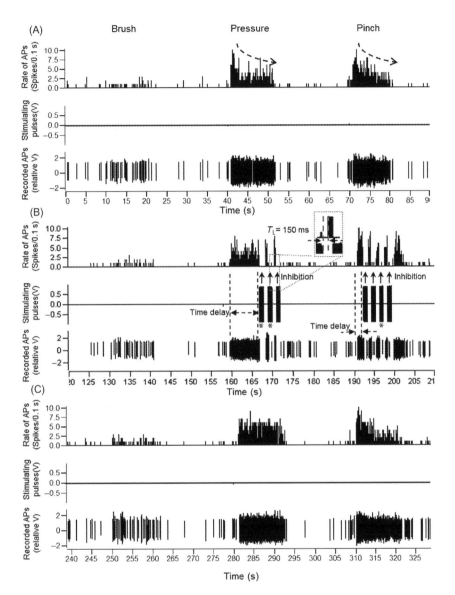

FIGURE 4.4

Representative responses from a dorsal horn WDR neuron with the closed-loop system. Neural activities during brush, pressure, and pinch stimuli are shown in terms of both individual action potentials (lower panel) and the rate histogram in bins of 0.1 s (upper panel). Three experiments are conducted for (A) Control. (B) Closed-loop; and (C) Recovery periods. During (A) and (C), the algorithm is turned off while the wireless

(Continued)

recovery periods during which the algorithm is turned off. The similar behaviors in these two periods show that the neuron is not injured from electrical stimulation. In Fig. 4.4(B), the algorithm is turned on with an interspike interval cluster size of 25 and a critical threshold of 0.7 seconds for nociception recognition. In Fig. 4.4(A), (B), and (C), the bottom traces show the raw data of spikes, and the middle ones show the stimulation delivered to the target. The spikes are counted every 0.1 seconds, and the rate of histograms in the upper traces illustrate the inhibitory effects. In each timeline, brush, pressure, and pinch stimuli are applied in sequence, with rest periods between stimuli. The suppression of neuronal activity during neurostimulation is obvious as indicated by arrows in Fig. 4.4(B). In this particular experiment, three and four trains of 1-V 100-Hz electrical stimulation pulses with a pulse duration of 0.1 ms are initiated for pressure and pinch stimuli, respectively. Each train lasts for 1 second, so 100 pulses are delivered during the stimulation.

When pressure is applied to the receptive field, the response increases to 43 spikes/seconds (at the 166th−167th s in Fig. 4.4(B)). The response reduces to 0 spikes/seconds when the first stimulating pulses are initiated (at the 167th−168th s in Fig. 4.4(B)), with an inhibition effect of 100%. One second after the neurostimulation, the rate of APs rises back to 28 spikes/seconds at the 168th−169th s in Fig. 4.4(B). Then a train of neurostimulation is triggered at the 169th−170th s in Fig. 4.4(B). It results in an inhibition of 88% compared with the one-second period before the stimulation. Comparing the histograms during control and recovery periods, the effect of neurostimulation is clear. When the mechanical stimuli are removed, the WDR neuron firing rates decrease dramatically and the neurostimulation trigger is stopped.

Without stimulation, the rates of APs in response to stimuli naturally decay during the ten-second period as indicated by the dashed arrows in Fig. 4.4(A). In some cases, the inhibitory effects of neurostimulation last after the neurostimulation is terminated. These cases are indicated by "*" in Fig. 4.4(B). The inset shows that the inhibition of APs lasts for 150 ms (T_L) beyond the completion of the stimulation. These are consistent with previous studies as neuron firing

recording continues. The histogram unit is spikes/0.1 seconds indicating the spike numbers in the bins with a duration of 0.1 seconds. We choose this unconventional unit to illustrate the activities with a higher resolution in the 0.1-s bins, instead of spikes/seconds, which is typically used in the Spike2 program to indicate activities in each 1-s bin. The decaying trends of neuron firing in response to constant mechanical stimuli of pressure and pinch in the control period are indicated by dashed curves in (A). The initiation of electrical stimulation and inhibition of neuronal firings are indicated by vertical arrows. Time delays between the initiation of mechanical stimuli and triggering of electrical stimulation are indicated by a horizontal dashed arrow. The lasting inhibitory effect of the electrical stimulation is indicated by "*". As an example, the duration of one of the lasting inhibitory effect (T_L) shown in the inset is measured as 150 ms.

naturally decays from the initial response [95]. This is a good indication that such a feedback loop can be helpful to prevent overstimulation.

The number of neurostimulations initiated during the brush stimuli is zero because the algorithm and user setting define it as an innocuous response. The average number of neurostimulations triggered during the pinch stimuli is significantly greater than that of the stimulations initiated during pressure, when the stimulating electrodes are in either the PAG or SC. A statistical analysis shows the average time delay for the pinch stimuli is significantly shorter than the one for the pressure stimuli during both PAG and SC stimulation. In addition, PAG stimulation has a significantly shorter average delay than SC stimulation during the pinch period.

In the PAG, neuronal responses (spikes/seconds) to different mechanical stimuli during the periods of control, neurostimulation, and recovery are (1) for brush: 9.45 ± 1.74, 9.41 ± 1.8, 9.59 ± 2.11, respectively; (2) for pressure: 28.45 ± 4.90, 18.89 ± 3.15, 30.75 ± 5.88, respectively; and (3) for pinch: 41.10 ± 6.13, 24.51 ± 4.34, 37.61 ± 5.61, respectively (Fig. 4.5(A)). The neuronal activities induced by the pressure stimuli are inhibited by electrical stimulation compared with those in the control and then return to an activity level similar to those in the control during the recovery period. There are no significant differences in AP rates between the control and recovery periods. The responses for pinch stimuli show a similar trend, as the activities are inhibited by neurostimulation and the control rates are similar to those during the recovery period. There are no differences when comparing any of the brush cases, as expected. The PAG is known for its descending inhibition [96–98], and our results show that the system is capable of inhibiting nociceptive activities by delivering electrical stimulation to the PAG. The average reduction rates are 28.22% and 40.98% for pressure and pinch stimuli, respectively. This shows that the algorithm triggers sufficient electrical stimulation intensities from the recording and pain signal recognition to suppress nociceptive activities.

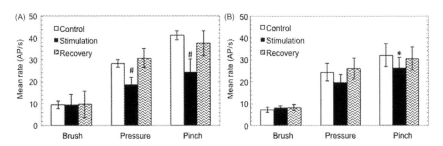

FIGURE 4.5

Mean of fired action potentials in dorsal horn neurons in response to graded mechanical stimuli with or without the closed-loop algorithm turned on in (A) PAG and (B) SC stimulation. *Note*: "*" $p < 0.05$, "#" $p < 0.001$.

In the SC, the responses (spikes/seconds) of WDR neurons to graded mechanical stimuli during the periods of control, neurostimulation, and recovery are: (1) for brush: 7.26 ± 1.15, 7.73 ± 1.12, 8.07 ± 1.39, respectively; (2) for pressure: 24.39 ± 3.95, 19.65 ± 3.76, 25.59 ± 4.83, respectively; and (3) for pinch: 32.18 ± 5.19, 26.30 ± 4.844, 30.69 ± 5.38, respectively, as shown in Fig 4.5(B). There is not a significant difference between the control and stimulation periods for pressure stimuli, suggesting no significant inhibition. The responses for pinch stimuli are inhibited by neurostimulation compared with controls. There are also no differences when comparing any of the brush conditions since the program defines brush as an innocuous stimulus. The SC is known for its prominent role in visual reflexes and for the coordination of head, neck, and eye movements, while some studies show antinociceptive effects of electrical stimulation in animals [82]. Our results indicate that SC stimulation has less antinociceptive effect than PAG stimulation. A possible descending pathway most likely exists through the SC connections to the PAG or outflow of electric fields from the electrode into the PAG area [94]. Although SC-stimulated inhibition is less efficient than PAG stimulation for pain relief, the deficiency can be improved by adjusting the electrical stimulation parameters such as amplitude, frequency, pulse duration, and/or the duration of the neurostimulation in the closed-loop system. This provides a better option for pain management as the electrodes can shift location over time and adapt to patients' conditions.

As our algorithm initiates stimulation based on the level of neuronal activities, it further provides flexibility in adjusting stimulation parameters such as voltage amplitudes, pulse frequencies, and pulse durations to optimize inhibition effects with the feedback mechanism. Currently, there is no such means for experiments or treatment options. The wireless closed-loop system can be a powerful tool to understand more about neural networks for neurological disorders and discover better means for treatment.

4.4.5 ELECTROCORTICOGRAPHY RECORDING

The same method can be applied for many different neural signal acquisitions and neurostimulations at different sites. For example, electrocorticography (ECoG) has been utilized to record brain activities from the surface of the dura in the brain [99,100]. It provides a better spatial resolution and signal characteristics such as fewer motion artifacts compared to the noninvasive electroencephalography [101]. The ECoG signals are different from individual neuronal APs with much lower frequency components. As a demonstration, the wireless module is retuned for the range of $1-150$ Hz with a sampling rate of 1 kSample/seconds. The electrodes are implanted with stainless steel screws in the somatosensory cortex as the working electrodes. Two electrodes over the cerebellum are the ground and reference electrodes. It is expected that the wirelessly recorded signals in

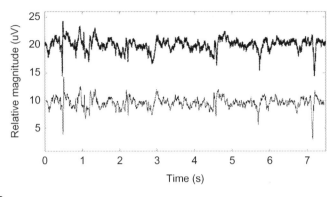

FIGURE 4.6

ECoG signals recorded by a conventional wired (top) and the wireless (bottom) systems. Only a section of 7.3 seconds was shown. The signals are identical.

animal are similar to the standard wired signals, as shown in Fig. 4.6, since the bandwidth is sufficient. The signals then can be utilized as the inputs for the closed-loop system.

4.5 ENDOLUMINAL APPLICATIONS

4.5.1 GASTRIC DYSMOTILITY DISORDERS

Another application example is to record gastric electrical activities (GEA) in the stomach. Gastric dysmotility disorders including GP cause chronic debilitating symptoms such as nausea, vomiting, early satiety, weight loss, and abdominal pain [102,103]. Gastric motility disorders can also cause gastroesophageal reflux disease (GERD) in which stomach content reflows into the esophagus from the stomach. It is estimated that GP affects more than 1.5 million Americans with over 100,000 suffering from severe forms of the disorder. A high percentage of type-1 and type-2 diabetic patients develop GP; thus it is estimated that the patient number will increase globally [104,105]. Gastroesophageal reflux disease affects approximately 15% of adult population in the United States. It is one of the most prevalent clinical conditions afflicting the gastrointestinal tract. Both the acid or nonacid reflux episodes damage esophagus and pharynx tissues [106–109]. GERD is the major risk factor for esophageal cancers which are the fastest growing cancers in developed countries [110]. Previously, a miniature batteryless wireless impedance and pH sensor capsule that can be implanted in the esophagus has been demonstrated to record the reflux episodes continuously for a long period of time [111–114]. The same passive signal transduction method was utilized as the one shown in Fig. 4.2. The sensors aim at detecting reflux symptom episode occurrences and the reflux pH values for diagnosis and prognosis during drug therapy.

4.5.2 GASTRIC ELECTRICAL STIMULATION

Gastric dysmotility disorders are characterized by the abnormal amplitude and frequency of the electric slow waves affecting myoelectrical activities in stomach [115]. Drug therapies with prokinetics have been used for symptomatic control, but only a portion of gastroparetic patients are tolerant due to side effects [116,117]. Gastric electrical stimulation (GES) has been shown to be effective for symptomatic relief [118]. GES alters intrinsic GEA by stimulating the stomach at a frequency generally higher than the normal electrophysical frequency of three cycles per minute [119]. The required electrical current for stimulation ranges from 2 to 6 mA. Simulation pulses having an on-time period more than 330 μs affect the interstitial cells of Cajal and smooth muscle cells to reactivate the motility of stomach [120].

Myoelectric signals with very low amplitudes and frequencies propagate as slow waves across the stomach [121−123]. The analog board for the wireless GEA recorder is designed to amplify and filter signals at a gain of 65 dB in the frequency range of 0.05−0.3 Hz. The sampling rate is reduced to 8 samples per second. The GEA signal is acquired from the serosal membranes of the stomach and compared with the one recorded through wired equipment. Sufficient bandwidth of the system allows reservation of the waveforms, as expected [124]. The in vivo GEA recording can be used to interpret the symptoms of gastric dysmotility. The signals can also be utilized to form the closed-loop feedback to determine gastrostimulation pulse parameters to be applied on mucosal or serosal tissues of the stomach in order to modulate the motility for individual patients.

GES is currently achieved by surgical implantation of a commercially available neurostimulators [44,125−127]. In some cases, temporary endoscopic stimulation is used to evaluate the efficacy of stimulation [128]. Despite promising results, the GES device has not been widely adopted in clinical applications due to the requirement of major implantation surgeries. The stimulator device containing a nonrechargeable battery has a size of $60 \times 55 \times 10 \text{ mm}^3$. The implantation requires a 1−3 hour operation under general anesthesia and several days of postoperative hospitalization. Other shortcomings of this method are the limited battery life and the requirement of repeated surgeries after the depletion of the battery. Even for temporary endoscopic stimulation, the bulky wires through the nostrils, throat, and esophagus hamper the patient's daily activities and cause pain and discomfort. Both methods are operated in the continuous-on mode with a fixed setting without the capability for closed-loop functions.

With the gastric electrical signals available as the feedback inputs, an implantable gastric stimulator in the wireless body network is needed to close the loop. Furthermore, while the wireless communication function can be added to the existing stimulators that are used for GES, the size and battery issues remain challenges as replacement surgeries are still required.

4.5.3 **BATTERYLESS WIRELESS GASTRIC STIMULATOR**

With the passive signal transduction system architecture, the myoelectrical signals can be load-modulated and transmitted through the tissues to the wearable module. At the same time, the electromagnetic energy can be harvested for the operation of the stimulator. In such a case, the battery then can be eliminated and replaced by a capacitor or a smaller rechargeable battery to reduce the device's size. The miniaturized wireless stimulators then can be implanted endoscopically through mouth, throat, and esophagus into the stomach without the need for surgery or external tethered wires. This method not only provides the possibility for closed-loop operation to treat gastric motility disorders according to the slow wave signals, but also enables an out-patient, easy-to-implant procedure with lower risk and costs. Patients will feel more comfortable with the small implant and not need to suffer from the surgery pain.

The wireless gastrostimulator is powered by inductive coupling at 1.27 MHz with a transmitting power of 4 W from the wearable module [129,130]. The wearable transmitter coil size is 11.5×11.5 cm^2, made of AWG (American wire gauge) metal wires. The coil is driven by a class-E amplifier with a biasing voltage of 5 V. Although the battery in the wearable module can be replaced easily, it is still more convenient to have lower power consumption for the battery to last longer. Therefore, it is important to have a higher quality factor by lowering the wire resistance, and to tune the coils in the implant and wearable module to reach resonance. Although it can be easily done with the coils in air, it is not straightforward for the implant system as the implant coil cannot be easily tuned after implantation. The abdominal tissues and liquid/food inside the stomach, all with complex permittivities, create an unpredictable environment for parasitic capacitances and resistance for the inductive coils. This means the dynamic tuning should be conducted for the external coil. The resonant frequency is also varied by the environment for the internal coil, which cannot be tuned. This presents a challenge as the operating frequency will not be optimized at the resonance for both coils. Thus, a trade-off strategy is to implement a lower quality factor for the internal coil, which means a wider frequency range to work with for the external coil.

The implant coil cross section has to be small enough to be enclosed in a capsule that can pass through the esophagus by the endoscope for implantation, while the wearable module coil cross section and weight need to be limited for wearability by different body types of patients. Since the implant coil cross section is small, the external coil should not be too large, or most of the electromagnetic energy cannot be captured and becomes wasted. However, the external coil needs to maintain a certain size to allow misalignment between two coils since the implant coil is not visible to the patients when they put on the wearable module. Because of the size difference, the mutual inductance is low between coils, by which the scenario falls nearly into the weakly coupling case. It gives convenience in circuit analysis since each coil can be independently analyzed. However, weakly coupling means a very low power transfer efficiency in the

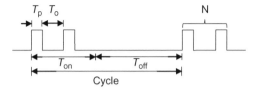

FIGURE 4.7

Pulse train definition.

Table 4.1 Pulse Specifications

Setting	T_p/T_o (μs/ms)	Pulse Frequency f (Hz)	Pulse Duty Cycle %	T_{on}/T_{off} (s)	Number of Pulses Per Cycle
Low	330/71.4	14	0.46	0.1/5.0	2
Medium	330/35.7	28	0.92	1.0/4.0	28
High	330/18.2	55	1.82	4.0/1.0	216

entire system. These factors present a significant challenge to the antenna and amplifier designs. Furthermore, the transmitting power should be limited or it may create excessive heat that can damage the skin or expose internal tissues to local heating caused by the electromagnetic energy.

The devices are made on miniature printed circuit boards wrapped with AWG-wire coils around the boards before packaging. The packaged prototypes have sizes of $12 \times 37 \times 9$ mm^3 for the first prototype and $7.2 \times 13 \times 8$ mm^3 for the second one. Once the implant coil couples the energy, a regulator regulates the voltage to charge a capacitor or a rechargeable battery. Charge pumps that consist of capacitors and diodes can be used to increase the operating voltage. A peripheral interface controller (PIC) generates stimulation pulses. The PIC is preprogrammed to generate pulse trains with specific frequencies and duty cycles. The pulse train parameters are defined in Fig. 4.7 and three different settings based on clinical works are shown in Table 4.1 [128,129].

The tissue thickness between the external transmitter coil and the implant coil is about 3 cm. The DC resistances of serosal and mucosal tissues are measured as 1179 Ω and 594 Ω, respectively, to calculate power delivered into tissues. The slow wave signals of the stomach are recorded at different stimulation settings. The recordings are analyzed by signal averaging for mean frequencies and amplitudes, as well as for frequency-to-amplitude ratios (FARs). Fig. 4.8 shows the stimulation waveforms at low, medium, and high settings in the tissues. The measured electrical currents into the tissues are 3.45 and 5 mA in the serosal and mucosal stimulations. The currents delivered to the serosal areas are lower due to the higher impedance. The instantaneous powers are 14.85 and 6.02 mW, respectively.

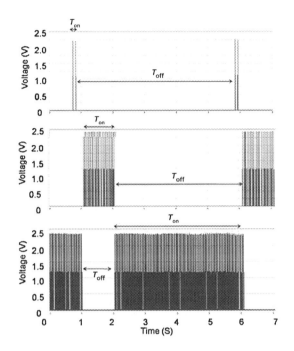

FIGURE 4.8

Stimulator outputs into the stomach tissues for low (top), medium, and high (bottom) settings.

The FARs confirm stimulation effects on gastric activities. The rhythm frequency ranges do not change significantly, but the average amplitudes of the myoelectrical signals vary noticeably. The control cases are measured, when the stimulation is turned off, with FARs of 22.92 and 33 bpm/mV at two different time points before and after a period of time for stimulation. This is typical as the stomach adapts to the electrical stimulation, so the slow waves change with the adaptation. The FAR changes to 44 bpm/mV at one case of medium dosage of stimulation and 21.43 bpm/mV after switching to the high dosage. The myoelectrical activities are modulated during stimulation as compared to the control. This indicates that the stimulation settings should be readjusted as the detected GEA vary to avoid overstimulation.

During the in vivo animal experiments, several issues are identified for the passive signal transduction and wireless power transfer to operate the closed-loop system devices. As mentioned before, misalignment issues are a main concern as the implants are small and their locations become unknown visually from outside the body after implantation. Searching the optimal location for wireless power transfer requires effort and it is not easy for daily operation since the effect of stimulation is not obvious until a certain period of time. The abdominal wall is measured as 3-cm thick on average after experiments, but the tissue thickness varies across the abdominal areas. As the electromagnetic fields decay quickly with distance, the available energy distribution inside the body is uneven at any cross section, making it difficult to find the optimal spot for the external coil. This calls for coil designs that have electromagnetic fields more uniformly distributed across the coil planes and less convergent or divergent with distance.

4.5.4 ELECTROMAGNETIC ENERGY TRANSFER

In our investigation, we map the effective coverage areas for each coil antenna designed. This is conducted by moving the implant coil two-dimensionally at the plane of interest, which is parallel to the external coil. It is difficult to conduct such experiments inside the body or in a phantom model; thus we propose to measure the distribution in air and with slabs of saline that mimic field behaviors in tissues. To separate the issues from the parasitic capacitance due to tissues, we first find an average power attenuation of 1.27 dB/cm in the tissue samples. Then, mapping experimental results can be scaled with the insertion losses. With a receiver coil cross section of 10×35 mm^2 and a transmitter coil with a size of 11.5×11.5 cm^2, Fig. 4.9 shows the maps of harvested power after taking the estimated tissue losses into account, within ± 10 cm from the center of the external coil. The currents are sufficient to deliver more than 2 mA into mucosal tissues at distances of 2, 3, 4, 5 cm within areas of 4.35, 4.93, 5.06, and 4.70 cm in radius,

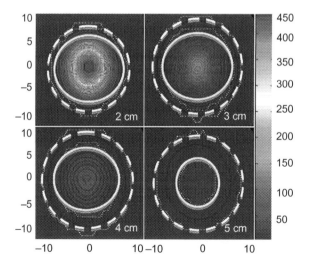

FIGURE 4.9

Received power distributions in the receiver planes at tissue thicknesses of 2, 3, 4, and 5 cm. The delivered currents are greater than 2 mA and 4 mA inside the circles with dashed lines and circles with solid lines, respectively. The color bar is power in mW.

respectively. When the current requirement is increased to 4 mA, the working areas reduce to radii of 3.93, 3.97, 3.61, and 2.11 cm, respectively, for tissue thicknesses of 2, 3, 4, and 5 cm. These estimations assume that the tissue attenuation is linear with thickness without considering anatomic variations.

When the coils are close to each other, the coupling become significant as the implant and wearable module circuits cannot be considered independent. The effective impedance from the implant side affects the impedances for the class-E amplifier. The overall power transfer efficiency then is affected by both the resonance and the amplifier efficiency. It can be observed that the input current into the class-E amplifier becomes dependent of the coil distance. The total power transfer efficiency in the system is defined as

$$n = n_{PA} \times n_e \qquad (4.1)$$

where n_{PA} is the class-E amplifier efficiency, and n_e is the link efficiency. The system efficiency is measured as

$$n = \frac{P_{out}}{P_{in}} \qquad (4.2)$$

where P_{out} is the total output power measured on the load, and P_{in} is the total DC power supplied to the circuit. Because the amplifier and coils are not separable, the results contain both the coupling effect and impedance matching effect for the amplifier. Therefore the distribution maps for the available power are different from the ones for power transfer efficiency. Fig. 4.10 shows the comparison.

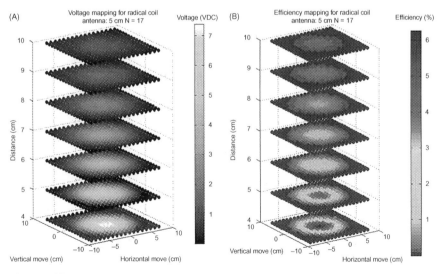

FIGURE 4.10

(A) Available power and (B) Wireless power transfer efficiency distribution (at the x–y plane) at various distances (z).

With a transmitter coil of 5 cm in radius and the 1×3.5 cm^2 internal coil, the average 3-dB (50%) power radii in each x$-$y plane are 3.7, 3.6, 3.75, 3.83, 4.15, 4.4, and 4.56 cm at the distances of 4, 5, 6, 7, 8, 9, and 10 cm, respectively. The efficiency maps, however, show the average 50% efficiency drop points at 3.46, 3.42, 3.67, 3.85, 4.09, 4.43, and 4.57 cm from the centers at the distances of 4, 5, 6, 7, 8, 9, and 10 cm, respectively. At a 4-cm distance, the produced current at the center is 14.78 mA, while it becomes 9.32 mA at a radius of 4 cm. At a 10-cm distance, the produced current at the center is 3.34 mA, while it becomes 2.52 mA at a radius of 4 cm. As the distance increases, the power and efficiency maps become similar because the coupling between coils becomes weaker. At near distances, special attention should be paid on the increase of power transfer efficiency in order to reduce the transmitting power requirement.

With single-radius coil antennas, the parameters for design variations are limited. Thus spiral coils are proposed to provide flexibility toward system requirements. With inner radius, outer radius, spacing between wires, and turn number, spiral coils give more degrees of freedom to tune the impedance, coupling coefficient and field distributions. For example, Table 4.2 shows two example coils with the main difference in the wire spacings resulting in different inner radii. Fig. 4.11 shows the normal components of magnetic field (H_z) distributions at the cross-section planes with spacings of 5 cm, 7 cm, and 10 cm to the spiral coils, obtained by finite element simulation. The colors indicate the magnitudes of the magnetic fields H when an electrical current of 1 A flows in the coil wires. The outer radius of both coils is 18 cm and the cross section of interest is ± 20 cm in both x and y directions aligned with the center of the transmitter spiral coil.

The normal component patterns indicate that the magnetic fields mainly focus near the center area. However, when the distance is 5 cm or less for coil #1, the peak power is not at the center but distributes around the center in a circular donut shape. This phenomenon manifests that the spiral coils can provide the flexibility to shape a desired power distribution at a specific plane. The spiral coil also provides another feature. As the wires generate both normal and tangential magnetic field components in space, concentric wires with different radii produce tangential

Table 4.2 Transmitter Coils

Parameters	TX Antenna	
	Coil #1	Coil #2
Inner radius (cm)	11	4
Outer radius (cm)	18	18
Turn number	10	10
Turn spacing (mm)	7.778	15.556
Average radius (cm)	14.5	11
Material	Litz wires	Litz wires

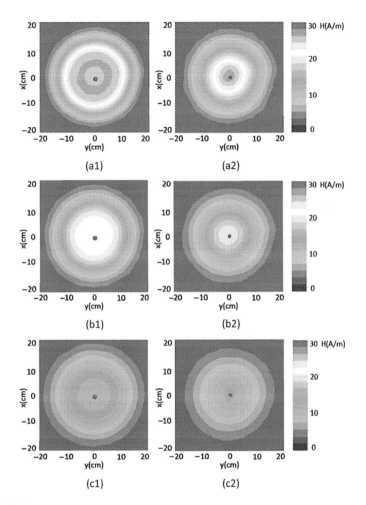

FIGURE 4.11

Field patterns of normal components (H_z) in two spiral antennas (coil #1 (a1,b1,c1) and coil #2 (a2,b2,c2)) at three planes parallel to the external coil with distances (z) of 5 cm (a1,a2), 7 cm (b1,b2), and 10 cm (c1,c2) to the implant coil.

components differently in the plane of interest. The patterns of the H_x component in Fig. 4.12 show that fields are mostly distributed off center at the outer areas. The H_y component patterns are similar to those for the H_x component, except that the patterns rotate by 90 degrees with the field nulls along the $y = 0$ axis instead of $x = 0$ axis. The tangential field patterns also depend on the distance between coils. As the divergence of field components depends on the wire spacing in the coil, it provides an opportunity to harvest power other than only from the normal component of the magnetic fields. As the implant has both position and angular

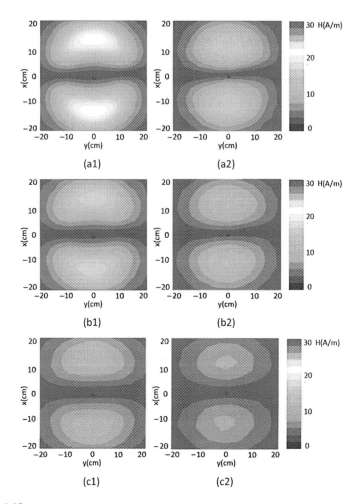

FIGURE 4.12

Field patterns for one of the tangential components (H_x) in two spiral antennas (coil #1 (a1,b1,c1) and coil #2 (a2,b2,c2)) at three planes parallel to the external coil with distances (z) of 5 cm (a1,a2), 7 cm (b1,b2), and 10 cm (c1,c2) to the implant coil.

arrangements other than being perfectly centered and parallel to the external coil, the implant coil produces the electrical current from both normal and tangential components. It is important in coil designs to determine the optimum position of the implant coil where the harvested energy can be maximized. The total received energy depends on the misalignment angle as well as the relative location of the receiver coil. This type of scenario often happens because it is difficult to maintain a perfect parallel arrangement between the implant coil, which is inside the body, and the transmitter coil, which is outside the body.

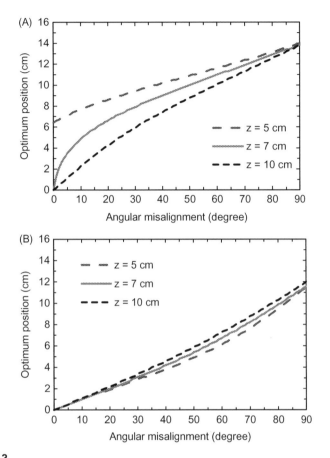

FIGURE 4.13

Optimal position to produce maximum harvested energy as a function of angular misalignment at three distances of 5, 7, and 10 cm between the transmitter and receiver coils. (A) Transmitter coil #1. (B) Transmitter coil #2. The optimal position is the distance from ($x = 0$, $y = 0$) where the implant coil aligns to the center of the transmitter coil, in the cross-sectional plane where the receiver coil locates.

Considering both normal and tangential components of fields, when the internal coil has a misalignment angle from the external coil, the optimal location to receive the maximum power, as shown in Fig. 4.13, moves toward the edge of the external coil in the plane. When the coils are in parallel, for coil #1 and a distance of 5 cm, the optimal location is 6.3 cm from the center as the peak power for the normal components is off center. For coil #1 and distances of 7 and 10 cm, the optimal location is in the center as the peak power for the normal components is at the center.

When the misalignment angle is 90 degrees, the optimal positions for coil #1 are 14, 13.8, and 13.8 cm from $(x = 0, y = 0)$ for the spacings of 5, 7, and 10 cm, respectively, as it only harvests energy from the tangential components, for which the maximum is at near ± 14 cm from the center. As the spacing between coils increases, the location of the maximum power stays at around ± 14 cm. For any misalignment angle, the optimal location to harvest the maximum energy exists. This can help to adjust the external coil location in order to get the best performance. When the spacing between coils is within the strongly-coupled region, the turn spacing between loop wires determines the field distribution patterns in the near-field. Fig. 4.14 shows the changes of the location of where the maximum

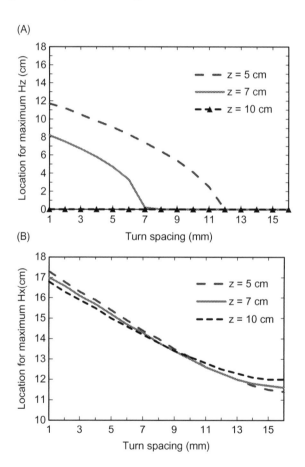

FIGURE 4.14

Location for the maximum field intensity of (A) H_z and (B) H_x components as a function of turn spacing between wires for a spiral transmitter coil with an 18-cm outer radius and 10 turns. The location is the distance from $(x = 0, y = 0)$ that aligns to the center of the transmitter coil.

field intensity is due to the wire spacing. As the spacing increases, the loop becomes smaller in the center and the optimal position to produce maximum H_z intensity moves toward the center, as shown in Fig. 4.14(A). The maximum tangential field location moves from the edge toward the inner area (around 11.5 cm) when the turn spacing increases from 1 to 16 mm. The maximum tangential field spot is not sensitive to the spacing between coils, as shown in Fig. 4.14(B).

The spiral coil provides more flexibility to shape the field distributions, so misalignment between the implant and the external coil can be used as a feature for power harvesting. The strategy to optimize power transfer efficiency in the system still depends on additional factors including the biological environments and physical design parameters, which are constrained by the size limitation in the implant and wearable module as well as the practical concerns in the patient's daily activities. One example is provided in [131], in which the constraints are given as clinical requirements and a strategy to optimize the coil designs is discussed.

4.6 CONCLUSION

In this chapter, two applications utilizing wireless closed-loop systems around the body to manage pain and gastric dysmotility by feedback-based neurostimulation and gastrostimulation have been presented. Neuronal APs are recorded and recognized for innocuous or noxious nociceptive signals by algorithms. The signals then are utilized by a process that combines the patient's inputs from his or her subjective feelings to generate commands for electrical stimulation in order to inhibit pain. Similarly, gastric myoelectric signals can be recorded for the analysis of stomach motility, which then can be modulated by electrical stimulation on stomach according to the signals. These digital methods demonstrate an effective means to manage symptoms and relieve suffering in patients. They are personalized medicine. The quantitative data of physiological signals indicating symptoms help caregivers to better understand patients' conditions during diagnosis and prognosis. The mechanism of wireless power transfer and signal transduction in a single miniaturized device allows the implants to be batteryless and wireless, enabling long-term wearability without constraining the patient's mobility and daily activities. It is clear that there are many potential clinical applications for such batteryless, wireless closed-loop systems to manage symptoms or provide therapy. The engineering efforts are disease-oriented as trade-off factors need to be considered according to the required functions, practical needs, and environmental parameters for a particular disorder. Some of the factors are discussed in this chapter include sensing modality, antenna design, and power transfer efficiency issues. Wireless closed-loop implant systems for autonomous management of diseases are promising solutions for the global grand challenge in healthcare. There is still much research needed to make them practical, deployable, and affordable.

REFERENCES

[1] Wenzel BJ, Boggs JW, Gustafson KJ, Grill WM. Closed loop electrical control of urinary continence. J Urol 2006;175(4):1559–63.

[2] Popovic DB, Stein RB, Jovanovic KL, Dai R, Kostov A, Armstrong WW. Sensory nerve recording for closed-loop control to restore motor functions. IEEE Trans Biomed Eng 1993;40(10):1024–31.

[3] Weinzimer SA, Steil GM, Swan KL, Dziura J, Kurtz N, Tamborlane WV. Fully automated closed-loop insulin delivery versus semiautomated hybrid control in pediatric patients with type 1 diabetes using an artificial pancreas. Diabetes Care 2008;31 (5):934–9.

[4] Macedo LG, Maher CG, Latimer J, McAuley JH. Feasibility of using short message service to collect pain outcomes in a low back pain clinical trial. Spine 2012;37 (13):1151–5.

[5] Luxton DD, McCann RA, Bush NE, Mishkind MC, Reger GM. mHealth for mental health: integrating smartphone technology in behavioral healthcare. Prof Psychol Res Pr 2011;42(6):505.

[6] Mann DL, Zipes DP, Libby P, Bonow RO. Braunwald's heart disease: a textbook of cardiovascular medicine. Elsevier Health Sciences; 2014.

[7] Ricci RP, Morichelli L, Santini M. Home monitoring remote control of pacemaker and implantable cardioverter defibrillator patients in clinical practice: impact on medical management and health-care resource utilization. Europace 2008;10 (2):164–70.

[8] Varma N, Stambler B, Chun S. Detection of atrial fibrillation by implanted devices with wireless data transmission capability. Pacing Clin Electrophysiol 2005;28(s1): S133–6.

[9] Yousef J, Lars A. Validation of a real-time wireless telemedicine system, using bluetooth protocol and a mobile phone, for remote monitoring patient in medical practice. Eur J Med Res 2005;10(6):254–62.

[10] Lee S, Su M, Liang M, Chen Y, Hsieh C, Yang C, et al. A programmable implantable microstimulator SOC with wireless telemetry: application in closed-loop endocardial stimulation for cardiac pacemaker. IEEE Trans Biomed Circuits Syst 2011;5(6):511–22.

[11] Medical Implant Communication Service (MICS) Federal Register. vol. 64; 1999: pp. 69926–134.

[12] North RB, Kidd DH, Zahurak M, James CS, Long DM. Spinal cord stimulation for chronic, intractable pain: experience over two decades. Neurosurgery 1993;32 (3):384–95.

[13] Song JJ, Popescu A, Bell RL. Present and potential use of spinal cord stimulation to control chronic pain. Pain Physician 2014;17(3):235–46.

[14] Picaud S, Sahel J. Retinal prostheses: clinical results and future challenges. C R Biol 2014;337(3):214–22.

[15] Maghami MH, Sodagar AM, Lashay A, Riazi-Esfahani H, Riazi-Esfahani M. Visual prostheses: the enabling technology to give sight to the blind. J Ophthalmic Vis Res 2014;9(4):494.

[16] Johnston JC, Durieux-Smith A, Angus D, O'Connor A, Fitzpatrick E. Bilateral paediatric cochlear implants: a critical review. Int J Audiol 2009;48(9):601–17.

[17] Wolfe J, Schafer E. Programming cochlear implants. Plural Publishing; 2014.

[18] Cameron T. Safety and efficacy of spinal cord stimulation for the treatment of chronic pain: a 20-year literature review. J Neurosurg: Spine 2004;100(3):254−67.

[19] Zeng F, Rebscher S, Harrison W, Sun X, Feng H. Cochlear implants: system design, integration, and evaluation. IEEE Rev Biomed Eng 2008;1:115−42.

[20] Kelly SK, Shire DB, Chen J, Doyle P, Gingerich MD, Cogan SF, et al. A hermetic wireless subretinal neurostimulator for vision prostheses. IEEE Trans Biomed Eng 2011;58(11):3197−205.

[21] Shire DB, Kelly SK, Chen J, Doyle P, Gingerich MD, Cogan SF, et al. Development and implantation of a minimally invasive wireless subretinal neurostimulator. IEEE Trans Biomed Eng 2009;56(10):2502−11.

[22] Li W, Rodger D, Tai Y. Integrated wireless neurostimulator. IEEE 22nd International Conference On Micro Electro Mechanical Systems, MEMS 2009; 2009.

[23] Li W, Rodger DC, Pinto A, Meng E, Weiland JD, Humayun MS, et al. Parylene-based integrated wireless single-channel neurostimulator. Sens Actuators A Phys 2011;166(2):193−200.

[24] Lee H, Park H, Ghovanloo M. A power-efficient wireless system with adaptive supply control for deep brain stimulation. IEEE J Solid-State Circuits 2013;48 (9):2203−16.

[25] Cho S, Cauller L, Rosellini W, Lee J. A MEMS-based fully-integrated wireless neurostimulator. In: 2010 IEEE 23rd International Conference on Micro Electro Mechanical Systems (MEMS); 2010.

[26] Ativanichayaphong T, Wang J, Huang W, Rao S, Tibbals H, Tang S, et al. Development of an implanted RFID impedance sensor for detecting gastroesophageal reflux. Proc. IEEE International Conference on RFID, Grapevine, TX, March 26−28, 2007.

[27] Ativanichayaphong T, Tang S.-J, Hsu L, Huang W, Seo YS, Tibbals HF, et al. An implantable batteryless wireless impedance sensor for gastroesophageal reflux diagnosis. Proc. IEEE MTT-S International Microwave Symposium, Anaheim May 23−28, 2010.

[28] Chang JY. Brain stimulation for neurological and psychiatric disorders, current status and future direction. J Pharmacol Exp Ther 2004;309(1):1−7.

[29] Kringelbach ML, Jenkinson N, Owen SL, Aziz TZ. Translational principles of deep brain stimulation. Nat Rev Neurosci 2007;8(8):623−35.

[30] Holm A, Staal M, Mooij JJ, Albers F. Neurostimulation as a new treatment for severe tinnitus: a pilot study. Otol Neurotol 2005;26(3):425−8.

[31] Tagliati M, Krack P, Volkmann J, Aziz T, Krauss JK, Kupsch A. Long-term management of DBS in dystonia: response to stimulation, adverse events, battery changes, and special considerations. Mov Disord 2011;26(S1):S54−62.

[32] Plow EB, Machado A. Invasive neurostimulation in stroke rehabilitation. Neurotherapeutics 2014;11(3):572−82.

[33] Charleston W, Deer TR, Inc C, Shurtleff M, Neuromodulation V. Intracranial neurostimulation for pain control: a review. Pain Physician 2010;13:157−65.

[34] Oujamaa L, Relave I, Froger J, Mottet D, Pelissier J. Rehabilitation of arm function after stroke. literature review. Ann Phys Rehabil Med 2009;52(3):269−93.

[35] Wu C, Sharan AD. Neurostimulation for the treatment of epilepsy: a review of current surgical interventions. Neuromodulation: Technology at the Neural Interface 2013;16(1):10−24.

[36] Saillet S, Langlois M, Feddersen B, Minotti L, Vercueil L, Chabardès S, et al. Manipulating the epileptic brain using stimulation: a review of experimental and clinical studies. Epileptic Disord 2009;11(2):100−12.

[37] Salanova V, Worth R. Neurostimulators in epilepsy. Curr Neurol Neurosci Rep 2007;7(4):315−19.

[38] Bronstein JM, Tagliati M, Alterman RL, Lozano AM, Volkmann J, Stefani A, et al. Deep brain stimulation for parkinson disease: an expert consensus and review of key issues. Arch Neurol 2011;68(2). pp. 165−165.

[39] Krack P, Batir A, Van Blercom N, Chabardes S, Fraix V, Ardouin C, et al. Five-year follow-up of bilateral stimulation of the subthalamic nucleus in advanced parkinson's disease. N Engl J Med 2003;349(20):1925−34.

[40] Coffey RJ. Deep brain stimulation devices: a brief technical history and review. Artif Organs 2009;33(3):208−20.

[41] Brazzelli M, Murray A, Fraser C. Efficacy and safety of sacral nerve stimulation for urinary urge incontinence: a systematic review. J Urol 2006;175(3):835−41.

[42] Patrick A, Epstein O. Review article: gastroparesis. Aliment Pharmacol Ther 2008;27(9):724−40.

[43] Lin Z, Forster J, Sarosiek I, McCallum RW. Review: treatment of gastroparesis with electrical stimulation. Dig Dis Sci 2003;48(5):837−48.

[44] Hasler W. Methods of gastric electrical stimulation and pacing: a review of their benefits and mechanisms of action in gastroparesis and obesity. Neurogastroenterol Motil 2009;21(3):229−43.

[45] Chen J. Mechanisms of action of the implantable gastric stimulator for obesity. Obes Surg 2004;14(1):S28−32.

[46] Weiland JD, Cho AK, Humayun MS. Retinal prostheses: current clinical results and future needs. Ophthalmology 2011;118(11):2227−37.

[47] Brown KD, Balkany TJ. Benefits of bilateral cochlear implantation: a review. Curr Opin Otolaryngol Head Neck Surg 2007;15(5):315−18.

[48] Balkany TJ, Hodges AV, Eshraghi AA, Butts S, Bricker K, Lingvai J, et al. Cochlear implants in children—a review. Acta Otolaryngol 2002;122(4):356−62.

[49] Lebedev MA, Nicolelis MA. Brain−machine Interfaces: past, present and future. Trends Neurosci 2006;29(9):536−46.

[50] Velliste M, Perel S, Spalding MC, Whitford AS, Schwartz AB. Cortical control of a prosthetic arm for self-feeding. Nature 2008;453(7198):1098−101.

[51] Marangell L, Martinez M, Jurdi R, Zboyan H. Neurostimulation therapies in depression: a review of new modalities. Acta Psychiatr Scand 2007;116(3):174−81.

[52] Mallet L, Polosan M, Jaafari N, Baup N, Welter M, Fontaine D, et al. Subthalamic nucleus stimulation in severe obsessive−compulsive disorder. N Engl J Med 2008;359(20):2121−34.

[53] Sidoti C, Agrillo U. Chronic cortical stimulation for amyotropic lateral sclerosis: a report of four consecutive operated cases after a 2-year follow-up: technical case report. Neurosurgery 2006;58(2):E384 discussion E384.

[54] George MS. Summary and future directions of therapeutic brain stimulation: neurostimulation and neuropsychiatric disorders. Epilepsy Behav 2001;2(3):S95−100.

[55] Gruber D, Kuhn A, Schoenecker T, Kopp U, Kivi A, Huebl J, et al. Quadruple deep brain stimulation in huntington's disease, targeting pallidum and subthalamic nucleus: case report and review of the literature. J Neural Transm 2014;121 (10):1303−12.

[56] Fassov JL, Lundby L, Laurberg S, Buntzen S, Krogh K. A randomized, controlled, crossover study of sacral nerve stimulation for irritable bowel syndrome. Ann Surg 2014;260(1):31–6.

[57] Gybels J. Thalamic stimulations in neuropathic pain: 27 years later. Acta Neurol Belg 2001;101(1):65–71.

[58] Blyth FM, Macfarlane GJ, Nicholas MK. The contribution of psychosocial factors to the development of chronic pain: the key to better outcomes for patients? Pain 2007;129(1):8–11.

[59] Blyth FM. Chronic pain—is it a public health problem? Pain 2008;137(3):465–6.

[60] Jordan KP, Thomas E, Peat G, Wilkie R, Croft P. Social risks for disabling pain in older people: a prospective study of individual and area characteristics. Pain 2008;137(3):652–61.

[61] Medical Data International, Market and Technology Reports, U.S. Markets For Pain Management Products, Report RP-821922. June, 1999.

[62] Melzack R. The mcgill pain questionnaire: major properties and scoring methods. Pain 1975;1(3):277–99.

[63] Price DD, McGrath PA, Rafii A, Buckingham B. The validation of visual analogue scales as ratio scale measures for chronic and experimental pain. Pain 1983;17 (1):45–56.

[64] Laureys S, Pellas F, Van Eeckhout P, Ghorbel S, Schnakers C, Perrin F, et al. The locked-in syndrome: what is it like to be conscious but paralyzed and voiceless? Prog Brain Res 2005;150:495–611.

[65] Price CI, Curless RH, Rodgers H. Can stroke patients use visual analogue scales? Stroke 1999;30(7):1357–61.

[66] Shields BJ, Cohen DM, Harbeck-Weber C, Powers JD, Smith GA. Pediatric pain measurement using a visual analogue scale: a comparison of two teaching methods. Clin Pediatr (Phila) 2003;42(3):227–34.

[67] Sandrine S, Guillaume C, Sadok G, Antoine D, Régis G, Olivier D. Closed-loop control of seizures in a rat model of absence epilepsy using the BioMEA™ system. In: 4th International IEEE/EMBS Conference On Neural Engineering, NER'09; 2009.

[68] Terry R. Jr. Vagus nerve stimulation: a proven therapy for treatment of epilepsy strives to improve efficacy and expand applications. Annual International Conference of the IEEE Engineering in Medicine and Biology Society, EMBC 2009; 2009.

[69] Breit S, Schulz JB, Benabid A. Deep brain stimulation. Cell Tissue Res 2004;318 (1):275–88.

[70] Butson CR, McIntyre CC. Differences among implanted pulse generator waveforms cause variations in the neural response to deep brain stimulation. Clin Neurophysiol 2007;118(8):1889–94.

[71] Chen AC. New perspectives in EEG/MEG brain mapping and PET/fMRI neuroimaging of human pain. Int J Psychophysiol 2001;42(2):147–59.

[72] Christmann C, Koeppe C, Braus DF, Ruf M, Flor H. A simultaneous EEG–fMRI study of painful electric stimulation. NeuroImage 2007;34(4):1428–37.

[73] Peng YB, Lin Q, Willis WD. Effects of GABA and glycine receptor antagonists on the activity and PAG-induced inhibition of rat dorsal horn neurons. Brain Res 1996;736(1):189–201.

[74] Price DD. Central neural mechanisms that interrelate sensory and affective dimensions of pain. Mol Interv 2002;2(6):392–403.

[75] Hosobuchi Y, Adams JE, Linchitz R. Pain relief by electrical stimulation of the central gray matter in humans and its reversal by naloxone. Science 1977;197 (4299):183–6.

[76] Millan MJ. Descending control of pain. Prog Neurobiol 2002;66(6):355–474.

[77] Hermann DM, Luppi P, Peyron C, Hinckel P, Jouvet M. Afferent projections to the rat nuclei raphe magnus, raphe pallidus and reticularis gigantocellularis pars A demonstrated by iontophoretic application of choleratoxin (Subunit B). J Chem Neuroanat 1997;13(1):1–21.

[78] Gerhart KD, Yezierski RP, Wilcox TK, Willis WD. Inhibition of primate spinothalamic tract neurons by stimulation in periaqueductal gray or adjacent midbrain reticular formation. J Neurophysiol 1984;51(3):450–66.

[79] Sandkühler J, Gebhart G. Relative contributions of the nucleus raphe magnus and adjacent medullary reticular formation to the inhibition by stimulation in the periaqueductal gray of a spinal nociceptive reflex in the pentobarbital-anesthetized rat. Brain Res 1984;305(1):77–87.

[80] Coimbra N, Brandão M. Effects of 5-HT 2 receptors blockade on fear-induced analgesia elicited by electrical stimulation of the deep layers of the superior colliculus and dorsal periaqueductal gray. Behav Brain Res 1997;87(1):97–103.

[81] Mensah-Brown E, Garey L. The superior colliculus of the camel: a neuronal-specific nuclear protein (NeuN) and neuropeptide study. J Anat 2006;208(2):239–50.

[82] Kandel ER, Schwartz JH, Jessell TM. Principles of neural science. New York: McGraw-Hill; 2000.

[83] Obeid I, Wolf PD. Evaluation of spike-detection algorithms fora brain-machine interface application. IEEE Trans Biomed Eng 2004;51(6):905–11.

[84] Wheeler BC, Nicolelis M. Automatic discrimination of single units. Methods for Neural Ensemble Recordings. CRC Press; 1999.

[85] Panzeri S, Petersen RS, Schultz SR, Lebedev M, Diamond ME. The role of spike timing in the coding of stimulus location in rat somatosensory cortex. Neuron 2001;29(3):769–77.

[86] Weidner C, Schmelz M, Schmidt R, Hammarberg B, Orstavik K, Hilliges M, et al. Neural signal processing: the underestimated contribution of peripheral human C-fibers. J Neurosci 2002;22(15):6704–12.

[87] Farajidavar A, Hagains CE, Peng YB, Chiao J-C. A closed loop feedback system for automatic detection and inhibition of mechano-nociceptive neural activity. IEEE Trans Neural Syst Rehabil Eng 2012;20(4):478–87.

[88] Farajidavar A, Hagains CE, Peng YB, Chiao J-C. Modulation of the mechano-nociceptive neural activities in the ventral posterolateral nucleus in thalamus. In: Proc. 6th International IEEE/EMBS Conference On Neural Engineering (NER), San Diego, November 6–8, 2013.

[89] Farajidavar A, Hagains CE, Peng YB, Behbehani K, Chiao J-C. Recognition and inhibition of dorsal horn nociceptive signals within a closed-loop system. Annual International Conference of the IEEE Engineering in Medicine and Biology Society (EMBC), Buenos Aires, Argentina, August 31, 2010.

[90] Kuncel AM, Grill WM. Selection of stimulus parameters for deep brain stimulation. Clin Neurophysiol 2004;115(11):2431–41.

[91] García-Larrea L, Sindou M, Mauguière F. Nociceptive flexion reflexes during analgesic neurostimulation in man. Pain 1989;39(2):145–56.

[92] Bartsch T, Paemeleire K, Goadsby PJ. Neurostimulation approaches to primary headache disorders. Curr Opin Neurol 2009;22(3):262−8.

[93] Mendell LM. Physiological properties of unmyelinated fiber projection to the spinal cord. Exp Neurol 1966;16(3):316−32.

[94] Beitz AJ. Periaqueductal gray. In: Paxinos G, editor. The rat nervous system. 2nd ed. San Diego: Academic Press; 1995. p. 173−82.

[95] Senapati AK, Huntington PJ, Peng YB. Spinal dorsal horn neuron response to mechanical stimuli is decreased by electrical stimulation of the primary motor cortex. Brain Res 2005;1036(1):173−9.

[96] Behbehani MM, Fields HL. Evidence that an excitatory connection between the periaqueductal gray and nucleus raphe magnus mediates stimulation produced analgesia. Brain Res 1979;170(1):85−93.

[97] Behbehani MM. Functional characteristics of the midbrain periaqueductal gray. Prog Neurobiol 1995;46(6):575−605.

[98] Peng YB, Lin Q, Willis WD. Involvement of alpha-2 adrenoceptors in the periaqueductal gray-induced inhibition of dorsal horn cell activity in rats. J Pharmacol Exp Ther 1996;278(1):125−35.

[99] Ritaccio A, Brunner P, Cervenka MC, Crone N, Guger C, Leuthardt E, et al. Proceedings of the first international workshop on advances in electrocorticography. Epilepsy Behav 2010;19(3):204−15.

[100] Leuthardt EC, Schalk G, Wolpaw JR, Ojemann JG, Moran DW. A brain−computer interface using electrocorticographic signals in humans. J Neural Eng 2004;1(2):63.

[101] Ball T, Kern M, Mutschler I, Aertsen A, Schulze-Bonhage A. Signal quality of simultaneously recorded invasive and non-invasive EEG. NeuroImage 2009;46(3):708−16.

[102] Rodriguez CRC. Gastrointestinal motility disorders. Top Palliat Care 1997;1:61.

[103] Vantrappen G, Janssens J, Coremans G, Jian R. Gastrointestinal motility disorders. Dig Dis Sci 1986;31(9):5−25.

[104] Wang YR, Fisher RS, Parkman HP. Gastroparesis-related hospitalizations in the United States: trends, characteristics, and outcomes, 1995−2004. Am J Gastroenterol 2008;103(2):313−22.

[105] Jung HK. The incidence, prevalence, and survival of gastroparesis in Olmsted County, Minnesota, 1996−2006 (Gastroenterology 2009;136:1225−1233). J Neurogastroenterol Motil 2010;16(1):99−100.

[106] Sifrim D, Holloway R, Silny J, Xin Z, Tack J, Lerut A, et al. Acid, nonacid, and gas reflux in patients with gastroesophageal reflux disease during ambulatory 24-hour pH-impedance recordings. Gastroenterology 2001;120(7):1588−98.

[107] Mainie I, Tutuian R, Shay S, Vela M, Zhang X, Sifrim D, et al. Acid and non-acid reflux in patients with persistent symptoms despite acid suppressive therapy: a multicentre study using combined ambulatory impedance-pH monitoring. Gut 2006;55(10):1398−402.

[108] DeVault KR, Castell DO, Practice Parameters Committee of the American College of Gastroenterology. Updated guidelines for the diagnosis and treatment of gastroesophageal reflux disease. Am J Gastroenterol 1999;94(6):1434−42.

[109] DeVault KR, Castell DO. Updated guidelines for the diagnosis and treatment of gastroesophageal reflux disease. Am J Gastroenterol 2005;100(1):190−200.

[110] Ofman JJ. The relation between gastroesophageal reflux disease and esophageal and head and neck cancers: a critical appraisal of epidemiologic literature. Am J Med 2001;111(8):124–9.

[111] Ativanichayaphong T, Huang W, Wang J, Rao SM, Tibbals H, Tang S, et al. A wireless sensor for detecting gastroesophageal reflux. SPIE International Smart Materials, Nano- & Micro-Smart Systems Symposium, Biomedical Applications of Micro- and Nanoengineering Conference, Adelaide, Australia, December 10–13, 2006.

[112] Ativanichayaphong T, Wang J, Huang W, Rao S, Chiao J-C. A simple wireless batteryless sensing platform for resistive and capacitive sensors. IEEE Sensors, Atlanta, Georgia, October 26–28, 2007.

[113] Cao H, Landge V, Tata U, Seo YS, Rao S, Tang S, et al. An implantable, batteryless, and wireless capsule with integrated impedance and pH sensors for gastroesophageal reflux monitoring. IEEE Trans Biomed Eng 2012;59(11):3131–9.

[114] Cao H, Rao S, Tang S, Tibbals HF, Spechler S, Chiao J-C. Batteryless implantable dual-sensor capsule for esophageal reflux monitoring. Gastrointest Endosc 2013;77(4):649–53.

[115] Corazziari E. Definition and epidemiology of functional gastrointestinal disorders. Best Pract Res Clin Gastroenterol 2004;18(4):613–31.

[116] Xu J, Ross RA, McCallum RW, Chen JD. Two-channel gastric pacing with a novel implantable gastric pacemaker accelerates glucagon-induced delayed gastric emptying in dogs. Am J Surg 2008;195(1):122–9.

[117] Eckhauser FE, Conrad M, Knol JA, Mulholland MW, Colletti LM. Safety and long-term durability of completion gastrectomy in 81 patients with postsurgical gastroparesis syndrome/discussion. Am Surg 1998;64(8):711.

[118] Zhang J, Chen J. Pacing the gut in motility disorders. Curr Treat Options Gastroenterol 2006;9(4):351–60.

[119] Familoni BO, Abell TL, Nemoto D, Voeller G, Johnson B. Efficacy of electrical stimulation at frequencies higher than basal rate in canine stomach. Dig Dis Sci 1997;42(5):892–7.

[120] Bortolotti M. The "electrical way" to cure gastroparesis. Am J Gastroenterol 2002;97(8):1874–83.

[121] Smout AJPM. Myoelectric activity of the stomach: gastroelectromyography and electrogastrography. DUP Science, Delft 1980.

[122] Parkman H, Hasler W, Barnett J, Eaker E. Electrogastrography: a document prepared by the gastric section of the American motility society clinical GI motility testing task force. Neurogastroenterol Motil 2003;15(2):89–102.

[123] Chen J, McCallum R. Electrogastrography: measurement, analysis and prospective applications. Med Biol Eng Comput 1991;29(4):339–50.

[124] Chiao J-C, Farajidavar A, Cao H, McCorkle P, Sheth M, Seo Y, et al. Wireless implants for in vivo diagnosis and closed-loop treatment. Proc. IEEE Annual Wireless and Microwave Technology Conference (WAMICON), Clearwater, FL, April 18–19, 2011.

[125] Forster J, Sarosiek I, Lin Z, Durham S, Denton S, Roeser K, et al. Further experience with gastric stimulation to treat drug refractory gastroparesis. Am J Surg 2003;186(6):690–5.

[126] Lin Z, Forster J, Sarosiek I, McCallum RW. Treatment of diabetic gastroparesis by high-frequency gastric electrical stimulation. Diabetes Care 2004;27(5):1071−6.

[127] Lin Z, Sarosiek I, Forster J, McCallum R. Symptom responses, long-term outcomes and adverse events beyond 3 years of high-frequency gastric electrical stimulation for gastroparesis. Neurogastroenterol Motil 2006;18(1):18−27.

[128] Abell T, McCallum R, Hocking M, Koch K, Abrahamsson H, LeBlanc I, et al. Gastric electrical stimulation for medically refractory gastroparesis. Gastroenterology 2003;125(2):421−8.

[129] Deb S, Tang S, Abell TL, Rao S, Huang W, To S, et al. An endoscopic wireless gastrostimulator (with Video). Gastrointest Endosc 2012;75(2):411−15.

[130] Rao S, Dubey S, Deb S, Hughes Z, Seo Y, Nguyen MQ, et al. Wireless gastric stimulators. Proc. Texas Symposium on Wireless and Microwave Circuits and Systems, Waco, TX; 2014.

[131] Nguyen MQ, Hughes Z, Woods P, Seo Y, Rao S, Chiao J-C. Field distribution models of spiral coil for misalignment analysis in wireless power transfer systems. IEEE Trans Microw Theory Tech 2014;62(4):920−30.

Human-aware localization using linear-frequency-modulated continuous-wave radars

J.-M. Muñoz-Ferreras[1], R. Gómez-García[1] and C. Li[2]

[1]*University of Alcalá, Madrid, Spain* [2]*Texas Tech University, Lubbock, TX, United States*

CHAPTER OUTLINE

C. Li, M. Tofighi, D. Schreurs and T-Z. J. Horng (Eds): Principles and Applications of RF/Microwave
in Healthcare and Biosensing.

5.1 INTRODUCTION

Nowadays, short-range radars are emerging as interesting devices for indoor and outdoor applications [1−6]. Non-contact human-aware localization can be attained through these radar sensors, which must be compact and have good performance [7−10]. Applications range from eldercare, patient monitoring, detection of surviving people after avalanches or earthquakes, and real-time healthcare to radar-based augmented reality [11−16].

To this purpose, two main architectures have traditionally been proposed in the literature: Doppler radars and impulse radio ultra-wideband (IR-UWB) radar systems. The former employ a single tone as the transmitting waveform, not possessing range resolution [11,17,18]. The latter transmit extremely narrow pulses which are usually difficult to acquire [12,19−22]. In both cases, many schemes and associated processing approaches have been suggested with interesting and promising results in the field of biosensing and healthcare [17−22].

This chapter is devoted to the recently suggested linear-frequency-modulated continuous-wave (LFMCW) radar architecture for human-aware localization applications [7,23−26]. By also adding the coherence feature, the system combines an excellent phase-based precision in the range measurements of targets with the disposition of range resolution [23−26]. Additionally, the acquisition of the echoes returned from the targets can be largely simplified, which can be exploited to develop inexpensive radar prototypes [23−26].

The chapter is written in a self-contained style, so that the reader can find all the important information to understand, construct, use, and even simulate a coherent LFMCW radar prototype for short-range applications. Given its unique features, it is the opinion of the authors that this kind of radar system will proliferate in the near future to improve our quality of life. Perhaps, the automotive sector is currently leading this race [27−29].

The next section describes the so-called deramping technique, which is the key concept for a low-cost LFMCW radar. It consists in mixing a replica of the transmitted signal with the echoes coming from the targets.

An easy-to-follow mathematical analysis to understand the signal waveforms is introduced in Section 5.3, which provides the reader with important concepts, such as fast-time, slow-time, range resolution, and so forth. Section 5.3 also details key aspects related to the coherence maintenance of the radar, so that the phase/Doppler history of the target scatterers can be exploited. Additionally, two simple algorithms to obtain the range evolution of targets are introduced. Respectively based on the amplitude and the phase of the slow-time signal, these algorithms can be employed to derive the range history of targets, which is the main output required in human-aware localization applications.

Signal processing issues are provided in Section 5.4, with the introduction of the conventional data formatting and the construction of important matrices, such as the range-profile matrix or the range-Doppler map. Also, a more formal description of the algorithms to track the range history of targets is provided in Section 5.4.

Section 5.5 reviews the important concepts of range resolution, precision, and accuracy in the context of coherent LFMCW radars. The authors have noted that sometimes these concepts are not adequately used in the literature, which usually leads the reader to confusion. The concepts are also briefly described for other dimensions, such as angle (azimuth or elevation) and Doppler.

On other hand, clutter is any unwanted return which may have a negative impact on the correct working of the radar. In a context of human-aware localization, Section 5.6 provides a mathematical framework for possibly arising clutter effects and the proposal of Doppler-based mitigation techniques for coherent LFMCW radars.

Simulated and experimental results are respectively given in Sections 5.7 and 5.8. Simulated examples enable corroboration of the principle of operation of the radar system, the implied mathematics, the indicated limitations, and the proposed algorithms. Experimental results also confirm the concepts and make the connection with reality. Additionally, a description of the constructed prototype is briefly provided.

The final section (Section 5.9) briefly suggests our future work, which is mainly concentrated on the construction of a millimeter-wave coherent LFMCW prototype. The advantages of increasing the operation frequency of the radar sensor are commented on in Section 5.9.

5.2 LFMCW RADAR ARCHITECTURE

A radar is a microwave/RF (radio frequency) system which incorporates a transmission (Tx) and a reception (Rx) chain. It is able to generate signals to be transmitted into the air and to capture the echoes returned from the targets. Fig. 5.1 shows a general scheme for any radar [30,31].

In the Tx part, an important block is the signal generation stage, which is responsible for the correct generation of the waveform to be transmitted. The generation block usually works at baseband and is conventionally made up of a direct

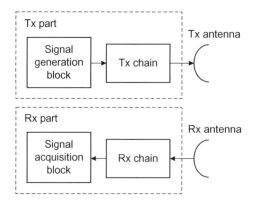

FIGURE 5.1

General two-antenna architecture for radar systems.

digital synthesizer (DDS), which generates the samples corresponding to the desired waveform. To avoid spectral aliasing, the output of the DDS is filtered with a low-pass profile with a cut-off frequency in the range of the sampling frequency. The Tx chain is responsible for upconverting the baseband signal to the RF range, in which the Tx to the air is possible. This signal conditioning is made through mixers and/or frequency multipliers. The Tx chain also possesses amplifiers and filters to further enhance the spectrum to be transmitted. One important element in the Tx chain is the usually-last high power amplifier, which increases the level of the output signal so that the range of the radar to detect far targets can be improved [30,31].

Fig. 5.1 shows that the transmitter uses a specific antenna, different from the one in the Rx part. This two-antenna scheme is usually preferred in continuous-wave (CW) radar prototypes, since for this case, the transmitter is operating all the time, and the Tx−Rx isolation must be large so that the high levels in the Tx part cannot damage the sensitive receiver. A scheme based on a single antenna (not shown in Fig. 5.1) must employ a bulky circulator and a limiter in the Rx chain to attain protective Tx−Rx isolation levels. In any case, the two-antenna architecture is preferred and easily found in many LFMCW radars [32,33].

Regarding the Rx part, the microwave chain is shaped by the cascading connection of amplifiers, mixers, and filters, which achieves an adequate conditioning of the received signal. A first element in the Rx chain is usually a low-noise amplifier (LNA), which permits the increase of the sensitivity of the radar for weak targets, since the noise level at Rx is reduced. After one or two downconversions from the RF stage to an intermediate frequency stage, it is usual to find an In-Phase/Quadrature (I/Q) demodulation, in which the signal is split into two branches and respectively mixed with two in-quadrature tones. The I and Q baseband channels are subsequently acquired by the signal acquisition block, which makes the corresponding sampling of these signals. The sampling process is made through an analog-to-digital converter (ADC), which is a costly device, especially if a very high sampling frequency (f_s) is required [30,31].

If a good range resolution is desired, radars usually transmit a large bandwidth (B) (say in the range of hundreds of megahertz) [32,33]. This leads to the fact that both the Tx and Rx parts must adequately handle this wideband signal. The ADC of the acquisition block should employ a very high sampling frequency to permit the correct handling of the signal bandwidth without spectral overlapping (aliasing) [30,31]. By supposing an I/Q demodulation, the sampling frequency must satisfy the following inequality:

$$f_s \geq B. \tag{5.1}$$

In other words, according to the architecture in Fig. 5.1, a very high sampling frequency must be used if a good range resolution is required. This means that very expensive acquisition blocks should be used.

Long-range radars are expensive equipment which use complex generation and acquisition blocks. Additionally, they have to transmit large powers. On the contrary, the intended human-aware localization scenarios belong to short-range applications. The LFMCW waveform is widely employed for short-range radars and its associated architecture, based on the analog deramping technique, can circumvent the restriction imposed by Eq. (5.1), which fortunately leads to cost-effective high-resolution radar solutions.

5.2.1 DERAMPING-BASED LFMCW RADAR ARCHITECTURE

For short-range applications, the most interesting architecture for an LFMCW radar is depicted in Fig. 5.2. The key difference with respect to Fig. 5.1 is found in the RF mixing stage, in which a replica of the transmitted signal is mixed with the returned echoes. This analog processing is called deramping or dechirping [32,33].

The signal after the mixing process is the so-called beat signal, which is subsequently processed by the Rx chain. The important point is that the bandwidth of the beat signal is much lower than the transmitted one B. This means that its acquisition can be achieved by a low-end ADC, which largely simplifies the whole radar system. In particular, it will be later shown in Section 5.3 that the sampling frequency of the ADC must now meet the following constraint [32,33]:

$$f_s \geq \frac{2\gamma}{c} R_{\max} \tag{5.2}$$

where γ is the chirp rate, c is the speed of light, and $R_{\max}$ is the maximum range from which echoes of targets are expected. For short ranges (say tenths of meters), the bandwidth of the beat signal (i.e., the right-hand term of Eq. 5.2) is much lower than the transmitted bandwidth B. This bandwidth reduction simplifies the radar prototype while maintaining the same range resolution.

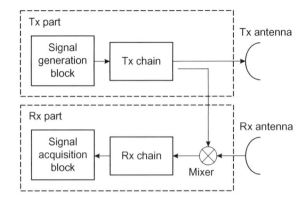

FIGURE 5.2

Scheme for an LFMCW radar based on the analog deramping technique.

This simplification of the ADC enables the transformation of these radars into quasi-commercial off-the-shelf prototypes, which are cost effective. Indeed, their development has been accomplished out of military scenarios. Examples can be found in the technical literature, such as the miniaturized Miniaturized Synthetic Aperture Radar (MiSAR) developed to fly on small unmanned aerial vehicles [34], the millimeter wave high-range-resolution Technical University of Madrid prototype [35], the Brigham Young University μSAR systems [36,37], or the X-band Delft University of Technology demonstrator [38]. Also, deramping-based LFMCW prototypes can be found for automotive radars [27−29].

5.3 LFMCW WAVEFORM

Fig. 5.3 shows a representation of the instantaneous frequency for the LFMCW waveform [25]. As can be seen, the frequency varies linearly in each period T. This is the typical behavior for a chirp, whose amplitude versus time plotting is also depicted in Fig. 5.3. The transmitted bandwidth is given by the parameter B, whereas the center transmitted frequency is f_c. The inverse of the period T is historically referred to as the pulse repetition frequency (PRF), although the LFMCW waveform is not a pulsed signal.

CW radars continuously transmit a signal. This implies that the receiver is opened for all of the time and can integrate more energy in comparison with pulsed radars, in which the energy is concentrated in short pulses and limited in Tx (i.e., it is very costly to obtain high peak powers in pulsed radars). This feature of CW radars leads to a better signal-to-noise ratio (SNR), which ultimately implies a better performance (e.g., in terms of range precision) [30,31].

The LFMCW waveform inherits this advantage from its CW nature, but additionally it provides range resolution due to the transmitted bandwidth B.

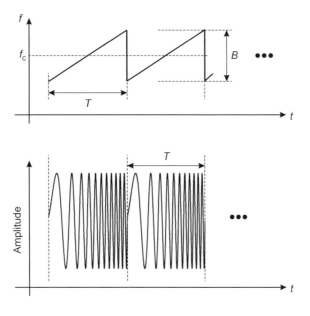

FIGURE 5.3

Transmitted waveform for an LFMCW radar. Instantaneous frequency versus time, and amplitude versus time.

The range resolution (which is higher when the transmitted bandwidth B is larger) confers interesting range-isolation properties upon LFMCW radar systems.

5.3.1 MATHEMATICAL ANALYSIS

The LFMCW waveform can be mathematically treated. Important concepts (such as range resolution, sampling requirements, phase history, Doppler, etc.) emerge from this analysis. Also, this mathematics is important for understanding the associated signal processing methods, such as the construction of the raw-data matrix or the phase-based range tracking of an isolated target [25].

In one waveform period T, the expression for the transmitted complex analytic signal can be expressed as

$$s_{\mathrm{Tx}}(t) = \exp\left(j(2\pi f_c t + \pi\gamma t^2)\right) \tag{5.3}$$

where f_c is the central frequency (refer to Fig. 5.3), γ is the chirp rate (i.e., the slope of the linear frequency-time curve in Fig. 5.3), and t is the "fast-time" that must be defined in the interval $[-T/2, T/2]$. The chirp rate γ can be easily shown to be

$$\gamma = \frac{B}{T} = B \cdot \mathrm{PRF}. \tag{5.4}$$

Note that the instantaneous frequency of the signal Eq. (5.3) can be calculated as

$$f_{\text{Tx}}(t) = \frac{1}{2\pi} \frac{d\phi_{\text{Tx}}(t)}{dt} = f_{\text{c}} + \gamma t \tag{5.5}$$

where the phase $\varphi_{\text{Tx}}(t)$ is the phase corresponding to Eq. (5.3). That is,

$$\phi_{\text{Tx}}(t) = 2\pi f_{\text{c}} t + \pi \gamma t^2. \tag{5.6}$$

According to Eq. (5.5) and to the fact that the fast-time t is defined in the interval $[-T/2, T/2]$, the instantaneous frequency linearly sweeps from $f_{\text{c}} - \gamma T/2$ to $f_{\text{c}} + \gamma T/2$. This speaks of the importance of defining the fast-time t in the interval $[-T/2, T/2]$ so that the parameter f_{c} is indeed the center frequency. Also, note the distinction between fast-time and time. The time varies continuously, whereas the fast-time t is always defined in the interval $[-T/2, T/2]$ for every waveform period T.

Suppose a target whose range to the radar is given by $R(\tau)$, where τ is the so-called "slow-time." The target is assumed to be a point scatterer and the slow-time τ is assumed to be discretized at T intervals. That means that the slow-time τ only takes values at 0, T, $2T$, and so forth, so that the slow-time naturally appears sampled at the rate PRF. In other words, the range $R(\tau)$ is assumed to be constant during each interval T, or the target is assumed to hop from its past position to its current location instantaneously. Obviously, this does not reflect a realistic smooth motion of the target, but this so-called "stop-and-go" or "stop-and-hop" approximation is excellent for slow targets (such as the ones considered in this chapter) and drastically simplifies the mathematical analysis of radar signals [30–33]. Furthermore, the stop-and-go hypothesis is intrinsic to every single piece of work in the field of radar signal processing, such as algorithms for synthetic aperture radar (SAR) imagery or conventional moving target indicator (MTI) processing for long-range surveillance radars [30–33]. In any case, works beyond the stop-and-hop assumption can be found in the literature when dealing with fast targets, and its violation leads to unexpected effects, such as the range broadening for the blades of a helicopter [39].

The signal reflected from the point scatterer in the interval T can be simply written as

$$s_{\text{Rx}}(t) = \sigma \cdot s_{\text{Tx}}\left(t - \frac{2R(\tau)}{c}\right) \tag{5.7}$$

where σ is the amplitude of the received signal with respect to the amplitude of the transmitted signal, which was assumed to be the unity. This amplitude obviously depends on the distance of the target to the radar, propagation losses, additional losses in the microwave chains, antenna pointing errors, and the radar cross section of the target.

At the output of the RF mixer in the deramping-based architecture in Fig. 5.2, the beat signal $s_b(t)$ arises. By combining Eqs. (5.3) and (5.7), the mathematical form for $s_b(t)$ can be written as

$$s_b(t) = s_{Tx}(t) \cdot s_{Rx}^*(t) = \sigma \cdot \exp\left(j\left(\frac{4\pi\gamma R(\tau)}{c} t + \frac{4\pi f_c R(\tau)}{c} + \phi_{RVP} \right) \right) \qquad (5.8)$$

where φ_{RVP} is the so-called residual video phase [32]

$$\phi_{RVP} = -\frac{4\pi\gamma R^2(\tau)}{c^2}. \qquad (5.9)$$

The term φ_{RVP} in Eq. (5.9) is negligible for short ranges (note that the speed of light appears squared in the denominator), and thus, it is usually disregarded (and assumed to be zero in the next equations) [32]. By taking a look at Eq. (5.8), one concludes that the beat signal for a point-scatterer target for the period T is a sinusoidal signal whose fast-time frequency (beat frequency f_b) is given by

$$f_b(\tau) = \frac{1}{2\pi} \frac{d\phi_b(t)}{dt} = \frac{2\gamma R(\tau)}{c} \qquad (5.10)$$

where the phase of the beat signal $\varphi_b(\tau)$, from Eq. (5.8), is

$$\phi_b(t) = \frac{4\pi\gamma R(\tau)}{c} t + \frac{4\pi f_c R(\tau)}{c}. \qquad (5.11)$$

That is, in each slow-time instant (for which the target is assumed to be static at the range position $R(\tau)$), the beat frequency is constant and given by Eq. (5.10). It comes to light that a Fourier transform of the beat signal (Eq. 5.8) gives rise to a spectrum with a single peak (corresponding to the single point-scatterer target) at f_b. According to Eq. (5.10), the frequency of the beat signal is proportional to the range, so that a simple scaling can transform the frequency axis to a range axis; hence the "range profile" can be obtained (one range profile for each period T of the LFMCW waveform). Note that this range profile provides the range positions of the illuminated scatterers, since the beat signal for many targets is simply the sum of the corresponding signals for each scatterer Eq. (5.8).

Regarding the acquisition of the beat signal, it comes to light that the maximum beat frequency, according to Eq. (5.10), can be expressed as

$$f_{b,max} = \frac{2\gamma R_{max}}{c} \qquad (5.12)$$

where R_{max} is the maximum range from which echoes are expected to come. Assuming that the beat signal is a complex-valued function (i.e., it comes from an I/Q demodulation), Eq. (5.2) is demonstrated. This means that the ADC for the deramping-based architecture (Fig. 5.2) can use a low sampling frequency, which largely simplifies the radar prototype. In other words, for short-range LFMCW radars, it happens that

$$f_{b,max} \ll B. \qquad (5.13)$$

As an example, consider that $R_{max} = 30$ m, $B = 1$ GHz, and $T = 1$ ms. The chirp rate (see Eq. 5.4) is $\gamma = 10^{12}$ Hz/seconds. From Eq. (5.12), the maximum beat frequency Eq. (5.12) is only $f_{b,max} = 200$ kHz, which is much lower than the transmitted bandwidth ($B = 1$ GHz).

The fast-time Fourier transform $S_b(f)$ of Eq. (5.8) (i.e., the mathematical expression for the range profile) can be easily shown to be

$$S_b(f) = \sigma T \cdot \exp\left(j \frac{4\pi f_c R(\tau)}{c}\right) \cdot \text{sinc}\left(T\left(f - \frac{2\gamma R(\tau)}{c}\right)\right) \tag{5.14}$$

where

$$\text{sinc}(x) = \frac{\sin(\pi \cdot x)}{\pi \cdot x}. \tag{5.15}$$

The frequency f is the Fourier variable associated with the fast-time t. As seen in Eq. (5.14), the sinc function appears centered at the beat frequency f_b of the target (see Eq. 5.10). Additionally, as is widely known, the 3-dB width of the sinc in the f-variable is given by

$$\Delta f = \frac{1}{T}. \tag{5.16}$$

After the aforementioned scaling to transform the f-axis into a range-axis, the width of the sinc (i.e., the width of the point spread function (PSF) of the target or the range resolution) can be written as

$$\Delta R = \frac{c}{2\gamma} \Delta f = \frac{c}{2\gamma} \cdot \frac{1}{T} = \frac{c}{2B}. \tag{5.17}$$

The right-hand term in Eq. (5.17) is general for the range resolution of any radar. It says that a higher resolution can be attained if the transmitted instantaneous bandwidth B is increased. As is further explained in Section 5.5.1, the range resolution is the ability of the radar to discriminate two targets closely spaced in range. For example, a high range resolution (HRR) radar transmits a large bandwidth so that the targets become "extended targets"; i.e., a single target appears in many range resolution cells [33]. A bandwidth B of 1 GHz implies a range resolution $\Delta R = 15$ cm. In the particular case of a deramping-based LFMCW short-range radar, the resolution is still given by Eq. (5.17). The clear advantage is that the acquisition block can be very simple and thus, the radar can be inexpensive, as previously anticipated in Section 5.2.

In an LFMCW radar, it is obviously necessary to apply a Fourier transform in order to obtain a range profile for each waveform period T. This Fourier transform operation is digitally performed (through a Fast Fourier Transform (FFT)) after the acquisition of the low-pass-filtered beat signal.

A clarifying statement is given next. Fig. 5.4 indicates that the beat signal Eq. (5.8) is not present during the complete T interval for all the targets. The useful interval T_u is the one which should be acquired and processed, which implies that the processed effective bandwidth B_{eff} is slighter lower than the transmitted

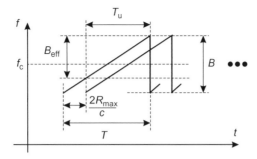

FIGURE 5.4

Definition of the useful interval and the effective bandwidth for an LFMCW radar.

bandwidth B. All the expressions in this section should be modified by writing T_u instead of T, and B_{eff} instead of B. In any case, note that the useful interval T_u is given by

$$T_u = T - \frac{2R_{max}}{c} \qquad (5.18)$$

which is only slightly lower than the period T for an LFMCW short-range radar. For example, if $T = 1$ ms and $R_{max} = 30$ m, the difference $T - T_u$ is only 0.2 μs.

5.3.2 COHERENCE MAINTENANCE

The Doppler frequency (f_{Dop}) of a moving target illuminated by a radar is well-known to be [30–33]

$$f_{Dop} = \frac{2v_{rad}}{\lambda} \qquad (5.19)$$

where v_{rad} is the radial velocity of the target (i.e., the speed component along the radar line-of-sight (LOS) vector) and λ is the wavelength associated with the center frequency, whose expression is

$$\lambda = \frac{c}{f_c}. \qquad (5.20)$$

A convention for the sign of the Doppler frequency is usually followed. It says that the Doppler is positive if the target is approaching the radar, and negative when the target is moving away.

Regarding the LFMCW waveform, if one takes a look at Eq. (5.14), it is noticeable that the range profile for a point scatterer is shaped by the multiplication of an exponential factor and a sinc factor. The latter has profoundly been explained in the previous section; remember that it is related to the range position of the target. From range profile to range profile, it becomes clear that the target can migrate through range resolution cells (depending on its dynamics and the value of the range resolution ΔR). In any case, this migration can be corrected by specific algorithms, such as the extended envelope correlation or global range alignment [40,41]. In relation to the former factor (the exponential one in Eq. 5.14), it turns out to represent a phase history of the target in the slow-time τ. Indeed, it becomes trivial that this phase $\varphi_t(\tau)$ is given by

$$\phi_t(\tau) = \frac{4\pi f_c R(\tau)}{c}. \tag{5.21}$$

Assume now that the range history of the target $R(\tau)$ can be written as

$$R(\tau) = R_0 + v_{\text{rad}} \cdot \tau \tag{5.22}$$

where R_0 is the initial range of the target and v_{rad} is its along-LOS speed. That is, the target is assumed to be uniformly moving. For this particular case and according to Eq. (5.21), the phase history of the target is

$$\phi_t(\tau) = \frac{4\pi f_c}{c}(R_0 + v_{\text{rad}} \cdot \tau). \tag{5.23}$$

In other words, the exponential factor of Eq. (5.14) is a sinusoidal signal in the slow-time τ with a frequency f_{Dop}, given by

$$f_{\text{Dop}} = \frac{1}{2\pi} \frac{d\phi_t(\tau)}{d\tau} = \frac{2f_c v_{\text{rad}}}{c}. \tag{5.24}$$

Eq. (5.24) is identical to Eq. (5.19), which means that the Fourier variable associated with the slow-time τ is the Doppler frequency.

A radar is said to be coherent if it preserves the phase history of the targets (see Eq. 5.21). If the radar is coherent, then the obtaining of the Doppler becomes trivial, after applying a Fourier transform (i.e., an FFT) to the slow-time signal. If a time-frequency transform (such as the spectrogram or the Wigner–Ville distribution) is applied to this signal, the Doppler frequency as a function of the slow-time τ (i.e., the Doppler history of the targets) can be reconstructed. An incoherent radar can only obtain range profiles and its performance is always poorer than that of a coherent radar [31]. The LFMCW radars here exposed are coherent. To preserve the phase information, a careful design of the radar becomes mandatory. As stated later, the most important rule is that the generation and acquisition blocks share the same clock [25].

Regarding the coherence preservation at the Tx part, the arrows shown in Fig. 5.5 highlight the initial phase for each transmitted waveform period. These initial phases must be equal (or at least known, to compensate for their effects). The generation block (normally based on DDS and voltage-controlled oscillator technologies) is normally capable of guaranteeing this phase control. Depending on the particular design, the irrelevant unprocessed interval T_r shown in Fig. 5.5 can be useful to match the aforementioned phases, since it can make easier the solution to possibly-arising locking-related issues.

More importantly, the generation and acquisition blocks must share a common clock or, at least, the clock shifts must be known. In other words, the samples taken by the ADC must be synchronized with the produced ramps in order to

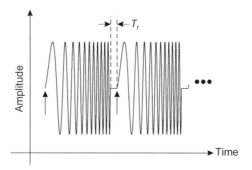

FIGURE 5.5

Control of the initial phase for each transmitted ramp to guarantee radar coherence.

guarantee that no additional unknown phase term is added to Eq. (5.21). If this synchronization is achieved, the number of samples of the beat signal for each waveform period T is constant.

The lack of clock-matching can mean a severe degradation for the radar coherence. For example, if the unwanted additional phase term in Eq. (5.21) is a white uniform stochastic process between $-\pi$ and π, the phase history of the target is completely destroyed and no Doppler reconstruction is possible. In practice, these things are not so bad, and the fact of using different clocks at the generation and acquisition stages leads to a slow phase-drifting, which is typically observed as a quasi-linear leaning of the always-present zero-Doppler returns. In that case, the data formatting becomes, however, uncomfortable, since the number of samples of the beat signal per each period T is not constant.

5.3.3 OBTAINING OF THE RANGE

Human-aware localization is a short-range application. Depending on the specific scenario, different information can be interesting to be extracted. For example, if the application consists in the detection of people trapped after a snow avalanche, the vital signs can be used as moving targets. In a healthcare scenario, an exact monitoring of the respiration and/or heartbeats can be of high interest. For gaming applications, it may be useful to analyze the human gait and monitor the different parts of the body (hands, arms, legs, etc.), which have different range histories, referred to as micro-Doppler signatures [42].

A common denominator for all these human-aware applications is the interest in obtaining the range histories (or, equivalently, Doppler histories) of different scatterers, while trying to avoid interferences among them. A short-range coherent LFMCW radar sensor is often preferred for this task and, furthermore, provides range resolution which can be very useful for the isolation of individual scatterers.

Eq. (5.14) gives the key to derive two independent algorithms to extract the range history $R(\tau)$ of the illuminated point-scatterer target:

- An amplitude-based tracking of the range, which uses the sinc factor of Eq. (5.14).
- A phase-based tracking of the range, which uses the exponential factor of Eq. (5.14).

In relation to the amplitude-based algorithm, the idea consists in tracking the amplitude peak from range profile to range profile. As will be seen later in Section 5.5.2, this algorithm is not very precise and usually suffers from the target-scintillation phenomenon. In practice, any target must be considered as the combination of many little scatterers, so that the received signal is the coherent sum of many single contributions. Little changes in the aspect angle of the target lead to amplitude fluctuations, which make the range profile have a noisy appearance, even when the SNR is high [30,31].

In relation to the phase-based algorithm, the range can be extracted by making use of Eq. (5.21). That is, the range $R(\tau)$ can be written as [25]

$$R(\tau) = \frac{c \cdot \phi_t(\tau)}{4\pi f_c}.\qquad(5.25)$$

This algorithm must first extract the phase history of the desired target. To that purpose, it is clear that (1) the radar must be coherent, and (2) the desired phase $\varphi_t(\tau)$ must be isolated from surrounding powerful interferences (these unwanted targets are usually referred to as clutter). Details will be given later in this chapter.

Additionally, the phase history $\varphi_t(\tau)$ must be unwrapped, because the measured phase always takes values between $-\pi$ and π. The phase-unwrapping algorithm works in a one-dimensional space, and thus it is usually simple, especially when the phase jump between any two consecutive slow-time instants is not greater than 2π. This implies that the along-LOS range motion of the target should not exceed $\lambda/2$ between any two consecutive periods. In practice, and especially for the slow targets here referred to, this limit is not overcome and the phase-based algorithm can be easily implemented.

The phase-based algorithm to extract the range history of scatterers has a better performance in terms of precision in comparison with the amplitude-based algorithm (see Section 5.5.2). Furthermore, it is more robust to the target-scintillation phenomenon.

5.4 SIGNAL PROCESSING ISSUES

The previous section provides the mathematics to understand important concepts around coherent LFMCW radars (such as fast-time, slow-time, range profile, beat signal, Doppler frequency, or range resolution), and the ideas for some algorithms. This section makes use of the previous results to justify the standard signal processing manipulations usually made for coherent LFMCW radar sensors. Aspects about data formatting, construction of important matrices, and algorithms for range tracking of targets are provided.

5.4.1 DATA FORMATTING

Fig. 5.2 indicates that the signal to be sampled is the beat signal, which is obtained after the mixing of the returned echoes with a replica of the transmitted signal (deramping technique). The conventional data formatting for a coherent LFMCW radar is shown in Fig. 5.6. As seen, the beat signal is sampled at the rate of the low-end ADC (sampling frequency f_s) during a long period of time (say some seconds). For a coherent radar, the total time to be sampled is referred to as the coherent processing interval (CPI).

After the acquisition of the beat signal during the total time CPI, a large number of samples is available (N_{CPI}). The usual data formatting consists in stacking the samples of the beat signal in rows of the so-called "raw-data matrix," so that the horizontal dimension of the matrix (along columns) is the fast-time. The number of samples in each row N corresponds to one waveform period T, and can be mathematically written as

$$N = \lfloor f_s \cdot T \rfloor \tag{5.26}$$

where the function $\lfloor x \rfloor$ refers to the lowest integer of x.

The vertical dimension of the raw-data matrix (along the rows) corresponds to the slow-time τ. The number of rows M (i.e., the number of slow-time instants) is given by

$$M = \left\lfloor \frac{\text{CPI}}{T} \right\rfloor = \frac{N_{CPI}}{N}. \tag{5.27}$$

Only a perfect coherence of the radar guarantees that the number of samples per row is constant with the time. The raw-data matrix clearly puts emphasis on the fact that the sampling frequency corresponding to the fast-time t is the

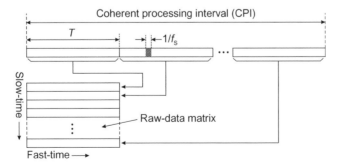

FIGURE 5.6

Conventional data formatting for a deramping-based coherent LFMCW radar.

sampling frequency of the ADC f_s, whereas the sampling frequency associated with the slow-time τ is the PRF.

5.4.2 CONSTRUCTION OF IMPORTANT MATRICES

The raw-data matrix contains the samples directly obtained by the ADC. Assuming an I/Q demodulation, the elements of the matrix are complex-valued. Obviously, the raw-data matrix contains contributions from the desired targets, thermal noise of the receiver, clutter (undesired targets), hardware mismatches, spurious signals, interferences, and so forth [30,31].

Given these $M \times N$ complex values, signal processing manipulations (i.e., algorithms) are required for multiple purposes, such as detection based on thresholding, target localization, construction of tracks based on a Kalman filtering, imaging, etc. For a short-range LFMCW radar and human-aware localization applications, the most interesting output of the intended algorithms is the range history of some scatterers, as previously mentioned. In this context, the reader must be familiar with the three following matrices in the context of a coherent LFMCW radar:

- The raw-data matrix $\mathbf{M}_r$.
- The range-profile matrix $\mathbf{M}_{rp}$.
- The range-Doppler matrix $\mathbf{M}_{r,Dop}$.

The previous section has profoundly explained the raw-data matrix. Regarding the range-profile matrix, one should note that a Fourier transform for each row of the raw-data matrix gives rise to it, since the beat frequency is proportional to the range for deramping-based LFMCW radars (see Section 5.3). Fig. 5.7 schematically shows the obtaining of this range-slow-time matrix $\mathbf{M}_{rp}$ (i.e., the range profiles) from the original raw-data matrix.

After noting that the Doppler is the frequency associated with the slow-time τ (see Section 5.3), it becomes trivial that a column-wise Fourier transform of the range-profile matrix leads to obtaining the range-Doppler matrix $\mathbf{M}_{r,Dop}$ (see Fig. 5.7). Since the slow-time is sampled at the PRF rate, it comes to light that the unambiguous interval for the Doppler is [−PRF/2, PRF/2], which means an unambiguous interval from −λ·PRF/4 to λ·PRF/4 for the radial along-LOS velocity of the targets.

The range-Doppler matrix is usually referred to as the inverse synthetic aperture radar (ISAR) image, since in ISAR the cross-range is proportional to the Doppler and thus, a scaled projection of the target shape may be reconstructed (in range and cross-range dimensions) [33].

5.4.3 ALGORITHMS TO EXTRACT THE RANGE HISTORY

Section 5.3.3 anticipated two algorithms to extract the range history of target scatterers. Their advantages and drawbacks were also given in Section 5.3.3. Here,

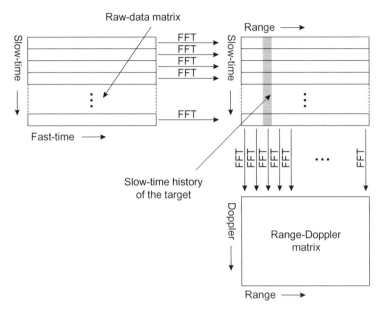

FIGURE 5.7

Construction of the range-profile and range-Doppler matrices from the raw-data matrix.

the algorithms are described under the signal processing framework of this section. Note that the information here provided is complementary to Section 5.6, where some ideas about clutter mitigation can lead to a better range tracking of short-range scatterers.

Assume that the raw-data matrix is called $\mathbf{M}_r[m, n]$ ($m = 1, 2, \ldots, M$; $n = 1, 2, \ldots, N$). M is the number of rows of the matrix (i.e., the number of transmitted ramps) and N is the number of samples for each waveform period T (i.e., the number of columns).

5.4.3.1 Amplitude-based algorithm

This algorithm to extract the range history of a target can be used with an incoherent radar, with the penalty of a reduced performance. The idea is simple: one must track the peak corresponding to the target in order to derive the range history. Moreover, to attain better precision in the determination of the peak position, the technique of "zero-padding" is conventionally employed. The details of the amplitude-based algorithm are given next:

- Apply an FFT of N_{FFT} points to each row of the matrix $\mathbf{M}_r[m, n]$. N_{FFT} is much larger than N. The resulting matrix is the range-profile matrix $\mathbf{M}_{\text{rp}}[m, k]$, where $k = 0, 1, \ldots, N_{\text{FFT}} - 1$.
- For each range profile m, obtain the value of k for which the magnitude of the range profile is maximum. Call to this value $k^*[m]$.

- Obtain the range evolution of the target as Eq. (5.28) indicates.

$$R(\tau) = k^*[m] \frac{f_s}{N_{\text{FFT}}} \cdot \frac{c}{2\gamma}. \tag{5.28}$$

The proposed method utilizes an interpolation to gain some precision in the determination of the peak position. Normally, large values of N_{FFT} are required, which leads to a computationally-costly Fourier transform. Simpler algorithms can use a parabolic interpolation only based on the neighboring samples surrounding the peak [30]. In any case, this method has a performance poorer than the phase-based algorithm and is less robust to the target-scintillation phenomenon.

5.4.3.2 Phase-based algorithm

If the coherence is properly preserved, this algorithm can be applied. Basically, it adapts the data to extract the slow-time phase history of the desired target, which is assumed to be situated in the same range bin during all the CPI (i.e., there is no range migration of the target). The range evolution can then be extracted from the unwrapped phase history, as suggested by Eq. (5.25).

The phase-based algorithm to extract the range history is divided into the following steps [25]:

- Apply an FFT to each row of the matrix $\mathbf{M}_r[m, n]$. The resulting matrix is the range-profile matrix $\mathbf{M}_{rp}[m, n]$.
- Select the range bin n* within which the target is contained. Pick up the corresponding column of $\mathbf{M}_{rp}[m, n]$ to construct the signal $s[m] = \mathbf{M}_{rp}[m, n^*]$.
- Obtain the phase of the signal $s[m]$ and proceed with its unwrapping.
- Apply Eq. (5.25) to derive the range evolution of the target.

As previously commented, the phase unwrapping should not be problematic. Also, if there is migration of the target through range cells during the CPI, specific algorithms (such as the extended envelope correlation or global range alignment) can be applied without affecting the desired phase history [40,41].

5.5 RANGE RESOLUTION, PRECISION, AND ACCURACY

An important measure of radars is the range (i.e., the along-LOS distance to the target). Moreover, the determination of the range history of scatterers is here identified as the key output product for short-range human-aware localization applications. Two algorithms to derive the track history of targets have been provided in Section 5.4.3. In this section, a big emphasis is put on the concepts of range resolution, precision, and accuracy [30–33], which are sometimes used confusedly in the literature. A brief subsection addresses the same concepts for other dimensions.

5.5.1 **RANGE RESOLUTION**

The range resolution of a radar refers to its ability to discriminate (i.e., to resolve) two close targets in range. Assuming a 3-dB criterion and a uniform weighting (i.e., a uniform window), the range resolution of a radar has the general expression [30–33]

$$\Delta R = \frac{c}{2B}, \tag{5.29}$$

where c is the speed of light, and B is the instantaneous transmitted bandwidth.

Eq. (5.29) is valid for an LFMCW radar, as demonstrated in Eq. (5.17), and is intimately connected with the fundamental limit of the Heisenberg's uncertainty principle (in the signal processing area, this limitation is usually addressed as the Gabor limit).

According to Eq. (5.29), if one looks for an improvement in the range resolution, the only way to achieve it is to transmit and to handle a larger bandwidth B. If the transmitted bandwidth is very large, one real target may be viewed as an extended target (i.e., the LOS dimension of the target is chopped in many pieces). If one does not have hardware and/or budget limitations, a finer range resolution is always pursued, because richer information (e.g., more interesting radar signatures) from the target can be extracted.

There are some techniques (such as MUSIC, Capon estimator, Autoregressive coefficients, and so forth), usually referred to as superresolution techniques, which can improve the conventional resolution. They are based on parametric models, which assume very specific forms for the signals. Indeed, they work properly for such signals, but can lead to important problems for real signals, such as the appearance of ghost targets and/or the distortion of the Fourier-based radar signatures of targets [43]. In the opinion of the authors, these techniques are controversial. Hence, any reference to resolution in this chapter must be understood as the conventional Heisenberg-limited resolution.

Fig. 5.8 shows the range profile corresponding to two simulated equal-amplitude point scatterers separated 50 cm in range, when illuminated by an LFMCW radar. The range resolution is 15 cm ($B = 1$ GHz), and consequently, the two targets can be easily discriminated. The windowing is uniform, which implies that the secondary lobes appear 13.5 dB below the main lobes. As is widely known, the application of a specific window (e.g., Taylor or Hamming) leads to an increase of the sidelobe rejection, but at the cost of a slight broadening of the main lobe.

When the two targets are very close, it is difficult to tell whether there are two point scatterers or only one. Fig. 5.8 also shows the range profile when the targets are spaced 15 cm (i.e., the range-resolution value). Two little peaks indicate the presence of the two targets. However, the addition of the always-existing noise or an approaching of the two targets would merge them into a single one.

The formula Eq. (5.29) for the range resolution must not be understood as a strong limit for the ability of discriminating two point scatterers. Indeed, depending on the relative phases of the two targets, the two peaks in Fig. 5.8 can be

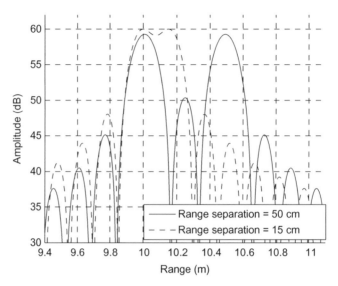

FIGURE 5.8

Range profiles corresponding to two point scatterers spaced 50 cm (continuous line) and 15 cm (dashed line) along LOS.

distinguished more easily. Eq. (5.29) speaks about the broadening of the target PSF and is perhaps the most famous formula for the radar community.

Additionally, the range resolution indicates the range size of the bins/cells of any radar. For example, for a conventional radar plan position indicator, the cells are angular sectors, in which the along-LOS width is given by the range resolution, and the azimuth width mainly depends on the antenna beamwidth.

5.5.2 RANGE PRECISION

When a radar is measuring the range of a point-scatterer target, it obtains an estimate $\hat{R}$. Obviously, the radar will not obtain an exact measurement of the actual range R. There is an error, which can be written as

$$\varepsilon_R = R - \hat{R}. \tag{5.30}$$

The estimate $\hat{R}$ is a function of the observations and thus, it is a random variable. Consequently, the error ε_R is also a random variable, whose probability density function may be very complex, depending on the nature of the observations (e.g., the observations can be the combination of the returns of a fluctuating target, complex clutter, and flicker and thermal noise).

The range precision is defined as the standard deviation of the error ε_R. Anytime that the experiment is repeated, the radar obtains a different value (i.e., a

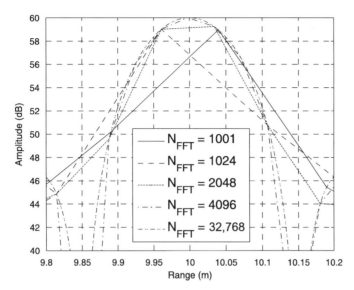

FIGURE 5.9

Range profiles corresponding to a point scatterer for different zero-padding levels.

different realization) for the range estimate. This fluctuating error around the ground-truth value is related to the range precision.

Fig. 5.9 plots the simulated range profile for a point scatterer situated at 10 m from the deramping-based LFMCW radar. The range resolution is $\Delta R = 15$ cm ($B = 1$ GHz), the central frequency is $f_c = 10$ GHz, the period is $T = 1$ ms, and the employed sampling frequency is $f_s = 1$ MHz. The different curves correspond to different points of the FFT (N_{FFT}) in the zero-padding technique (see Section 5.4.3).

As observed, an increase in the N_{FFT} parameter leads to a better estimation of the range position of the point-scatterer target. This is commented in Section 5.4.3 for the amplitude-based algorithm. Table 5.1 details the error as a function of N_{FFT} and confirms the previous asseveration. It comes to light that the amplitude-based algorithm can be very precise (i.e., the error can be very small) for a very high SNR and a nonfluctuating target.

Unfortunately, noise is always present and degrades the precision of any radar. It can be shown that the range precision σ_R of a radar is inversely proportional to the square root of the SNR (the noise is assumed to be white and Gaussian) [30]:

$$\sigma_R = \frac{K}{\sqrt{SNR}}. \tag{5.31}$$

Fig. 5.10 depicts three range profiles for three different realizations and a high number of FFT points ($N_{FFT} = 32768$). The SNR in the beat-frequency domain is considered to be 10 dB. As can be seen, the peak fluctuates around the ground-truth value of 10 m.

Table 5.1 Absolute and Relative Error in the Amplitude-Based Range Measurement of a Simulated Point Scatterer as a Function of the Zero-Padding Level

N_{FFT}	Range of Peak (m)	Absolute Error (cm)	Relative Error (%)
1001	10.04	4	0.4
1024	9.9609	−3.91	−0.391
2048	10.0342	3.42	0.342
4096	9.9976	−0.24	−0.024
32768	10.0021	0.21	0.021

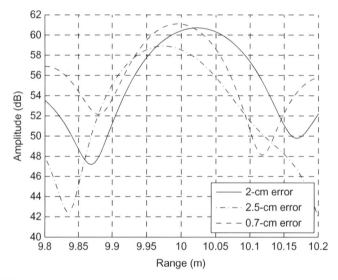

FIGURE 5.10

Realizations of range profiles corresponding to a point scatterer for a SNR of 10 dB.

It is important to highlight the big differences between the concepts of range resolution and precision. The range resolution is related to the width of the target PSF and the ability to discriminate two targets closely spaced in range, whereas the range precision is related to the process of measuring the distance and the associated fluctuating errors.

Regarding the two algorithms provided in Section 5.4.3, both of them have the purpose of estimating the range history of targets. However, their performance is quite different. It can be shown that the range precision for the amplitude-based algorithm is given by [30]

$$\sigma_{R,\text{amp}} = K_1 \frac{\Delta R}{\sqrt{\text{SNR}}}, \tag{5.32}$$

whereas the corresponding equation for the phase-based algorithm can be written as [44]

$$\sigma_{R,\text{phase}} = K_2 \frac{\lambda}{\sqrt{\text{SNR}}}. \tag{5.33}$$

In Eqs. (5.32) and (5.33), the constants K_1 and K_2 are similar. It comes to light that the range precision is usually better for the phase-based algorithm, especially for high-frequency radars. Furthermore, the phase-based method is more robust against the target-scintillation phenomenon, as verified in Section 5.7.4. These advantages for the phase-based method are gained from the effort of constructing a coherent radar.

5.5.3 RANGE ACCURACY

The range precision is related to the standard deviation of the error in Eq. (5.30). The range accuracy refers to the mean (i.e., the expected value) of the error in Eq. (5.30). In other words, it is said that the range measurement is accurate if the range estimate $\hat{R}$ has a small bias [30]. The expected value of $\hat{R}$ can be generally written as

$$E[\hat{R}] = R + R_{\text{bias}} \tag{5.34}$$

where $E[x]$ is the expectation operator for the random variable x, and R_{bias} is the range bias from the ground-truth range R.

The bias observed in a measure is also referred to as a systematic error, since it is repeated every time the same experiment is conducted. Contrary to the range precision, the range accuracy is not related to fluctuating errors, but to a systematic error.

Possible causes for systematic errors can be very complex and even difficult to detect. In an LFMCW radar, one typical origin for a range bias is an uncertainty in the delay of the replica of the transmitted signal at the input of the deramping mixer. In this case, a simple shifting of the range axis can be enough to improve the accuracy of the radar.

The procedures to reduce systematic errors are called calibration processes, which can be very simple (such as the aforementioned shifting of the range axis) or very sophisticated. For example, the multipath phenomenon and/or hardware mismatches can lead to large range biases (or even to ghost echoes).

In any case, it is sometimes difficult to tell in practice (in specific real experiments) whether the measurements are accurate or not. The point is that to calculate the bias R_{bias}, it is necessary to know the ground-truth range R (see Eq. 5.34). This range R has been defined for a point-scatterer target, which is not a real entity. Real targets are always extended targets, for which it is difficult to tell their real and single range R. For example, a metal plate illuminated by a radar can have returns from its surface and from its edges, but the radar singly observes the coherent combination of all these returns.

This leads to the fact that little variations on the experiment configuration may require different calibration procedures. There are sometimes additional available measurements (e.g., an infrared marker to monitor vital signs) and, of course, calibration procedures could be applied to match the radar measures to the available ones. In any case, these reference measures are not necessarily identical to the ground truth.

The tolerance to systematic errors largely depends on the specific application, and normally, the calibration procedures are designed to be simple. All in all, the important aspect is to note that the range accuracy is different from the range precision and that a better SNR does not necessarily imply having a more accurate radar sensor.

5.5.4 OTHER DIMENSIONS

The concepts of resolution, precision, and accuracy can be applied to other dimensions beyond the aforementioned range. In the case of radar systems, the Doppler and the angle (e.g., azimuth and elevation) are to be considered.

In relation to the Doppler frequency, a good Doppler precision requires a high SNR, whereas systematic Doppler errors require calibration procedures [30,33]. On the other hand, a good Doppler resolution requires a large CPI, but note that target scatterers usually change their Doppler frequencies during long times. In ISAR contexts, big efforts are devoted to compensating for the motion of the targets [33].

Regarding the angle, a good angular precision again requires a high SNR, and calibration may be required to improve the angular accuracy. Different techniques to precisely mark the azimuth/elevation of the targets have been proposed, such as the well-known monopulse technique or the moving-window technique. On the other hand, the angular resolution is limited by the beamwidth of the radar antenna. Hence, if a better angular resolution is desired, a larger antenna, a phased array, or a synthetic array (e.g., SAR techniques) may be used [30,33].

5.6 CLUTTER MITIGATION

In a general sense, clutter is any returned echo which affects the correct estimation of the parameters of wanted targets. For human-aware localization scenarios, the desired output has been here identified as the range history of the target scatterers. Any clutter return which may have an influence on the correct extraction of this range history should be sought to be mitigated.

LFMCW radars have range resolution, and thus it becomes trivial that clutter returns in range bins different from the range cell in which the desired target is located will not have a big influence on the correct estimation of the range history of the target (note that problematic secondary lobes of clutter may be mitigated by simple windowing). This clutter-isolation property is a clear advantage for LFMCW radars in comparison with conventional Doppler radars, which do not transmit an instantaneous bandwidth. An incoherent LFMCW radar can also exploit this clutter-isolation property [45−47].

A more complex scenario is the one in which the clutter returns are located at the same range bin of the wanted target. For example, in a vital sign monitoring scenario, the stationary limbs of the breathing subject (e.g., the legs and the arms) may influence the correct range tracking of respiration and heartbeats [45−47].

If the clutter echoes are situated at the same range cell of the wanted target (and, of course, both of them are within the radar antenna beamwidth), and the radar is incoherent, the only difference between them is the amplitude of the corresponding signals. Only if the signal-to-clutter ratio (SCR) is high, a correct tracking of the wanted range history will be possible. On the contrary, if the radar is coherent, the Doppler domain is available to try to filter/mitigate the clutter echoes.

This clutter filtering will be effective if the Doppler signatures of the clutter and the target can be distinguished. In other words, if the Doppler-domain spectrums of clutter and target are quite different, the application of a suitable filter may mitigate the degrading effects of clutter. In the case of the vital sign monitoring application, the implied signals (i.e., respiration and heartbeats) are very slow and weak, so that the mitigation of the stationary clutter (e.g., legs, arms, and head) becomes very challenging. In the case that the clutter is powerful and moving (e.g., when trying to monitor the vital signs of a sleeping patient who is moving in the bed), the radar-based noncontact tracking of the respiration and heartbeats becomes almost an impossibility.

The next subsections provide the mathematics and algorithms to understand the ideas underlying the Doppler-based clutter-mitigation methods for a coherent LFMCW radar. The application of the methods to simulated and experimental data is provided in Sections 5.7.5 and 5.8.3.

5.6.1 MATHEMATICAL FORMULATION

From Section 5.4, and assuming that there are no range migrations, it comes to light that the slow-time signal for the range bin in which the target is located (deramping-based coherent LFMCW radar) can be written as [46]

$$s(\tau) = K \cdot \exp\left(j\frac{4\pi f_{\mathrm{c}}}{c} R(\tau)\right) \tag{5.35}$$

where K is the amplitude of the desired signal, and $R(\tau)$ is the wanted range history. From Eq. (5.35) and also according to Eq. (5.21), the phase history for the desired target is given by

$$\phi_{\mathrm{d}}(\tau) = \frac{4\pi f_{\mathrm{c}}}{c} R(\tau). \tag{5.36}$$

Suppose the existence of a stationary point-scatterer clutter echo situated at the same range bin. The clutter-plus-target signal in that range bin can be expressed as

$$s_r(\tau) = K_d \cdot \exp(j\phi_d) + K_c \cdot \exp(j\phi_c) \tag{5.37}$$

where K_d and φ_d are the amplitude and phase of the desired target (see Eq. 5.36), whereas K_c and φ_c are the amplitude and phase of the clutter scatterer, respectively. The clutter phase φ_c can be written as

$$\phi_c = \frac{4\pi f_c}{c} R_c \tag{5.38}$$

where R_c is the range of the clutter scatterer. Note that the clutter phase Eq. (5.38) is constant with the slow-time τ, since the clutter scatterer has been assumed to be stationary.

By noting Eq. (5.37), the SCR is defined as

$$\mathrm{SCR} = \frac{|K_d|^2}{|K_c|^2}. \tag{5.39}$$

For the particular case of a low SCR (SCR $\ll$ 1), Fig. 5.11 schematically shows the complex phasor representation of Eq. (5.37).

Geometrical observations in Fig. 5.11 permit the conclusion that the slow-time signal Eq. (5.37) can be written as follows for a low SCR

$$s_r(\tau) = K_r \cdot \exp(j\phi_r) \tag{5.40}$$

where K_r is approximately equal to K_c, and the phase φ_r must fulfill the following equation:

$$\tan(\phi_r - \phi_c) = \frac{K_d \sin(\phi_d - \phi_c)}{K_c + K_d \cos(\phi_d - \phi_c)}. \tag{5.41}$$

For a low SCR, note that

$$\tan(\phi_r - \phi_c) \approx \phi_r - \phi_c, \tag{5.42}$$

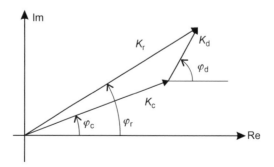

FIGURE 5.11

Complex phasor representation of the slow-time signal in Eq. (5.37) for a low SCR.

$$K_c >> K_d\cos(\phi_d - \phi_c), \tag{5.43}$$

which leads to the fact that the phase extracted by the phase-based algorithm in Section 5.4.3 would be [46]

$$\phi_{r,\text{ low SCR}} = \phi_c + \frac{K_d\sin(\phi_d - \phi_c)}{K_c}. \tag{5.44}$$

As concluded from Eq. (5.44) and after noting that $K_c >> K_d$ (SCR $<< 1$), the reconstructed phase φ_r is almost identical to the clutter phase φ_c, and no information from the desired phase can be recovered. This necessitates a clutter-mitigation technique.

On contrary, the phase φ_r for a high SCR can simply be written as [46]

$$\phi_{r,\text{ high SCR}} = \phi_d + \frac{K_c\sin(\phi_c - \phi_d)}{K_d}. \tag{5.45}$$

In the case of a high SCR, the reconstructed phase is nearly the desired one. Nevertheless, the second term in Eq. (5.45) indicates that clutter can have some influence on the exact extraction of the wanted range history even for a high SCR.

5.6.2 DOPPLER-BASED CLUTTER MITIGATION

Depending on the relative level of the clutter return and the desired target, the previous section has demonstrated that clutter can completely obscure the wanted range history of the target. A HRR is always preferred, because it confers range-isolation capabilities upon the radar system. However, it is sometimes unavoidable to have clutter returns at the same range bin of the desired target. If the radar is coherent, a Doppler-based filtering can still alleviate the effects of the clutter [45–47].

As demonstrated in Section 5.3.2, the Fourier variable associated with the slow-time τ is the Doppler frequency, and hence the PRF is the sampling frequency for these slow-time signals, such as the one in Eq. (5.37). The second term of Eq. (5.37) corresponds to a stationary-assumed clutter scatterer, which means a zero-Doppler component; whereas the first summand refers to the moving target and gives rise to a broader lowpass Doppler spectrum. For example, if the desired range is assumed to vary sinusoidally (e.g., a respiration signal may be modeled as a tone), the Doppler spectrum for the vital sign is the well-known spectrum of a signal frequency-modulated by a modulating sinusoid—i.e., some Bessel-weighted deltas separated by the sinusoid frequency with an overall bandwidth estimated by Carson's rule.

Thus, for a coherent radar, clutter appears as a zero-Doppler component, and the desired target has a broader low-pass-type spectrum. Fig. 5.12 schematically emphasizes these features. As observed, overlapping of both signal spectrums is a fact. However, the clutter can be mitigated if a high-pass digital filter is applied

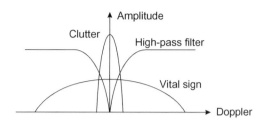

FIGURE 5.12

Schematic plotting of the Doppler spectrums corresponding to a stationary clutter echo and a slow-moving target.

to the slow-time signal. Conventional coherent surveillance radars use these filters (usually called cancelers) to reduce the levels of surface clutter (e.g., land and crops) [30−31]. The technique extrapolated here to short-range radars belongs to the so-called MTI Doppler-mitigation approaches. However, the differences between both scenarios are big, because for human-aware localization applications and especially for vital-sign monitoring, the implied signals are very slow and the bandwidth of the desired spectrum is usually small.

The suggested high-pass filter in Fig. 5.12 attenuates the very low-frequency Doppler components of the wanted slow-time signal. Consequently, the distortion on the desired signal is inevitable. The design of the high-pass digital system must search for a tradeoff between a proper mitigation of the clutter and a tolerable distortion of the extracted range evolution of the desired target. Examples will be given in Section 5.7.

In conclusion, when having a coherent radar, it is always mandatory to take a look at the possibilities of exploiting the Doppler dimension to mitigate clutter. For example, in a gaming scenario where some people are moving in a room, it may be interesting to isolate the range tracks of each person. Fig. 5.13 schematically shows a processing to isolate the range histories of two people (one moving away from the radar and the other approaching the radar). Fig. 5.13 (top left) shows the range-profile matrix, where two leaned echoes can be seen. After applying an FFT in the slow-time dimension, two returns from the two people are isolated in the range-Doppler domain (Fig. 5.13 (bottom left)). An image-processing windowing (or clustering) can be used to isolate one of the returns, as Fig. 5.13 (bottom right) shows for the receding person. An inverse FFT (IFFT) returns the isolated echo to the range-slow-time domain (Fig. 5.13 (top right)). In this domain, a range-bin-alignment method together with the phase-based algorithm of Section 5.4.3 can be applied to extract the range history of the isolated person.

An important conclusion is that, contrary to incoherent radars, the Doppler dimension can be exploited for clutter-mitigation purposes in coherent short-range deramping-based LFMCW radars.

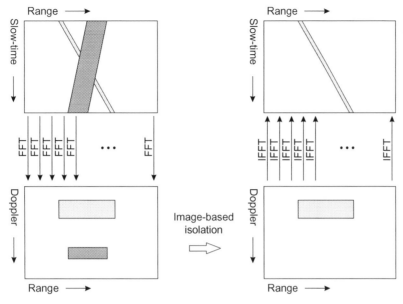

FIGURE 5.13

Bi-dimensional Doppler-based clutter/interference-mitigation technique.

5.7 SIMULATIONS

In previous sections, many important concepts and algorithms regarding coherent deramping-based LFMCW radars have been introduced. In this section, some simulated results are shown with the aim of emphasizing the main aspects of this kind of radar for short-range human-aware localization applications. The code for the simulations has been developed under the signal processing environment Matlab.

5.7.1 PROOF-OF-CONCEPT SIMULATION

For this first simulation, consider the scenario shown in Fig. 5.14. A coherent LFMCW radar is illuminating a point scatterer situated at a mean range of R_c. The scatterer is sinusoidally vibrating along the LOS. The peak-to-peak motion of the scatterer is given by the parameter R_{pp}, whereas the motion frequency is given by f_t. In this example, R_c is 5.625 m, R_{pp} is 30 cm, and the frequency of the sinusoidal motion is $f_t = 0.5$ Hz.

In relation to the simulation, the important parameters of the radar are summarized as follows. An S-band radar working at $f_c = 5.8$ GHz is assumed. The transmitted bandwidth is 160 MHz, which provides the radar with a range resolution of $\Delta R = 93.75$ cm, according to Eq. (5.17). Given the considered very slow motion (which is typical for human-centered applications), the induced Doppler is

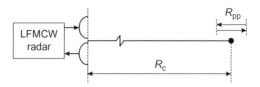

FIGURE 5.14

Simulated scenario for a point-scatterer target sinusoidally vibrating along LOS.

small, and thus the PRF is not chosen to be very high (PRF = 200 Hz). The CPI is 12 seconds, so that it is long enough to observe some periods of the wanted slow signal (the period of the sinusoidal motion is 2 seconds). The sampling frequency in the simulation is assumed to be 8 kHz, which implies a maximum range from which echoes are assumed to come of 37.5 m, according to Eq. (5.12). This low sampling frequency means a simple and inexpensive architecture for the radar prototype. From the simulation viewpoint, a low sampling frequency implies a reduced computational burden, since the number of samples is not very large. Indeed, the number of samples of the beat signal associated with the waveform period T ($T = 5$ ms) is $N = 40$ according to Eq. (5.26), and the number of transmitted ramps is $M = 2400$ (see Eq. 5.27). This means that the raw-data matrix is not very large and has 2400 rows and 40 columns.

Fig. 5.15 shows the range-profile matrix $\mathbf{M}_{rp}[m, n]$, after applying a row-wise FFT to the raw-data matrix (refer to Fig. 5.7). As seen, the target appears situated at the seventh range bin during the CPI, and no range migrations are observed for it, because its sinusoidal motion is not greater than the range resolution $\Delta R = 93.75$ cm (i.e., it is not greater than the range size of the radar cell). The secondary lobes can be observed next to the main vertical strip in Fig. 5.15. Remember that the level of secondary lobes can be reduced through simple weighting (e.g., Hamming or Taylor windowing), at the cost of a width increase of the main lobe.

The result of applying the phase-based algorithm in Section 5.4.3 is shown in Fig. 5.16, which shows the range history extracted for the sinusoidally-moving point-scatterer target. As seen, the range history can be perfectly recovered, since no noise (and thus, an infinite SNR) has been considered in the simulation. Fig. 5.16 also shows that the stage of phase unwrapping is of high importance if a proper tracking of the target motion is desired, because the original phase extracted for the seventh column of the matrix $\mathbf{M}_{rp}[m, n]$ in Fig. 5.15 is wrapped between $-\pi$ and π.

One important point to emphasize is the fact that the range-tracking ability shown in this section is not related to an absolute-range tracking. Indeed, as far as an absolute range marking of the target is concerned, one should simply conclude that the scatterer is situated at the seventh range bin. This is the normal output after the detection stage for a conventional radar: the target is in a specific radar cell/bin. In any case, it is obvious that, depending on the real SNR and the target-scintillation phenomenon, a more-precise measure of the absolute range might be gained after applying, e.g., the amplitude-based method in Section 5.4.3.

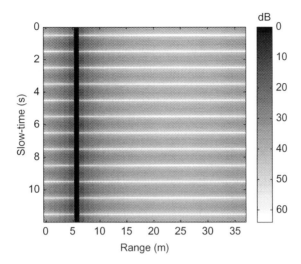

FIGURE 5.15

Simulated range-profile matrix for a vibrating scatterer illuminated by an LFMCW radar.

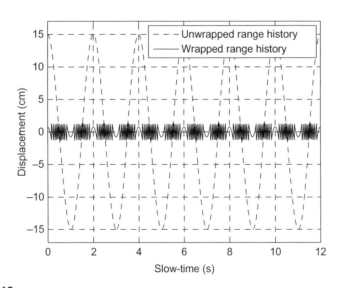

FIGURE 5.16

Wrapped and unwrapped range histories for a vibrating scatterer after applying the phase-based method.

5.7.2 **CLUTTER IN OTHER RANGE BINS**

Conventional coherent Doppler radars can obtain the range history of the wanted target by making use of the phase-based algorithm. However, these radar sensors do not transmit bandwidth and thus, they do not have range resolution (see Section 5.1). As a consequence, echoes coming from surrounding clutter may have a great influence on the correct derivation of the desired range evolution. Contrary to Doppler radars, LFMCW radars do possess range resolution and hence, they can benefit from this range-isolation capability.

To show this potentiality, a simulation of a wanted scatterer together with a moving clutter return has been conducted. The simulation parameters are the same as those in Section 5.7.1 for the desired point-scatterer target, whereas the unwanted target (clutter) is assumed to uniformly move along LOS from a range of 1 m to a range of 15 m during the 12-second CPI. Fig. 5.17 shows the range-profile matrix (i.e., the range-slow-time map) for this simulation scenario. The wanted target remains in its seventh range bin, whereas the clutter scatterer is subject to a linear motion through resolution cells. The range crossing of both histories happens at around 4 seconds.

Fig. 5.18 shows the result of applying the phase-based algorithm in Section 5.4.3 to the seventh range bin in which the desired target is located. Around the time interval in which the two scatterers are situated at the same range cell, the interference is manifest. At other slow-time intervals, the effect of

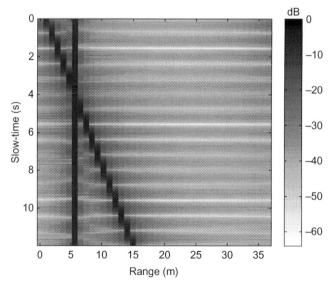

FIGURE 5.17

Simulated range-profile matrix for a desired vibrating point-scatterer target and an unwanted moving clutter scatterer.

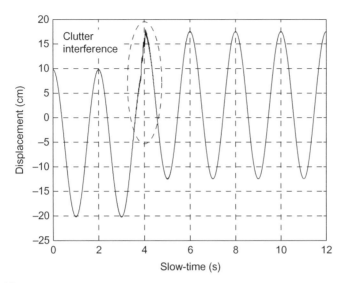

FIGURE 5.18

Range history extracted for a desired vibrating scatterer. The clutter influence is manifest.

the clutter scatterer on the desired range history is minor and the correct sinusoidal pattern with the adequate periods and amplitudes can be retrieved.

This section highlights a big advantage of having a good range resolution: clutter returns at other range bins usually have a minor effect on the range tracking of the wanted scatterer. This is contrary to the Doppler radars. In any case, it is clear that clutter returns situated at the desired range bin may have a big influence on the correct results, as seen in Fig. 5.18 and also anticipated in Section 5.6. Results around the application of clutter-mitigation techniques for in-cell clutter are given later.

5.7.3 PERFORMANCE AGAINST NOISE

Fig. 5.16 has shown that an infinite SNR leads to a perfect reconstruction of the range history of the desired point scatterer. It becomes clear that a poorer SNR should mean an imperfect tracking of the range evolution. The mean square error (MSE) is defined as

$$\text{MSE} = \frac{1}{M} \sum_{m=1}^{M} (R_t[m] - \hat{R}_t[m])^2 \tag{5.46}$$

where $R_t[m]$ is the ground-truth slow-time range history of the target, and $\hat{R}_t[m]$ is its estimate.

Fig. 5.19 shows the MSE as a function of the beat-signal SNR for the scatterer of Section 5.7.1 when centered on the range cell ($R_c = 5.625$ m) and at a range-bin extreme ($R_c = 5.15625$ m). The Monte Carlo method is employed so that each

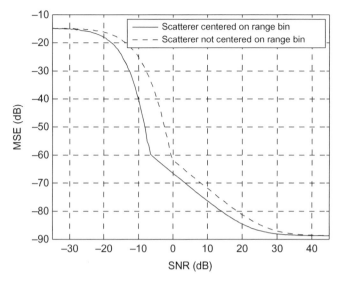

FIGURE 5.19

Performance of the phase-based method against white and Gaussian noise. The MSE is plotted as a function of the SNR.

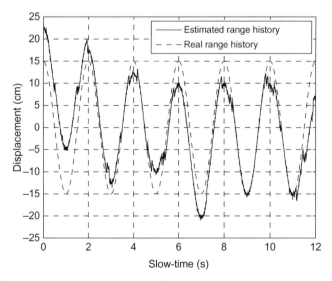

FIGURE 5.20

Estimated range history of a vibrating scatterer for an SNR of −15 dB.

graph point is the average of one thousand realizations, where the noise has been considered to be white and Gaussian. It comes to light that a higher SNR leads to a better reconstruction of the range history in terms of the MSE. The MSE saturation observed for very poor SNRs comes from the fact that the corrupted phase oscillates in the interval $[-\pi, \pi]$, whereas the MSE saturation for high SNRs is explained by numerical negligible errors.

As an example, Fig. 5.20 depicts the range-history reconstruction for one realization with an SNR of -15 dB. In spite of the imperfect estimate, the periods of the sinusoidally-vibrating scatterer can still be identified.

5.7.4 MULTI-SCATTERER SIMULATION

Section 5.7.1 refers to a single scatterer which is sinusoidally vibrating. In practice, it is difficult to only have one predominant point scatterer. As an example, the signal coming from the moving chest of a breathing patient may be considered as a combination of different contributions which do not have the same motion amplitude.

Consider one hundred scatterers which are moving sinusoidally in phase, but with different along-LOS range amplitudes. The motion amplitude (R_{pp}) is drawn from a uniform distribution from 0 to 1 cm. Also, the signal amplitude is drawn from an independent uniform random variable from 0 to 1. Fig. 5.21 details the range history extracted from the desired range bin after applying the phase-based method. For comparison purposes, the maximum 1-cm peak-to-peak motion is also indicated.

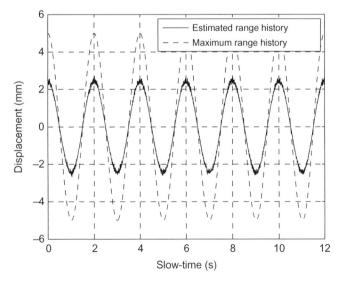

FIGURE 5.21

Range evolutions of multiple in-phase vibrating scatterers. The target-scintillation phenomenon is manifest.

Four important conclusions can be derived from Fig. 5.21. First, the sinusoidal pattern of the motion can be identified, which is encouraging, since the signal is the complex coherent combination of 100 in-phase contributions. Second, the maximum 1-cm peak-to-peak amplitude is not recovered by the algorithm, because many of the scatterers do not fluctuate so much (indeed, the reconstructed amplitude has an approximate value of $R_{pp} = 0.5$ cm). In any case, the desired result is to detect the presence of a motion and hence, the amplitude is unimportant. Third, some negligible noise is observed. Taking into account that the simulated SNR is infinite, the origin of that noise must be found in the aforementioned target-scintillation phenomenon. However, the noise is little, which comes from the fact that the used phase-based method is more robust to this phenomenon than the amplitude-based algorithm, as previously stated in Section 5.3.3. Fourth, the amplitude decrease in Fig. 5.21 corroborates the fact that an energetic stationary clutter scatterer in the same range bin may lead to the impossibility of detecting the wanted motion, as highlighted by Eq. (5.44) and further verified in the next section.

5.7.5 **CLUTTER IN THE SAME RANGE BIN**

LFMCW radar sensors can easily deal with unwanted targets situated in range bins in which the desired targets are not located. This is why an improvement in the range resolution is normally pursued, so that this range-isolation capability can be further enhanced. Unfortunately, depending on the application, it is sometimes unavoidable to have clutter in the same desired range bin (e.g., if the vital signs of a person are to be monitored, it is almost impossible not to receive echoes from undesired parts, such as head or limbs). If the radar is coherent, the new dimension Doppler can be exploited to devise specific methods which try to mitigate this in-cell unwanted clutter (see Section 5.6).

Consider the scenario shown in Fig. 5.22, which is analogous to the one depicted in Fig. 5.14, with the inclusion of a stationary clutter scatterer at the same distance of the wanted sinusoidally-vibrating target. The radar prototype is assumed to have a center frequency of $f_c = 5.8$ GHz. The transmitted bandwidth is $B = 160$ MHz and the PRF is 500 Hz, whereas the CPI is 12 seconds. The

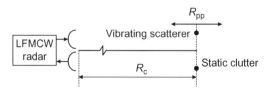

FIGURE 5.22

Simulated scenario for a point-scatterer target sinusoidally vibrating along LOS and a stationary clutter return.

mean range to the targets is $R_c = 5.625$ m, whereas the sinusoidal motion of the desired target is defined through its range amplitude (peak-to-peak $R_{pp} = 50$ cm) and its frequency ($f_t = 0.5$ Hz). The amplitude of the desired target is $K_d = 1$, whereas the amplitude of the clutter scatterer is $K_c = 1.2$.

The slow-time signal for the range bin in which the targets are present can be written as

$$s_b(\tau) = K_d \cdot \exp\left(j\frac{4\pi f_c}{c}\left(R_c + \frac{R_{pp}}{2}\sin(2\pi f_t \tau)\right)\right) + K_c \cdot \exp\left(j\frac{4\pi f_c}{c}R_c\right). \tag{5.47}$$

The second term in Eq. (5.47) is the zero-Doppler clutter component, whereas the first term corresponds to the wanted signal, which has a Doppler frequency, according to Eq. (5.24), given by

$$f_{Dop}(\tau) = -\frac{2\pi f_c f_t R_{pp}}{c}\cos(2\pi f_t \tau). \tag{5.48}$$

For the simulation example in this section, the maximum Doppler frequency can be derived from Eq. (5.48) and turns out to be 30.4 Hz. This value corresponds to the result in Fig. 5.23, which shows the spectrogram for the slow-time signal in the corresponding range bin. The sinusoidal Doppler history associated with the moving target (see Eq. 5.48), together with the zero-Doppler component corresponding to the clutter scatterer, can be seen in Fig. 5.23.

The Fourier transform of the slow-time signal Eq. (5.47) is depicted in magnitude in Fig. 5.24. This figure permits the corroboration of the spectrum distribution shown in Fig. 5.12. It becomes clear that a high-pass filter in the slow-time τ

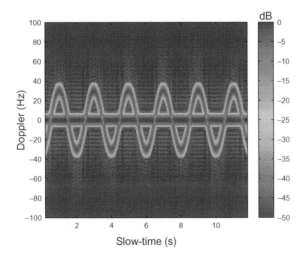

FIGURE 5.23

Spectrogram (Doppler vs slow-time) for a vibrating target and a stationary clutter scatterer.

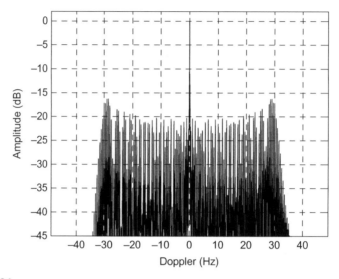

FIGURE 5.24

Doppler spectrum of the simulated slow-time signal for the scenario in Fig. 5.22.

permit the mitigation of the clutter zero-Doppler component at the cost of some distortion added to the desired Doppler spectrum.

The application of the mitigation technique (a fifth-order Chebyshev type-I filter with a 0.5-dB ripple in the passband and a 40-dB attenuation in the 2-Hz-edge-frequency stopband has been employed in this example) permits the recovery of the desired sinusoidal motion of the vibrating scatterer, as shown in Fig. 5.25. Before the mitigation approach of Section 5.6.2, the clutter influence on the desired range history is enormous.

A more compromised simulation considers that the peak-to-peak amplitude is reduced. Assume now that $R_{pp} = 10$ mm. For this situation, the maximum Doppler is only 0.61 Hz, which leads to a desired Doppler spectrum almost identical to DC. Fig. 5.26 shows the range history extracted before and after applying a high-pass filter with cut-off frequency of 0.2 Hz. The ground-truth motion of the wanted target is also shown as a reference. It is true that the sinusoidal pattern is observed before the mitigation technique. However, the Doppler filtering amplifies the displacement amplitude, which may help with the detection of the motion. In any case, due to the added distortion, the amplitude of the estimated range evolution after clutter mitigation does not correspond with the ground-truth 10-mm peak-to-peak amplitude.

As a conclusion, the detection of slow targets when competing with strong stationary clutter is of great difficulty, and any radar in that scenario (e.g., maritime radars trying to detect small and slow boats under heavy clutter) sees its performance degraded.

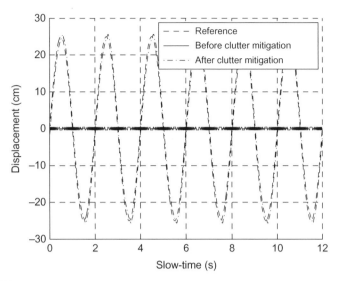

FIGURE 5.25

Tracked motion pattern for a desired vibrating scatterer before and after applying a clutter-mitigation high-pass filter ($R_{pp} = 50$ cm).

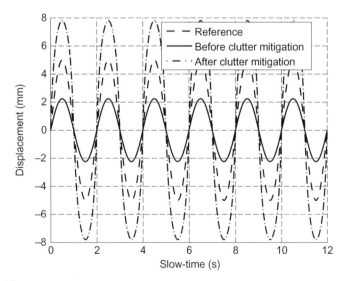

FIGURE 5.26

Tracked motion pattern for a desired vibrating scatterer before and after applying a clutter-mitigation high-pass filter ($R_{pp} = 10$ mm).

5.8 EXPERIMENTAL RESULTS

The suitability of coherent LFMCW radars to short-range human-aware localization applications has also been verified by experimental results. In this section, the radar prototype and some experiments are described to further emphasize the concepts and features around these kinds of systems.

5.8.1 RADAR PROTOTYPE

A block diagram of the developed radar prototype is shown in Fig. 5.27. As can be seen, it uses a deramping-based architecture so that the acquisition block is simple (it works at a reduced sampling frequency of 10 kHz). Some details are given next.

The generation of the ramps is accomplished through the PXIe 5450, which is a baseband vector signal generator able to control the initial phase of each period and to guarantee a high ramp linearity. Its output is upconverted with the PXIe 5611 and PXIe 5652, and is amplified with the PXIe 5691. The final ramp sweeps its frequency from 5.72 to 5.88 GHz, which leads to a transmitted bandwidth of 160 MHz. The waveform period is 2 ms (i.e., a PRF of 500 Hz).

The output of the PXIe 5691 is split through a Wilkinson-type power divider. One output is the signal to be transmitted after a 30-dB gain block, whereas the other one is used as a replica of the transmitted signal in the mixing stage (deramping technique).

In the Rx chain, the signal returned from the targets is amplified. To that purpose, two LNAs and a gain block with a total gain of 47.5 dB are used. The mixer used for the analog deramping is a quadrature mixer, so that the I and Q components of the beat signal are provided at its output. As previously mentioned, the beat signal is acquired with a low-end acquisition block working with a sampling frequency of 10 kHz. The samples of the beat signal are stored for further off-line processing.

The architecture uses a two-antenna scheme, which consists of two 2×2 planar patch arrays, so that the isolation between them can be small. The inclusion of two delay lines permits keeping the desired beat frequency far from the low-frequency baseband spectrum, which is usually contaminated with strong flicker noise. The extra delay added by the lines can be easily compensated for through calibration.

Finally, the coherence of the radar is guaranteed by sharing a common clock between the generation and acquisition blocks, as expounded in Section 5.3.2. Two photographs of the prototype are provided in Fig. 5.28, with the interconnection details in Fig. 5.28(B).

Fig. 5.29 schematically depicts the experimental setups made with the radar prototype, which are respectively associated with the next sections.

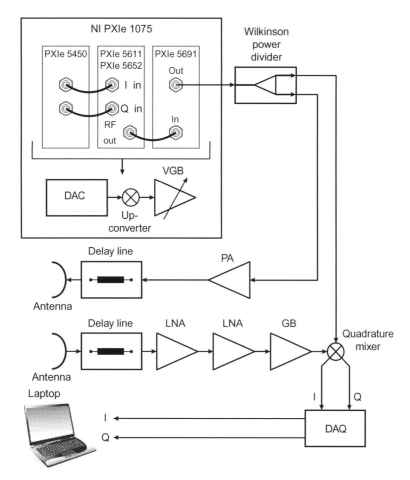

FIGURE 5.27

Block diagram for the coherent C-band LFMCW radar prototype (power amplifier: PA; low-noise amplifier: LNA; digital-to-analog converter: DAC; data acquisition: DAQ; gain block: GB; variable gain block: VGB).

5.8.2 ABSOLUTE RANGE MEASURES OF METAL PLATES

In this experiment, a metal plate on a small cart is perpendicularly located at different ranges from the radar (see Fig. 5.29(A)). The range position of the plate is estimated through the amplitude-based method, which obtains the position of the corresponding peak observed in the frequency domain of the beat signal. The calibration distance was set at 4.5 ft (137.16 cm). Fig. 5.30 shows the spectrum of the beat signal when the metal plate was situated at 4.5 ft and at 8.5 ft. In addition to the echo associated with the desired target, some peaks corresponding to clutter

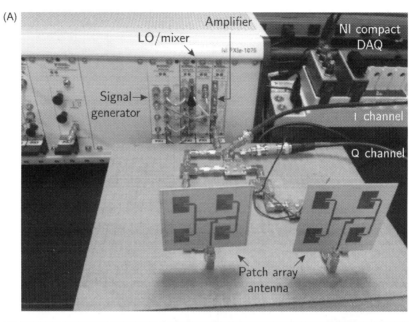

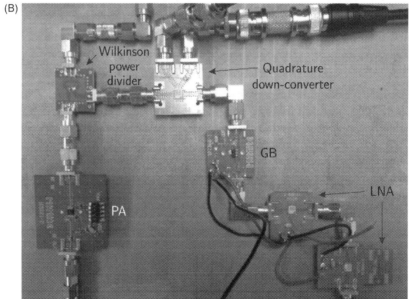

FIGURE 5.28

Photograph of the coherent C-band LFMCW radar prototype (power amplifier: PA; low-noise amplifier: LNA; digital-to-analog converter: DAC; data acquisition: DAQ; gain block: GB; local oscillator: LO). (A) Overall view. (B) RF detail.

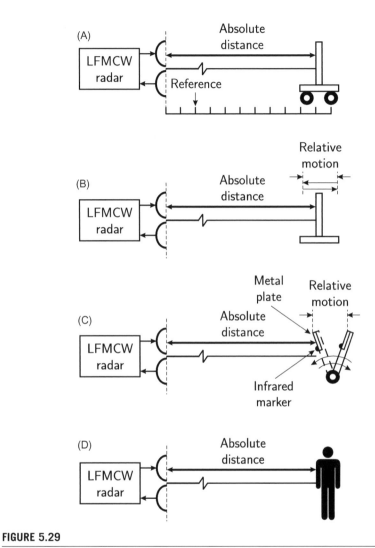

FIGURE 5.29

Schematic representations of the experimental setups. (A) Absolute ranging of a metal plate. (B) Tracking of the range history of a sinusoidally-vibrating metal plate. (C) Tracking of the range history of a phantom-borne metal plate with stop periods. (D) Monitoring of vital signs.

are observed. The Tx−Rx coupling requires special mention, since it means a powerful return, which may be even greater than that associated with the metal plate (especially when the target is far from the radar, and thus the wanted received power is lower). This powerful Tx−Rx coupling means large sidelobes, which may mask desired weak targets, and calls for a redesign of the antennas so that the Tx−Rx isolation may be improved.

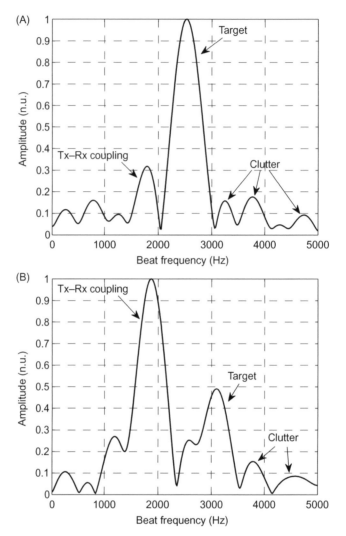

FIGURE 5.30

Beat-signal spectrum magnitude for a metal plate situated at the calibration distance of 4.5 ft (A), and at a range of 8.5 ft (B).

Table 5.2 details the absolute range measures of the metal plate at different locations. As seen, a systematic error is observed for every position, which limits the accuracy of the system. Related with the discussions in Section 5.5.3, the origin of this biasing may be found in the high interference of the Tx−Rx coupling, or in the fact that a metal plate is a good scatterer, but cannot be considered a point scatterer.

Table 5.2 Experimental Absolute Range Measurements and Associated Errors [25]

Reference Distance (ft/cm)	Beat Frequency (Hz)	Measured Distance (ft/cm)	Associated Error (ft/cm)
3.0/91.44	2305.8	3.11/94.79	+0.11/ + 3.35
3.5/106.68	2401.1	3.69/112.47	+0.19/ + 5.79
4.0/121.92	2456.1	4.03/122.83	+0.03/ + 0.91
4.5/137.16	2531.3	4.50/137.16	0.00/0.00
5.0/152.40	2606.5	4.96/151.18	−0.04/ − 1.22
5.5/167.64	2656.6	5.27/160.63	−0.23/ − 7.01
6.0/182.88	2731.8	5.73/174.65	−0.27/ − 8.23
6.5/198.12	2807.0	6.19/188.67	−0.31/ − 9.45
7.0/213.36	2882.2	6.66/203.00	−0.44/ − 13.41
7.5/228.60	2957.6	7.12/217.02	−0.43/ − 13.11
8.0/243.84	3022.4	7.52/229.21	−0.48/ − 14.63
8.5/259.08	3105.3	8.03/244.75	−0.47/ − 14.32
9.0/274.32	3222.2	8.75/266.70	−0.25/ − 7.62
9.5/289.56	3272.0	9.16/279.20	−0.34/ − 10.36
10.0/304.80	3339.1	9.57/291.69	−0.43/ − 13.11
10.5/320.40	3406.0	10.08/307.24	−0.42/ − 12.80
11.0/335.28	3523.0	10.60/323.09	−0.40/ − 12.19
11.5/350.52	3606.0	11.11/338.63	−0.39/ − 11.89
12.0/365.76	3673.0	11.62/354.18	−0.38/ − 11.58

5.8.3 RANGE TRACKING OF A VIBRATING METAL PLATE

The metal plate is now situated in an actuator which generates a sinusoidal motion with a peak-to-peak amplitude of 1.5 cm and a period of 4.762 seconds (see Fig. 5.29(B)). After applying the phase-based method in Section 5.4.3 to track the motion of the metal plate, the results are depicted in Fig. 5.31. As can be checked, the tracking accuracy is less than 1 mm, although it must again be emphasized than the metal plate is not a point scatterer and the real actuator does not have an infinite precision.

Fig. 5.32 shows the same scenario of Fig. 5.29(B), with the inclusion of a stationary metal plate which emulates strong clutter returns at the same range bin. Referring to the nomenclature in Fig. 5.32, the mean range to the desired target is $R_d = 2$ m. Now, the actuator generates a sinusoidal motion with a peak-to-peak amplitude $R_{pp} = 2$ cm and a period of 3 seconds. The range separation between the vibrating metal plate and the stationary one is $R_{d-c} = 20$ cm.

The tracked motion of the vibrating metal plate before and after applying a high-pass filter with a cut-off frequency of 0.2 Hz is shown in Fig. 5.33. For this

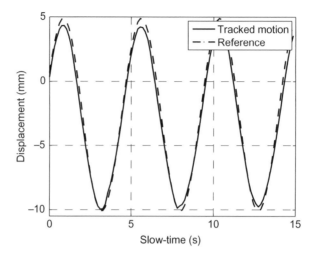

FIGURE 5.31

Experimental range evolution extracted for an along-LOS sinusoidally-vibrating metal plate.

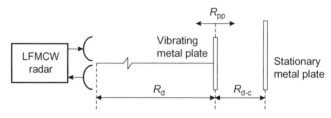

FIGURE 5.32

Experimental scenario for a sinusoidally-vibrating metal plate and a clutter-assumed stationary metal plate.

particular example, the clutter-mitigation technique is effective and enables a correct tracking of the desired motion.

As mentioned, Doppler radars usually require complex architectures to follow DC components of the desired signal. These DC components are present in practical signals, such as vital signs. The proposed coherent LFMCW architecture is simple and also permits the maintaining of the DC structure of the wanted signal. To experimentally verify this feature, the metal plate was mounted on a professional phantom which mimics the chest movement (see Fig. 5.29(C)). Fig. 5.34 shows the detected motion pattern of the metal plate. As seen, the motion is tracked even during the stop periods of the phantom motion. Fig. 5.34 also shows a reference motion based on an infrared marker. A fair agreement between both motions is observed.

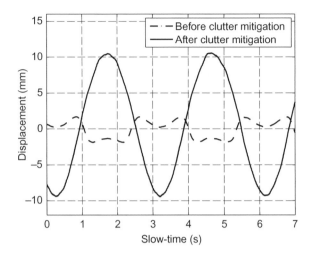

FIGURE 5.33

Range evolution extracted for an along-LOS sinusoidally-vibrating metal plate before and after applying the clutter-mitigation high-pass filtering.

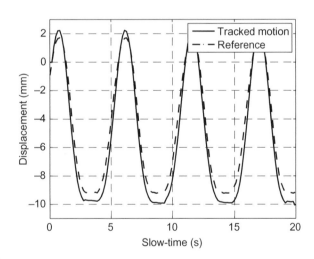

FIGURE 5.34

Tracked motion and infrared-based reference for a phantom-borne metal plate.

5.8.4 MONITORING OF VITAL SIGNS

The coherent short-range LFMCW radar prototype was also used to illuminate a breathing subject (see Fig. 5.29(D)). The patient was sitting in front of the radar and was normally breathing.

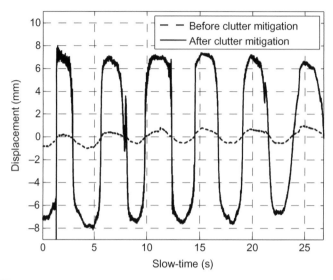

FIGURE 5.35

Range tracking of vital signs before and after applying the clutter-mitigation approach.

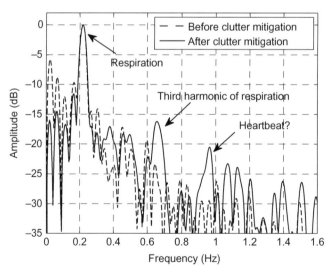

FIGURE 5.36

Fourier transform of the range histories corresponding to the experimental vital sign monitoring scenario before and after clutter-mitigation high-pass filtering.

Fig. 5.35 shows the tracked signal before and after applying a high-pass filter with a cut-off frequency of 0.1 Hz. Note that the vital sign is extremely slow. Indeed, by supposing an approximate breathing period of 4.5 seconds and a peak-to-peak amplitude of the respiration signal of 15 mm, the maximum Doppler frequency is only 0.4 Hz. Additionally, the vital sign is weak, when compared with the echoes coming from the metal plates. This makes the tracking of vital signs extremely challenging.

As shown in Fig. 5.35, the mitigation technique permits the enhancement of the amplitude of the vital sign. After applying a Fourier transform to that signal, the frequency components of the vital sign can be distinguished (see Fig. 5.36). The results in Fig. 5.36 indicate that the Doppler high-pass filtering is effective in cleaning the spectrum, and thus even the heartbeat signal appears to be detectable.

5.9 FURTHER WORK

Coherent LFMCW radar systems are interesting devices for short-range human-aware localization applications. From our point of view, moving to high-frequency devices (e.g., millimeter-wave radars) is the next step. Reasons are indicated in the following:

- A higher absolute bandwidth can be transmitted. This means that a higher range resolution can be achieved, and thus better range-isolation capabilities can be accomplished.
- A higher range precision for the phase-based method can be obtained, according to Eq. (5.33). A look should be taken at the phase-unwrapping process, which may be more complex.
- The antennas can have a greater directivity. This leads to a better Tx−Rx isolation and to a better concentration of the radiation on the desired target (thus diminishing the clutter contributions).
- The Doppler frequencies would be increased. This means that the under-study slow signals would be filtered more easily in the Doppler domain.

ACKNOWLEDGMENTS

The work of José-María Muñoz-Ferreras and Roberto Gómez-García was financially supported by the Spanish Ministry of Economy and Competiveness under project TEC2014-54289-R and by the University of Alcalá under project CCG2015/EXP-017.

REFERENCES

[1] Charvat GL, Williams J, Zeng S, Nickalaou J. Small and short-range radar systems. Taylor & Francis; 2014.

[2] Muñoz-Ferreras JM, Pérez-Martínez F, Calvo-Gallego J, Asensio-López A, Dorta-Naranjo BP, Blanco-del-Campo A. Traffic surveillance system based on a high resolution radar. IEEE Trans Geosci Remote Sens 2008;46(6):1624−33.

[3] Ayhan S, Scherr S, Pahl P, Kayser T, Pauli M, Zwick T. High-accuracy range detection radar sensor for hydraulic cylinders. IEEE Sens J 2014;14(3):734−46.

[4] Armbrecht G, Zietz C, Denicke E, Rolfes I. Antenna impact on the gauging accuracy of industrial radar level measurements. IEEE Trans Microw Theory Techn 2011;59 (10):2554−62.

[5] Felguera-Martín D, González-Partida JT, Almorox-González P, Burgos-García M. Vehicular traffic surveillance and road lane detection using radar interferometry. IEEE Trans Veh Technol 2012;61(3):959−70.

[6] Raj RG, Chen VC, Lipps R. Analysis of radar human gait signatures. IET Signal Processing. 2010;4(3):234−44.

[7] Wang FK, Horng TS, Peng KC, Jau JK, Li JY, Chen CC. Detection of concealed individuals based on their vital signs by using a see-through-wall imaging system with a self-injection-locked radar. IEEE Trans Microw Theory Techn 2013;61 (1):696−704.

[8] Nanzer JA, Anderson MG, Josser TM, Li K, Olinger GA, Buhl DP, et al. Detection of moving intruders from a moving platform using a Ka-band continuous-wave Doppler radar. In: Proceedings IEEE Antennas Propagation Society International Symposium, Charleston, SC, USA, June 2009, pp. 1−4.

[9] Mercuri M, Schreurs D, Leroux P. SFCW microwave radar for in-door fall detection. In: Proceedings of IEEE Topical Conference Biomedical Wireless Technologies, Networks, Sensing Systems, Santa Clara, CA, USA, January 2012, pp. 53−56.

[10] Amin MG, Ahmad F. Change detection analysis of humans moving behind walls. IEEE Trans Aerosp Electron Syst 2013;49(3):1410−25.

[11] Li C, Lubecke VM, Boric-Lubecke O, Lin J. A review on recent advances in Doppler radar sensors for noncontact healthcare monitoring. IEEE Trans Microw Theory Techn 2013;61(5):2046−60.

[12] Schleicher B, Nasr I, Trasser A, Schumacher H. IR-UWB radar demonstrator for ultra-fine movement detection and vital-sign monitoring. IEEE Trans Microw Theory Techn 2013;61(5):2076−85.

[13] Hafner N, Mostafanezhad I, Lubecke VM, Boric-Lubecke O, Host-Madsen A. Non-contact cardiopulmonary sensing with a baby monitor. In: Proceedings of 29th Annual International Conference IEEE Engineering in Medicine and Biology Society, Lyon, France, August 2007, pp. 2300−2.

[14] Gu C, Li R, Zhang H, Fung AYC, Torres C, Jiang SB, et al. Accurate respiration measurement using DC-coupled continuous-wave radar sensor for motion-adaptive cancer radiotherapy. IEEE Trans Biomed Eng 2012;59(11):3117−23.

[15] Li C, Lin J, Xiao Y. Robust overnight monitoring of human vital signs by a non-contact respiration and heartbeat detector. In: Proceedings of 28th Annual International Conference IEEE Engineering in Medicine and Biology Society, New York, NY, USA, August 2006, pp. 2235−8.

[16] Morinaga M, Nagasaku T, Shinoda H, Kondoh H. 24-GHz intruder detection radar with beam-switched area coverage. In: Proceedings of IEEE MTT-S International Microwave Symposium, Honolulu, HI, USA, June 2007, pp. 389−92.

[17] Li C, Gu C, Li R, Jiang SB. Radar motion sensing for accurate tumor tracking in radiation therapy. In: Proceedings of IEEE 12th Annual Wireless Microwave Technology Conference, 2011, pp. 1−6.

[18] Wang J, Wang X, Zhu Z, Huangfu J, Li C, Ran L. 1-D microwave imaging of human cardiac motion: an AB-initio investigation. IEEE Trans Microw Theory Techn 2013;61(5):2101−7.

[19] Zito D, Pepe D, Mincica M, Zito F, Tognetti A, Lanata A, et al. SoC CMOS UWB pulse radar sensor for contactless respiratory rate monitoring. IEEE Trans Biomed Circuits Syst 2011;5(6):503−10.

[20] Gezici S. Theoretical limits for estimation of periodic movements in pulse-based UWB systems. IEEE J Sel Top Signal Process 2007;1(3):405−17.

[21] Bernardi P, Cicchetti R, Pisa S, Pittella E, Piuzzi E, Testa O. Design, realization, and test of a UWB radar sensor for breath activity monitoring. IEEE Sens J 2014;14 (2):584−96.

[22] Zhang C, Kuhn MJ, Merkl BC, Fathy AE, Mahfouz MR. Real-time noncoherent UWB positioning radar with millimeter range accuracy: theory and experiment. IEEE Trans Microw Theory Techn 2010;58(1):9−20.

[23] Wang G, Gu C, Inoue T, Li C, Hybrid FMCW-interferometry radar system in the 5.8 GHz ISM band for indoor precise position and motion detection. In: Proceedings of IEEE MTT-S International Microwave Symposium, Seattle, WA, USA, June 2013, pp. 1−3.

[24] Wang G, Muñoz-Ferreras JM, Gu C, Li C, Gómez-García R. Linear-frequency-modulated continuous-wave radar for vital-sign monitoring. In: Proceedings of IEEE Topical Conference Biomedical Wireless Technologies, Networks, Sensing Systems, Newport Beach, CA, USA, January 2014, pp. 1−3.

[25] Wang G, Muñoz-Ferreras JM, Gu C, Li C, Gómez-García R. Application of linear-frequency-modulated continuous-wave (LFMCW) radars for tracking of vital signs. IEEE Trans Microw Theory Techn 2014;62(6):1387−99.

[26] Wang G, Gu C, Inoue T, Li C. A hybrid FMCW-interferometry radar for indoor precise positioning and versatile life activity monitoring. IEEE Trans Microw Theory Techn 2014;62(11):2812−22.

[27] Belyaev E, Molchanov P, Vinel A, Koucheryavy Y. The use of automotive radars in video-based overtaking assistance applications. IEEE Trans Intell Transp Syst 2013;14(6):1035−42.

[28] Feger R, Pfeffer C, Scheiblhofer W, Schmid CM, Lang MJ, Stelzer A. A 77-GHz cooperative radar system based on multi-channel FMCW stations for local positioning applications. IEEE Trans Microw Theory Techn 2013;61(1):676−84.

[29] Luo T-N, Wu C-HE, Chen Y-JE. A 77-GHz CMOS FMCW frequency synthesizer with reconfigurable chirps. IEEE Trans Microw Theory Techn 2013;61(7):2641−7.

[30] Richards MA, Scheer JA, Holm WA. Principles of modern radar: basic principles. 2nd ed. Edison, NJ: SciTech Publishing; 2010.

[31] Skolnik MI. Introduction to radar systems. 3rd ed. New York, NY: McGraw-Hill; 2002.

[32] Carrara WG, Goodman RS, Majewski RM. Spotlight synthetic aperture radar: signal processing algorithms. Boston, MA: Artech House; 1995.

[33] Wehner DR. High-resolution radar. 2nd ed. Boston, MA: Artech House; 1995.

[34] Edrich M. Design overview and flight test results of the miniaturised SAR sensor MISAR. Proceedings of 1st European Radar Conference. Amsterdam, The Netherlands; 2004. p. 205−8.

[35] Almorox-Gonzalez P, González-Partida JT, Burgos-García M, de la Morena-Álvarez-Palencia C, Arche-Andradas L, Dorta-Naranjo BP. Portable high resolution LFM-CW radar sensor in millimeter-wave band. In: Proceedings of International Conference Sensor Technologies and Applications, Valencia, Spain, October 2007, pp. 5−9.

[36] Zaugg EC, Long DG. Theory and application of motion compensation for LFM-CW SAR. IEEE Trans Geosci Remote Sens 2008;46(10):2990−8.

[37] Zaugg EC, Hudson DL, Long DG. The BYU μSAR: a small, student-built SAR for UAV operation. In: Proceedings of IEEE International Geoscience Remote Sensing Symposium, Denver, CO, USA, August 2006, pp. 411−14.

[38] Meta A, de Wit JJM, Hoogeboom P. Development of a high resolution airborne millimeter wave FM-CW SAR. In: Proceedings of 1st European Radar Conference, Amsterdam, The Netherlands, October 2004, pp. 209−12.

[39] Muñoz-Ferreras JM, Pérez-Martínez F, Burgos-García M. Helicopter classification with a high resolution LFMCW radar. IEEE Trans Aerosp Electron Syst 2009;45 (4):1373−84.

[40] Wang J, Kasilingam D. Global range alignment for ISAR. IEEE Trans Aerosp Electron Syst 2003;39(1):351−7.

[41] Muñoz-Ferreras JM, Pérez-Martínez F. Subinteger range-bin alignment method for ISAR imaging of noncooperative targets. EURASIP J Adv Signal Processing 2010;2010(1):1−16.

[42] Chen VC. The micro-doppler effect in radar. Norwood, MA: Artech House; 2011.

[43] Muñoz-Ferreras JM, Pérez-Martínez F. Superresolution versus motion compensation-based techniques for radar imaging defense applications. EURASIP J Adv Signal Processing 2010;2010(1):1−9.

[44] Brennan PV, Lok LB, Nicholls K, Corr H. Phase-sensitive FMCW radar system for high-precision Antarctic ice shelf profile monitoring. IET Radar Sonar Navig. 2014;8 (7):776−86.

[45] Wang G, Muñoz-Ferreras JM, Gómez-García R, Li C. Clutter interference reduction in coherent FMCW radar for weak physiological signal detection. In: Proceedings of IEEE MTT-S International Microwave Symposium, Tampa Bay, FL, USA, 2014, pp. 1−3.

[46] Muñoz-Ferreras JM, Wang G, Li C, Gómez-García R. Mitigation of stationary clutter in vital-sign-monitoring linear-frequency-modulated continuous-wave radars. IET Radar Sonar Navig 2015;9(2):138−44.

[47] Muñoz-Ferreras JM, Peng Z, Gómez-García R, Wang G, Gu C, Li C. Isolate the clutter: pure and hybrid linear-frequency-modulated continuous-wave (LFMCW) radars for indoor applications. IEEE Microw Mag 2015;16(4):40−54.

Biomedical radars for monitoring health

F.-K. Wang[1], M. Mercuri[2], T.-S. J. Horng[1] and D.M.M.-P. Schreurs[3]

*[1]National Sun Yat-Sen University, Kaohsiung, Taiwan [2]Holst Centre, IMEC, Eindhoven,
Netherlands [3]KU Leuven, Leuven, Belgium*

6.1 INTRODUCTION

The biomedical radar is a sensor device that uses radio waves to detect or monitor biological objects, providing physiological, motion-related, and location-related information [1−3]. It is based on the detection of a reflected signal that is Doppler-shifted by movements of the target body or organ. The main benefits of biomedical radar sensing are that it is remote and contactless and that it penetrates obstacles. All of these characteristics are important for long-term human health monitoring because the examinee is not restrained or confined during the test and is unlikely to be concerned about the invasion of privacy.

In the 1970s, biomedical radar technology was first used for the remote detection of vital signs, such as rates of respiration and heartbeat [4,5]. Subsequently, life detectors based on this technology were developed to search for living victims

C. Li, M. Tofighi, D. Schreurs and T-Z. J. Horng (Eds): *Principles and Applications of RF/Microwave
in Healthcare and Biosensing.*

in earthquake rubble [6,7]. In more recent decades, biomedical radars with lower power, smaller sizes, and higher functionality have been developed, and these are becoming widely used for the contactless sensing of vital signs, such as those related to sudden infant death [8,9], sleep apnea [10,11], cardiopulmonary activity [12,13], human tumors [14], and the occupancy of buildings [15]. Some radars have been developed to locate or track people, even behind walls, without requiring those people to carry any devices, forming the basis of surveillance systems that detect falls [16] or intruders [17] in the home.

6.1.1 UWB PULSE RADAR VERSUS CW RADAR

The two main categories of biomedical radar are the ultra-wideband (UWB) pulse radar [18−20] and the continuous-wave (CW) radar [21−23], whose architectures are shown in Fig. 6.1(A) and (B), respectively. The former uses a pulse generator to transmit a sequence of short pulses of radio frequency (RF) energy, and the distance to the target is estimated by measuring the delay of the reflected pulses. The clutter effect on the vital sign signal is removed by switches with programmable delay controls. However, an important issue concerns how, in the detection process, the receiver identifies the pulses that are reflected from the target, as doing so usually involves considerable effort to

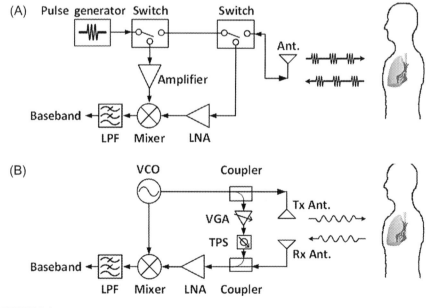

FIGURE 6.1

Two main categories of biomedical radar architectures. (A) UWB pulse radar. (B) CW radar.

filter out environmental clutter. In particular, the UWB pulse radar faces significant challenges that are related to system complexity and cost of integration.

The CW radar uses a voltage-controlled oscillator (VCO) to transmit continuously a CW signal. The receiver is always on to detect the echo signal. The CW radar is much simpler than the UWB pulse radar and so is easier to integrate into personal mobile devices. However, its major shortcomings include its lack of ranging capability and the need for isolation between the transmitted (Tx) and received (Rx) signals using either a circulator or separate Tx and Rx antennas. Additionally, its sensitivity is vulnerable to environmental clutter if no cancellation is applied. In Fig. 6.1(B), an auxiliary path with a variable gain amplifier and a tunable phase shifter yields a signal with the same amplitude as, but an opposite phase to that of, the received clutter, canceling it out. Nevertheless, any manner of clutter cancellation in a CW radar inevitably increases the complexity and cost of the system.

6.1.2 SIL RADAR AND SFCW RADAR

The idea of the self-injection-locked (SIL) radar is motivated by an empirical observation of a free-running oscillator [24]. When a human hand is waved up and down close to a free-running oscillator, the oscillation frequency shifts up and down accordingly, as illustrated in Fig. 6.2, because the oscillator is SIL by its leakage signal, which bounces back from the hand. When the motion of the hand is within a range that is much smaller than the free-space wavelength, the shift in oscillation frequency is proportional to the displacement of the hand. Therefore, a frequency demodulator can be utilized at the output of the oscillator to detect the motion of the hand.

Based on this observation, an SIL radar for vital sign detection was developed [25]. In its basic operation, a VCO transmits a CW signal, which is partially reflected by a distant human subject, and then injected into the same VCO to

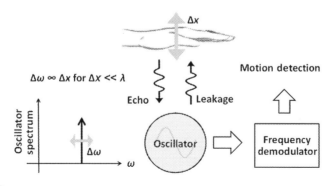

FIGURE 6.2

SIL effect on a free-running oscillator and its application to motion detection.

form an SIL state. The Doppler phase shift that is related to the subject's cardio-pulmonary movement can be simply extracted by frequency demodulation of the oscillator output. SIL radar-based vital sign sensors have been verified to exhibit a high sensitivity with low complexity [26]. Furthermore, their sensitivity is highly resistant to clutter [27]. With the help of frequency-modulated CW (FMCW) technology, the SIL radar can not only detect the distance to a target, but also overcome the null point problem [17,26]. Thanks to these advances, SIL radar has great potential for use in real-time vital sign monitors with a low-complexity single or array architecture. Section 6.2 elucidates the operating principles of the SIL radar and demonstrates some promising results concerning its applications in a through-wall life detector and a chest-worn health monitor.

The FMCW radar is a well-known extension of the CW radar that can range short-distance objects with clutter rejection. The stepped-frequency CW (SFCW) radar is basically a discretized version of the FMCW radar that measures distance to a target by sweeping frequency in discrete steps [28], as depicted in Fig. 6.3. The output of the SFCW radar receiver can be written in the discrete-time domain as

$$S(n) = \cos\left(\phi_0 + \frac{4\pi R}{c} \cdot n\Delta f\right). \tag{6.1}$$

where n is the index of data sequences, ϕ_0 specifies the phase at the starting frequency of the sweep, R represents the distance to the target, c denotes the speed of light, and Δf is the frequency step. Assume that N is the number of frequency step's and the sweep bandwidth B equals $N \cdot \Delta f$. The discrete Fourier transformation of Eq. (6.1) yields a peak at an integer point m_p, from which is determined the distance $R = m_p \cdot c/2B$. The difference between two transformed points is 1, which gives a range resolution of $c/2B$.

The major advantages of the SFCW radar over the FMCW radar are a reduced need for high sweep linearity and a lower analog-to-digital converter (ADC) sampling rate [29], making the former more cost-effective than the latter. The disadvantage of the SFCW radar is the longer duration of the frequency sweep, which

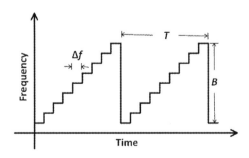

FIGURE 6.3

Time-frequency relation of an SFCW radar waveform.

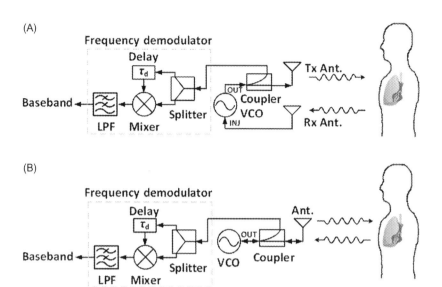

FIGURE 6.4

SIL radar architectures with (A) Two antennas and (B) A single antenna.

affects the accuracy of detecting moving targets because the Doppler frequency shift caused by the motion of the target perturbs the peak location in the Fourier transform spectrum. To improve the accuracy, a hybrid ranging system that combines the SFCW and CW radar waveforms can be utilized [16]. Section 6.3 presents more details concerning the design of the hardware for this system and its expansion to a wireless sensor network for home fall-detection applications.

6.2 POPULARIZING HEALTH MONITORING WITH CUTTING-EDGE VITAL SIGN SENSORS

6.2.1 SIL RADAR-BASED CARDIOPULMONARY SENSOR

As a crucial advantage, the SIL radar is inherently immune to stationary clutter, such as that produced by background reflection and antenna coupling. Fig. 6.4(A) and (B) presents a two-antenna and single-antenna SIL radar, respectively. The former radar mainly comprises a Tx antenna, an Rx antenna, a VCO with an injection terminal, a coupler, and a frequency demodulator that is composed of a mixer, a delay unit, and a lowpass filter (LPF). Fig. 6.5 shows the schematic and photograph of the VCO circuit which has a tuning range from 2.3 to 2.65 GHz and an output power of 5 dBm. Notably, the design of the VCO uses the Clapp configuration along with an injection port connected to the gate of the transistor.

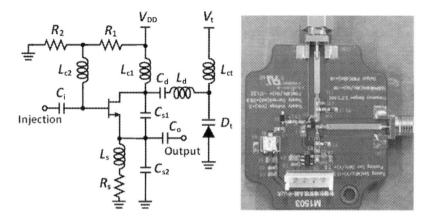

FIGURE 6.5

Schematic and photograph of the S-band VCO circuit with an injection terminal.

Fig. 6.4(B) shows the latter radar, with the same components as in Fig. 6.4(A), but without the Rx antenna or the VCO's injection terminal. It uses a single antenna to transmit the CW output signal toward the subject and to receive the reflected Doppler signal without the need to isolate both signals.

Assume that the VCO in Fig. 6.4(A) and (B) has an inherent oscillation frequency ω_{osc}, a constant oscillation amplitude V_{osc}, and a tank quality factor Q_{tank}. When a reflected signal with an instantaneous frequency $\omega_{inj}(t)$ and constant amplitude V_{inj} is injected into the VCO, the instantaneous VCO output frequency $\omega_{out}(t)$ can be obtained from Adler's theory [30], as follows.

$$\omega_{out}(t) = \omega_{osc} - \omega_{LR}\sin\alpha(t). \tag{6.2}$$

where

$$\omega_{LR} = \frac{\omega_{osc}}{2Q_{tank}}\frac{V_{inj}}{V_{osc}}. \tag{6.3}$$

is the locking range, and $\alpha(t)$ is the instantaneous phase difference between the injection and oscillation signals. Under SIL conditions, $\alpha(t)$ equals the propagation phase delay between the transmitted signal and the received signal, which is given by

$$\alpha(t) = \frac{2\omega_{osc}}{c}(R + x(t)) \tag{6.4}$$

where R and $x(t)$ are the distance to the target and the instantaneous displacement of the target, respectively. Substituting Eq. (6.3) and Eq. (6.4) into Eq. (6.2) and assuming that $x(t)$ is much smaller than the free-space wavelength yields the frequency-demodulated baseband signal $V_{bb}(t)$, which is approximated as

$$V_{bb}(t) \approx -\frac{\omega_{osc}^2 \tau_d}{Q_{tank}c} \cdot \frac{V_{inj}}{V_{osc}} \cdot \cos\left(\frac{2\omega_{osc}R}{c}\right) \cdot x(t) \tag{6.5}$$

where τ_d is the delay time of the delay unit in the frequency demodulator. Importantly, Eq. (6.5) indicates that the optimum and null points for small displacement detection are at a distance of $\cos(2\omega_{osc}R/c) = 0$ and ± 1, respectively, from the radar. The detected displacement $x(t)$ is proportional to the square of the operating frequency ω_{osc}; this relationship differs from that for a conventional CW radar, which is linear. A Laplace-domain analysis reveals that the SIL radar can be modeled as a first-order delta-sigma modulator with a noise shaping ability that increases the signal-to-noise ratio of the baseband signal by more than 100 dB at sub-Hz frequencies [26]. Moreover, from Eq. (6.2) and Eq. (6.4), if the clutter is stationary, meaning $x(t) = 0$, then its effect on an SIL radar is to cause a small, constant-frequency offset in the output signal of the oscillator, rather than a large DC offset, which is caused in the baseband signal of the conventional CW radar.

Fig. 6.6(A)−(D) presents four experiments to demonstrate the anti-clutter and anti-null-point capabilities of the SIL radar in detecting vital signs. These experiments are conducted under a combination of conditions, such as one or two antennas, optimum or null points, and the presence of metallic objects. In the experiments, a subject who is breathing normally is seated approximately one meter from the radar that is operated at 2.4 GHz. Fig. 6.6(A) illustrates the case using the two-antenna SIL radar while Fig. 6.6(B)−(D) shows the cases with the single-antenna SIL radar. In Fig. 6.6(C), a $60 \times 60 \text{ cm}^2$ metal plate is placed close to the subject. Moreover, in Fig. 6.6(A)−(C) the subject is at the optimum point position but in Fig. 6.6(D), the subject is at the null point position.

In the above experimental setups, the coupling coefficient between Tx and Rx antennas in Fig. 6.6(A) is around 0.03, and the reflection coefficient of the single antenna in Fig. 6.6(B) and (C) is around 0.2. The SIL radar has only one active component, which is the VCO circuit shown in Fig. 6.5. The rest of its components are all passive circuits and commercial, off-the-shelf products. The output baseband signals from the radar were processed by a digital filter with a passband of 0.1−10 Hz. As a result, the root-mean-square (RMS) values of the baseband signal waveforms in Fig. 6.6(A), (B), and (C) are 0.825 V, 0.823 V, and 0.822 V, respectively, indicating that the antenna and metal clutters have little influence on the detection outcome when the subject is at an optimum point.

Fig. 6.7 shows the Fourier-transformed spectra of the waveforms in Fig. 6.6. In Fig. 6.7(A)−(C), the spectral peak voltages associated with respiration are 1.074 V, 1.061 V, and 1.042 V, respectively. Those associated with heartbeat are 0.189 V, 0.183 V, and 0.169 V, respectively. All three sets of outputs have sufficient amplitudes to provide information about vital signs: the breathing rate is about 18 breaths/minutes and the heart rate is about 78 beats/minutes. Based on the above theory and experiments, the SIL radar is strongly resistant to stationary clutter that is caused by the reflection from the antenna and metallic plate. Therefore, the two-antenna SIL radar, which seeks to mitigate antenna clutter, can be replaced by the single-antenna SIL radar, since antenna clutter has a negligible impact on SIL radars that are used for CW Doppler detection.

Fig. 6.7(D) compares the baseband signal spectra of the single-antenna SIL radar: the results that are represented by the gray line were obtained at a null point

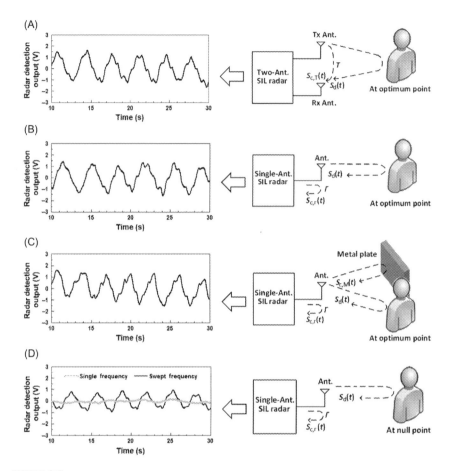

FIGURE 6.6

Four experiments of using SIL radar to detect vital signs with (A) The use of two antennas. (B) The use of single antenna. (C) The use of single antenna and the presence of a metal plate; and (D) The use of single antenna and at a null point.

with a single frequency (where $\cos(2\omega_{osc}R/c) = 0$), and the improved results, represented by the black line, were obtained using a swept frequency. Clearly, frequency sweeping effectively solves the null point problem. Notably, the swept frequency method is also effective for detecting vital signs in the presence of various forms of interference in the 2.4 GHz industrial, scientific and medical band [26].

6.2.2 THROUGH-WALL LIFE DETECTOR

Optical image sensors fail to provide visual information through obstacles and often arouse concerns about privacy infringement when used in monitoring homes

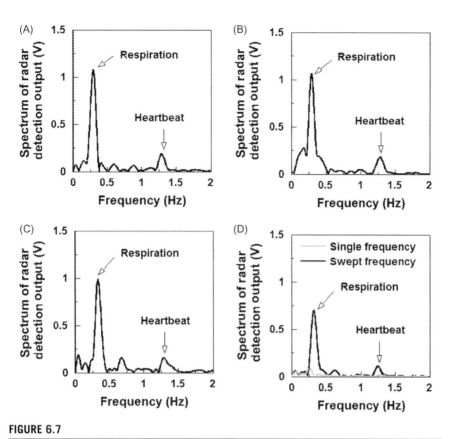

FIGURE 6.7

Spectra of the vital sign signals shown in (A) Fig. 6.6(A), (B) Fig. 6.6(B), (C) Fig. 6.6(C) and (D) Fig. 6.6(D).

for healthcare purposes. Real-time through-wall radars for use in monitoring health in the home have therefore drawn increasing attention because they can be concealed to detect, locate, and monitor individuals using electromagnetic waves. Since the early 1990s, FMCW technology has been adopted in imaging radars to provide real-time images in areas with obscured lines of sight [31]. It can locate a moving subject behind a wall by subtracting the output spectra from others in a short period of time to remove the clutter signals that are produced by stationary objects. However, tiny body movements, such as those caused by cardiopulmonary activity, cannot easily be identified in continuous range information from an FMCW radar because the noise sideband of a transmitter often swamps the Doppler signals associated with vital signs. However, some FMCW radars detect vital signs by demodulating low−frequency information from the modulation of the carrier frequency [32], but they fail to provide synchronous range information.

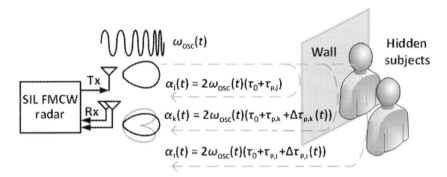

FIGURE 6.8

Through-wall life detector based on an SIL FMCW radar with sum and difference antenna patterns in the receiver.

Owing to the high sensitivity of the SIL radar to the detection of vital signs, an SIL FMCW radar has been used to discover the individuals hidden behind a wall by detecting their vital signs [17]. This radar system incorporates a sum−difference antenna architecture to construct a wall-transparent image for locating the individuals behind a wall, as shown in Fig. 6.8. Fig. 6.9 presents a block diagram with timing waveforms for this radar system that contains a VCO with an injection input and a differential output, a low-noise amplifier (LNA) with a power gain of 14 dB, a power amplifier (PA) with a power gain of 8 dB, a frequency demodulator composed of a mixer, a delay line, and an LPF, a Tx antenna, and an Rx antenna array using two antennas, a phase shifter, and a combiner to form the sum and difference beam patterns for azimuth detection. Both the Tx and the Rx antenna elements are horn antennas with a high gain of more than 10 dBi from 2 to 18 GHz. The VCO has a tuning range of 2.3 to 2.8 GHz, and it delivers an output power of approximately 0 dBm.

The SIL FMCW radar shown in Fig. 6.9 performs two time-division detection processes. Firstly, in the FMCW detection process, the scanning input voltage $V_s(t)$ is constant to fix the beam direction of the Rx antenna array. In contrast, the tuning input voltage $V_t(t)$ outputs a linear ramp waveform to frequency-modulate the VCO during the duration t_1. The positive output signal of the VCO $S_{out} + (t)$ is amplified by the PA and then emitted by the Tx antenna toward the targets hidden behind a wall. In Fig. 6.9, the signals that are reflected from human subjects and stationary objects (such as wall and table) are received by the Rx antenna array and are then amplified as the injection signals by the LNA. These injection signals enter the VCO via its injection input and cause the VCO to enter an SIL state.

When the linear ramp waveform is applied to the tuning terminal of the VCO, the oscillation frequency of the VCO is given as

$$\omega_{osc}(t) = \omega_{osc} + 2\pi K_v V_t(t) \tag{6.6}$$

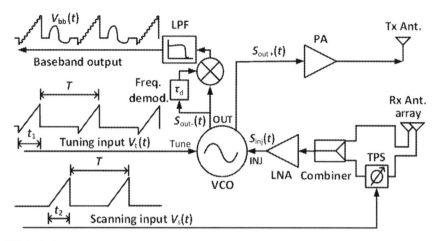

FIGURE 6.9

Block diagram of the SIL FMCW radar system with timing information.

where K_v is the tuning sensitivity and ω_{osc} is the initial oscillation frequency. Assume that the VCO has a small tuning range, and consider a situation in which the injection signal is a summation of multiple echo signals from various targets. Thus, Eq. (6.2) is rewritten as

$$\omega_{out}(t) = \omega_{osc}(t) - \sum_i \omega_{LR,i}\sin\alpha_i(t) \tag{6.7}$$

where $\omega_{LR,i}$ is the locking range that is associated with the echo signal from the ith target. The phase difference $\alpha_i(t)$ under an SIL condition equals the propagation phase delay of each injection signal, and can be expressed as

$$\alpha_i(t) = 2\omega_{osc}(t)(\tau_0 + \tau_{p,i} + \Delta\tau_{p,i}(t)) \tag{6.8}$$

where τ_0 and $\tau_{p,i}$ are the time delays within the radar system and between the system and the ith target, respectively, and where the target may be a human subject or a stationary object. Additionally, $\Delta\tau_{p,i}(t)$ represents the delay fluctuation that is caused by the cardiopulmonary movement of the ith target who is a human subject, and equals zero if the target is a stationary object. Frequency-demodulating the VCO output signal yields the following baseband signal.

$$V_{bb}(t) = G_c V_{osc}\tau_d(\omega_{out}(t) - \omega_{osc}) = G_c V_{osc}\tau_d\left(2\pi K_v V_t(t) - \sum_i \omega_{LR,i}\sin\alpha_i(t)\right) \tag{6.9}$$

where G_c and τ_d represent the conversion loss of the mixer and the delay time of the delay line, respectively, in the frequency demodulator.

As displayed in Fig. 6.9, the linear ramp waveform applied to the VCO has a period of T. The resultant baseband signal is processed by subtracting from it the initial period signal, which is approximated for small values of $\Delta\tau_{p,i}(t)$ as

$$\Delta V_m(n, t) = V_{bb}(t + nT) - V_{bb}(t) \approx -2G_c V_{osc}\omega_{osc}(t)\tau_d \sum_i \omega_{LR,i}\cos(2\omega_{osc}(t)(\tau_0 + \tau_{p,i}))$$
$$\times (\Delta\tau_{p,i}(t + nT) - \Delta\tau_{p,i}(t))$$

$$(6.10)$$

Since the period of the ramp signal T is significantly shorter than the period of the vital signs, $\Delta\tau_{p,i}(t)$ can be treated as a constant within each period T. Based on this approximation, the Fourier transform of $\Delta V_m(n,t)$ yields the following FMCW spectra:

$$\Delta V_m(n, \omega) = F(w(t)\Delta V_m(n, t)) \approx -G_c V_{osc}\omega_{osc}\tau_d \sum_i \omega_{LR,i}$$
$$\times (W(\omega - \omega_{s,i})e^{j\omega_{osc}\tau_{s,i}} + W(\omega + \omega_{s,i})e^{-j\omega_{osc}\tau_{s,i}}) \cdot (\Delta\tau_{p,i}(nT) - \Delta\tau_{p,i}(0))$$

$$(6.11)$$

where $w(t)$ and $W(\omega)$ represent the Fourier transform pair of a rectangular window function, $\omega_{s,i}$, which is the center frequency of each tone in the FMCW spectrum, equals $2\pi S_R\tau_{s,i}$, S_R is the sweep rate (rate of change of frequency per second), and $\tau_{s,i}$ equals $2\tau_0 + 2\tau_{p,i}$. Therefore, the distance to the targets within the sensing range can be calculated using

$$R_{p,i} = c\tau_{p,i} = c\left(\frac{\omega_{s,i}}{4\pi S_R} - \tau_0\right).$$

$$(6.12)$$

Moreover, according to Eq. (6.11), the vital sign signals of a human subject can be found from the voltage variation of the corresponding tone spectra in discrete time.

Secondly, the radar system is switched to the CW detection process by setting $V_t(t)$ to a constant voltage V_{t0} and $V_s(t)$ to a linear ramp waveform. This ramp waveform enters the voltage-controllable phase shifter during duration t_2 to enable scanning of the sum and difference beam patterns. Similarly, in this process, the signals that are reflected from the targets are injected into the VCO, causing the VCO to enter an SIL state. The resultant baseband signal that is obtained by frequency-demodulating the VCO output signal is expressed as

$$V_{bb}(t) = G_c V_{osc}\tau_d\left(2\pi K_v V_{t0} - \sum_i \omega_{LR,i}\sin\alpha_i(t)\right) = V_{dc} + \sum_i \Delta V_{m,i}(t)$$

$$(6.13)$$

where

$$V_{dc} \approx G_c V_{osc}\tau_d\left[2\pi K_v V_{t0} - \sum_i \omega_{LR,i}\sin(2\omega'_{osc}(\tau_0 + \tau_{p,i}))\right]$$

$$(6.14)$$

$$\Delta V_{m,i}(t) = -2G_c V_{osc}\omega'_{osc}\tau_d\omega_{LR,i}\cos(2\omega'_{osc}(\tau_0 + \tau_{p,i})) \cdot \Delta\tau_{p,i}(t)$$

$$(6.15)$$

where $\omega'_{osc} = \omega_{osc} + 2\pi K_v V_{t0}$. The DC level V_{dc}, given by Eq. (6.14), is the so-called DC offset, which commonly degrades the sensitivity of detection of a vital sign signal that is given by Eq. (6.15). Therefore, V_{t0} can be tuned to eliminate the dc offset according to Eq. (6.14). Following subtraction of the initial background, the vital sign signal can be written in the discrete-time domain as

$$\Delta V_{\mathrm{m}}(n) = \sum_i (\Delta V_{\mathrm{m},i}(nT) - \Delta V_{\mathrm{m},i}(0))$$
$$= -2G_{\mathrm{c}} V_{\mathrm{osc}} \omega'_{\mathrm{osc}} \tau_{\mathrm{d}} \sum_i \omega_{\mathrm{LR},i} \cos(2\omega'_{\mathrm{osc}}(\tau_0 + \tau_{\mathrm{p},i})) \cdot (\Delta \tau_{\mathrm{p},i}(nT) - \Delta \tau_{\mathrm{p},i}(0)) \tag{6.16}$$

As shown in Fig. 6.10, the radar system utilizes one Tx and two Rx antennas that are separated by a distance d. All of the elements are horn antennas with aperture width a and height b. The phase shifter in the system provides a voltage-controllable phase shift ξ between two Rx antennas. The Rx antenna array pattern can thus be written as

$$E(\phi, \xi) = F(\phi)A(\phi, \xi) \tag{6.17}$$

where the element factor $F(\phi)$, which is the horn antenna pattern function in this case, is given by [33]

$$F(\phi) = \frac{\pi^2}{2}(1 + \sin\phi)\frac{\cos\left(\frac{a \cdot \omega'_{\mathrm{osc}}}{2c}\cos\phi\right)}{\pi^2 - 4\left(\frac{a \cdot \omega'_{\mathrm{osc}}}{2c}\cos\phi\right)^2} \tag{6.18}$$

and the array factor $A(\phi, \xi)$ in Eq. (6.17) is

$$A(\phi, \xi) = \cos\left[\frac{1}{2}\left(\frac{\omega'_{\mathrm{osc}}d}{c}\cos\phi + \xi\right)\right] \tag{6.19}$$

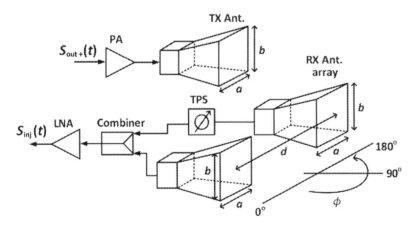

FIGURE 6.10

Tx and Rx antennas in the SIL FMCW radar system.

The tracking pattern $T(\phi_0,\xi)$ for a target that is located at an azimuth angle of ϕ_0 is defined as the ratio of the sum pattern to difference pattern as the phase shift ξ is varied at the azimuth angle of $\phi = \phi_0$. The pattern thus obtained is

$$T(\phi_0, \xi) = \frac{E(\phi_0, \xi)}{E(\phi_0, \xi + 180 \text{ degrees})} = \cot\left[\frac{1}{2}\left(\frac{\omega'_{\text{osc}}d}{c}\cos\phi_0 + \xi\right)\right] \qquad (6.20)$$

The tracking azimuth angle ϕ_t, at which the tracking pattern has an infinite peak amplitude, is found using

$$\frac{\omega'_{\text{osc}}d}{c}\cos\phi_t + \xi = 0 \qquad (6.21)$$

Substituting Eq. (6.21) into Eq. (6.20) yields

$$T(\phi_0, \phi_t) = \cot\left[\frac{\omega'_{\text{osc}}d}{2c}(\cos\phi_0 - \cos\phi_t)\right] \qquad (6.22)$$

and the output baseband signal in the CW detection process has an amplitude that is proportional to $|T(\phi_0,\phi_t)|$.

In the experiment, two subjects are seated between two wooden walls that are separated by 5.7 m and are surrounded by various objects that generate stationary clutter, as shown in Fig. 6.11. The azimuth angles of Subjects 1 and 2 are 65 and 95 degrees, respectively. In the FMCW detection process, the VCO frequency of the SIL radar is swept from 2.4 to 2.7 GHz at a rate of 30 MHz/ms, and in the CW detection process, the tracking azimuth angle is scanned from 55 to 125

FIGURE 6.11

Experimental scene for the through-wall life detection using an SIL FMCW radar.

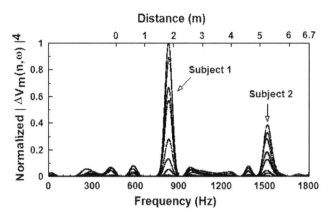

FIGURE 6.12

Normalized fourth power of the initial background-subtracted FMCW spectra.

Copyright © [2013] IEEE. Reprinted, with permission, from Wang F-K, Horng T-S, Peng K-C, Jau J-K, Li J-Y, Chen C-C. Detection of concealed individuals based on their vital signs by using a see-through-wall imaging system with a self-injection-locked radar, IEEE Trans Microw Theory Techn 2013; 61(1): 696–704.

degrees. The durations t_1 and t_2 are both set to 0.01 seconds and the period T is chosen to be 0.1 seconds.

Fig. 6.12 presents the normalized fourth power of the initial background-subtracted FMCW spectra for the experimental scene shown in Fig. 6.11. Clearly, the largest two peaks occur at frequencies that represent the distances of Subjects 1 and 2 from the radar because they are mainly caused by the subjects' cardiopulmonary movements in a 1s measurement time. Fig. 6.13(A) and (B) displays the radar-to-subject distance fluctuations that are converted from the frequency variations of the two peaks over a period of 30 seconds and the results of the Fourier transform for Subjects 1 and 2, respectively. It can be estimated from the spectra shown in these two figures that the movements associated with respiration and heartbeat are within 7.132 mm and 1.324 mm, respectively, for Subject 1, and within 8.47 mm and 1.766 mm, respectively, for Subject 2. The spectra also demonstrate a breathing rate of 18 breaths/minutes and a heart rate of 76 beats/minutes for Subject 1, and a breathing rate of 17 breaths/minutes and a heart rate of 72 beats/minutes for Subject 2.

Fig. 6.14(A) presents the normalized tracking pattern for the same scene as in Fig. 6.11. Since the antenna beam is not narrow enough to distinguish between the echo signals that originate from the two subjects, the received signal at various tracking azimuth angles contains the Doppler shifts produced by the two subjects. In Fig. 6.14(A), the tracking pattern is dominated by the echo signal from Subject 1, who is closer to the radar system, at an azimuth angle of 65 degrees, so identifying the other subject at an azimuth angle of 95 degrees is difficult. To solve this problem, a time-domain correction method is used to decouple the

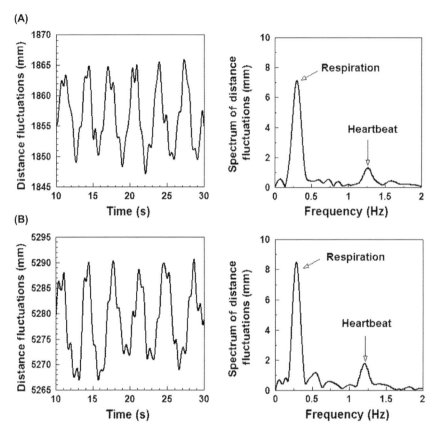

FIGURE 6.13

Radar-to-subject distance fluctuations due to the subjects' cardiopulmonary movements and the results of the Fourier transform. (A) The case of Subject 1. (B) The case of Subject 2.

contributions of the two subjects to the tracking pattern. In the CW detection process, the vital sign signals that are detected using sum and difference patterns can be described by

$$V_{\text{sum}}(\xi, t) = E(\phi_{01}, \xi)V_{\text{vs},1}(t) + E(\phi_{02}, \xi)V_{\text{vs},2}(t) \tag{6.23}$$

$$V_{\text{diff}}(\xi, t) = E(\phi_{01}, \xi + 180 \text{ degrees})V_{\text{vs},1}(t) + E(\phi_{02}, \xi + 180 \text{ degrees})V_{\text{vs},2}(t) \tag{6.24}$$

where $V_{\text{vs},1}(t)$ and $V_{\text{vs},2}(t)$ represent the vital sign signals of Subjects 1 and 2 at azimuth angles $\phi_{01} = 65$ degrees and $\phi_{02} = 95$ degrees, respectively. In

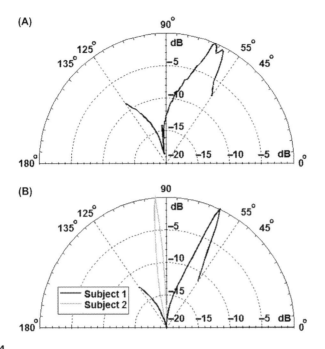

FIGURE 6.14

Tracking patterns for determining the azimuth angles of the hidden subjects.
(A) Uncorrected (coupled) pattern. (B) Corrected (decoupled) pattern.

Eqs. (6.23) and (6.24), $E(\phi_{01,2},\xi)$ and $E(\phi_{01,2},\xi+180$ degrees) denote the scanning sum and difference patterns obtained in the detection of these two subjects. Since $V_{vs,1}(t)$ and $V_{vs,2}(t)$ in Eqs. (6.23) and (6.24) are obtained as known signals of vital signs in the FMCW detection process, such as those shown in Fig. 6.13(A) and (B), Eqs. (6.23) and (6.24) can be solved for $E(\phi_{01,2},\xi)$ and $E(\phi_{01,2},\xi+180$ degrees) by sampling at two different times. Thereafter, the tracking patterns associated with these two subjects, $T(\phi_{01},\phi_t)$ and $T(\phi_{02},\phi_t)$, can be found as in Fig. 6.14(B), yielding a peak amplitude at azimuth angles of 65 degree and 95 degree, respectively. These two angles of peak amplitude agree with the azimuth of the two subjects.

The through-wall image for locating the subjects can be constructed from Figs. 6.12 and 6.14(B), which present the range and azimuth information, respectively, using the following equation.

$$I_{TW}(R,\phi_t) = \left|\Delta V_m(n,\omega(R))\right|^4 \cdot \left|T(\phi_0,\phi_t)\right|. \tag{6.25}$$

Fig. 6.15 displays the constructed through-wall image, in which the light-colored spots closely correspond to the positions of the seated subjects in Fig. 6.11. In this figure, the spot of Subject 1 has a greater brightness than that of

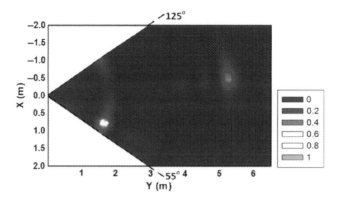

FIGURE 6.15

Through-wall image for localizing the hidden subjects.

Copyright © [2013] IEEE. Reprinted, with permission, from Wang F-K, Horng T-S, Peng K-C, Jau J-K, Li J-Y, Chen C-C. Detection of concealed individuals based on their vital signs by using a see-through-wall imaging system with a self-injection-locked radar, IEEE Trans Microw Theory Techn 2013; 61(1): 696–704.

Subject 2 because the former is closer to the radar system and so results in a stronger echo signal. Notably, the brightness of the spots varies with time according to the cardiopulmonary movements produced by the subjects.

6.2.3 **CHEST-WORN HEALTH MONITOR**

Wearable devices have become very popular for healthcare purposes owing to their effectiveness in the long-term monitoring of human posture and physiological parameters. However, most currently available devices integrate a wireless device with various sensors such as an accelerometer, an oxygen saturation meter and an electrocardiogram monitor [34], resulting in high complexity and high power consumption. Furthermore, some sensors must be in contact with the skin, so they cannot be worn on clothes. Doppler radars have great potential as wearable health monitoring devices because they use radio waves that easily penetrate clothes. The two major classes of Doppler radars are pulse radar and CW radar. A pulse radar transmits a stream of short RF pulses and determines the motion of a target by measuring the delay before the pulses that are reflected from the target are received in relation to their round-trip propagation time. However, a pulse radar cannot receive while it transmits, creating a blind interval that limits the short-range detection of Doppler signals and causes difficulties in its use in wearable monitors. A CW radar continuously transmits an RF signal, and its receiver is always on to detect the reflected signal. Therefore, it has no blind interval. However, its primary shortcoming is its need to isolate the transmitted from the received signals using either a circulator or separate Tx and Rx antennas. Hence, reducing the size and cost of integration of CW radar for wearable applications is challenging.

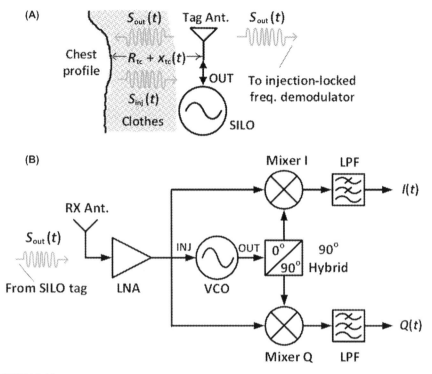

FIGURE 6.16

Block diagram of the bistatic SIL radar system for wearable health monitoring applications.
(A) SILO tag. (B) Injection-locked frequency demodulator.

Copyright © [2015] IEEE. Reprinted, with permission, from Wang F-K, Chou Y-R, Chiu Y-C, Horng T-S. Chest-worn health monitor based on a bistatic self-injection-locked radar. IEEE Trans Biomed Eng 2015; 62(12): 2931–40.

As mentioned earlier, the SIL radar can exhibit high sensitivity with low complexity, and its sensitivity is highly resistant to Tx—Rx coupling, so the radar can be operated with a single antenna. Moreover, the SIL radar can be made bistatic by spatially separating it into two parts—an SIL oscillator (SILO) and a frequency demodulator—each of which has its own antenna. These parts constitute the basic building block of a wearable monitoring system [35]. More explicitly, as shown in Fig. 6.16(A) and (B), the system has two main components: an SILO tag that is attached to the chest of the subject and an injection-locked frequency demodulator that can be simply integrated into the RF circuitry in a mobile gadget. Based on the bistatic configuration, the SIL radar can be developed as an economic low-power wearable device for tracking health and fitness. As an example, illustrated by Fig. 6.17, a chest-worn health monitor provides information concerning respiration and heart rates, exercise intensity, and the effect thereof on

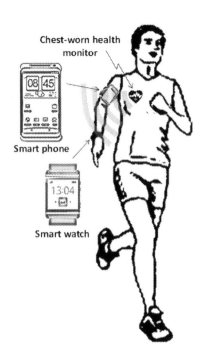

FIGURE 6.17

Chest-worn health monitor with wireless connection to mobile gadgets.

heart rate. Additionally, automatic wireless connection to nearby mobile gadgets does not excessively increase the power consumption overhead.

With reference to Fig. 6.16(A), when the SILO tag receives an echo signal $S_{inj}(t)$, the SILO outputs a signal $S_{out}(t)$ whose frequency is modulated with the Doppler phase shift in the signal $S_{inj}(t)$, according to Eqs. (6.2−6.4). Consequently, the waves that are emitted and received by the antenna of the SILO tag contain the Doppler modulation. Therefore, the injection-locked frequency demodulator can wirelessly receive signal $S_{out}(t)$ via its own antenna to allow demodulation of the Doppler phase shift at flexible locations. This feature of the bistatic SIL radar is unique.

In the SIL state, the output signal of the SILO tag is frequency-modulated by the time-varying phase $\alpha(t)$, as described by Eqs. (6.2) and (6.3), where $\alpha(t)$ specifies the propagation phase delay between the transmitted signal and the injected signal, as follows.

$$\alpha(t) = \frac{2\omega_{osc}}{c}(R_{tc} + x_{tc}(t)) \tag{6.26}$$

where R_{tc} represents the distance from the tag to the chest, and $x_{tc}(t)$ is the relative instantaneous displacement between the tag and the chest. When the body is stationary, $x_{tc}(t)$ is determined by the cardiopulmonary movement. However, when the body moves, a displacement must be added to $x_{tc}(t)$ to account for the effect of inertia on the SILO tag. Owing to the elasticity of clothes, the radar system acts as an accelerometer to detect this displacement as a means of measuring the acceleration of the body. Mathematically, $x_{tc}(t)$ is given by

$$x_{tc}(t) = x_v(t) + \frac{m}{k_s} a_b(t) \tag{6.27}$$

where $x_v(t)$ denotes the displacement of the chest due to respiration and heartbeat, m is the mass of the SILO tag, k_s represents the spring constant of the worn clothes, and $a_b(t)$ is the acceleration of the body. Importantly, the reflection from the tag antenna in proximity to the subject's chest causes a small constant frequency shift of the SILO.

The injection-locked frequency demodulator, displayed in Fig. 6.16(B), is constructed from an LNA, a VCO with a dedicated injection terminal, a pair of in-phase (I) and quadrature-phase (Q) mixers, and two LPFs. The Doppler-modulated signals that are emitted from the SILO tag are captured by the Rx antenna and amplified by the LNA. Then, the output signals from the LNA are fed into the RF terminals of the in-phase and quadrature (IQ) mixers and the injection terminal of the VCO, which enters an injection-locked state upon receiving the injected signals. The injection-locked VCO acts as a bandpass amplifier with a time delay τ_{IL}, and with the IQ mixers, it performs frequency demodulation using an arctangent function as follows.

$$\phi_{bb}(t) = \tan^{-1}\frac{Q(t)}{I(t)} \approx \omega_{osc}\tau_{IL} - \omega_{LR}\tau_{IL}\sin\left(\frac{2\omega_{osc}R_{tc}}{c}\right)$$
$$- \frac{2\omega_{osc}\omega_{LR}\tau_{IL}}{c}\cos\left(\frac{2\omega_{osc}R_{tc}}{c}\right)x_{tc}(t) \tag{6.28}$$

where $I(t)$ and $Q(t)$ are lowpass-filtered baseband I and Q signals, respectively. The time delay τ_{IL} is estimated as

$$\tau_{IL} = \frac{1}{\sqrt{\omega_{LR,ILO}{}^2 - \Delta\omega_{osc}{}^2}} \tag{6.29}$$

where $\omega_{LR,ILO}$ is the locking range that is associated with the injection-locked VCO, and $\Delta\omega_{osc}$ is the difference between the inherent VCO frequency and the frequency of the injection signal. Obviously, the chest motion $x_{tc}(t)$ can be extracted from the ac component in Eq. (6.28). Notably, since the VCO output signal is locked to the frequency of the signal that is emitted by the SILO tag, the problem of the aforementioned frequency shift in the SILO due to antenna reflection is solved.

In the experimental setup, an SILO tag is attached to the left chest of a subject on the front of his sleeveless shirt, as shown in Fig. 6.18. This tag, which was implemented using discrete components, delivers an RF power of -5.2 dBm with

FIGURE 6.18

Chest-worn SILO tag.

Copyright © [2015] IEEE. Reprinted, with permission, from Wang F-K, Chou Y-R, Chiu Y-C, Horng T-S. Chest-worn health monitor based on a bistatic self-injection-locked radar. IEEE Trans Biomed Eng 2015; 62(12): 2931–40.

a low-power consumption of 4.4 mW. It transmits a Doppler-modulated signal in the SIL state. An injection-locked frequency demodulator with an Rx antenna receives and demodulates this signal. The distance between the SILO tag and the injection-locked frequency demodulator is about 30 cm. A digital storage oscilloscope (DSO) is then utilized to digitize the baseband quadrature signals for further processing of the time-frequency spectrograms. The sampling rate of the DSO is set to 20 Hz. Herein, four experiments were conducted, as shown in Fig. 6.19(A)−(D), and are discussed as follows.

6.2.3.1 Experiment A: standing with fidgeting

In the first experiment, the respiration and heart rates of the subject were monitored as the subject stood still and breathed normally. Fig. 6.20(A) plots the arctangent demodulation data. The RMS value that is calculated from the results in Fig. 6.20(A) is 4.4 degrees, and it can be used as an indicator of the intensity of physical activity. In Fig. 6.20(A), the irregular fluctuation in the period between 32 and 38 seconds arises from the fidgeting of the subject, producing some artifacts in the Fourier-transformed spectra, degrading the obtained cardiopulmonary information. To combat this effect, a time-frequency spectrogram of the arctangent output was computed using a Kaiser window with a length of 400 points, a sampling rate of 20 Hz over an interval of 20 seconds, and a side lobe attenuation of 40 dB. Notably, the window length was chosen as a trade-off between frequency and time resolution. The window overlap was set to 398 points to ensure the smoothness of the frequency lines in the spectrogram.

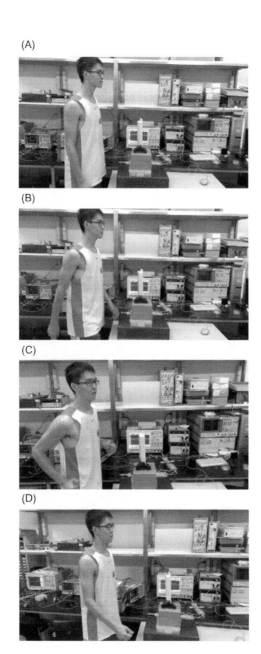

FIGURE 6.19

Experiments for a subject using a chest-worn health monitor. (A) Experiment A: standing with fidgeting. (B) Experiment B: walking with deep breathing. (C) Experiment C: jogging with rapid breathing; and (D) Experiment D: walking with gasping.

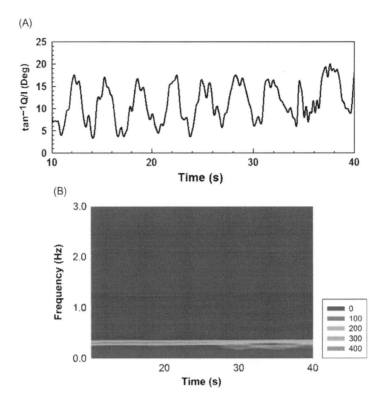

FIGURE 6.20

Monitoring results of Experiment A. (A) Arctangent demodulation data. (B) STFT-based spectrogram.

Copyright © [2015] IEEE. Reprinted, with permission, from Wang F-K, Chou Y-R, Chiu Y-C, Horng T-S. Chest-worn health monitor based on a bistatic self-injection-locked radar. IEEE Trans Biomed Eng 2015; 62(12): 2931–40.

Fig. 6.20(B) shows the spectrogram that is obtained from Fig. 6.20(A) using the Short-Time Fourier Transformation (STFT) with the DC offset removed. The recognizable frequency lines from bottom to top are associated with the fundamental (0.31 Hz), second harmonic (0.62 Hz), and third harmonic (0.93 Hz) of the respiration signal, and the heartbeat signal (1.25 Hz). In Fig. 6.20(B), the fundamental respiration frequency line is diffuse between 28 and 40 seconds because of the artifacts caused by fidgeting. A larger window overlap was empirically observed to yield a longer period of diffusion.

6.2.3.2 Experiment B: walking with deep breathing
The second experiment concerns monitoring the health of the subject when walking in place and breathing deeply. Fig. 6.21(A) plots the result of the arctangent

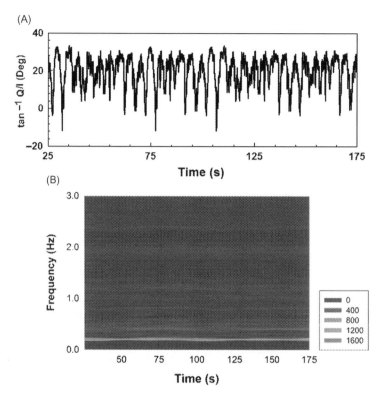

FIGURE 6.21

Monitoring results of Experiment B. (A) Arctangent demodulation data. (B) STFT-based spectrogram.

Copyright © [2015] IEEE. Reprinted, with permission, from Wang F-K, Chou Y-R, Chiu Y-C, Horng T-S. Chest-worn health monitor based on a bistatic self-injection-locked radar. IEEE Trans Biomed Eng 2015; 62(12): 2931–40.

demodulation. The obtained RMS value is 8.4 degrees. To find more distinguishable frequencies that are related to the physiological and physical activities of an exercising subject, the window and overlap lengths are changed to 1000 and 998 points, respectively, to increase the frequency resolution at the cost of reduced time resolution of the spectrograms. The same window conditions are applied in the next two experiments in which the same subject performs different exercises.

In the spectrogram in Fig. 6.21(B), the fundamental and harmonic respiration frequency lines, up to the fifth harmonic, can be seen clearly at up to five multiples of 0.21 Hz because the subject took deep breaths while walking. In addition to the fundamental and harmonic respiration frequency lines, the heartbeat frequency line is observed at 1.39 Hz. The frequency lines at 1.19 and 2.38 Hz correspond to half of the frequency and the full frequency of the signal that is

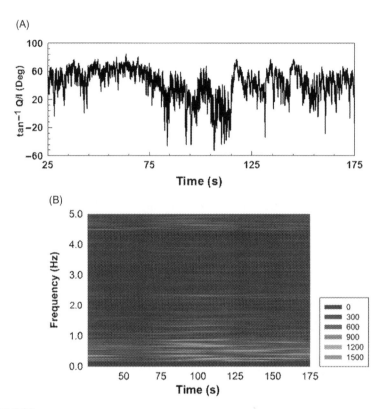

FIGURE 6.22

Monitoring results of Experiment C. (A) Arctangent demodulation data. (B) STFT-based spectrogram.

Copyright © [2015] IEEE. Reprinted, with permission, from Wang F-K, Chou Y-R, Chiu Y-C, Horng T-S. Chest-worn health monitor based on a bistatic self-injection-locked radar. IEEE Trans Biomed Eng 2015; 62(12): 2931–40.

associated with walking. Since walking causes one side of the chest to move forward and then backward as each pair of consecutive steps is taken, and the other side of the chest to move in the opposite direction, the half and full frequencies that are associated with walking are the fundamental and second harmonic frequencies of this periodic chest movement.

6.2.3.3 Experiment C: jogging with rapid breathing

In the third experiment, health monitoring was carried out while the subject was jogging in place and breathing rapidly. The RMS value of the arctangent demodulation data in Fig. 6.22(A) is 20.1 degrees, which markedly exceeds those in the previous two experiments because the intensity of motion is greater in this experiment. In Fig. 6.22(A), the waveform changes very rapidly from 75 to 125 seconds because the subject stumbles in this period. Fig. 6.22(B) presents the spectrogram that is obtained from Fig. 6.22(A), and some frequency artifacts are observed in the period when the subject stumbled.

In this experiment, the fundamental respiration and full jogging rate signals are identified by their having the largest signal amplitude in the frequency range 0.1−0.75 Hz and 1.75−5 Hz, respectively; their frequencies are 0.35 and 4.63 Hz, respectively. Since the frequency artifacts that are caused by the stumble appear only briefly, the heartbeat signal is identified as associated with a time-continuous 1.54 Hz frequency line in the range 0.75−1.75 Hz.

6.2.3.4 Experiment D: walking with gasping

In this experiment, the subject walked in place to cool down after the jogging exercise. Fig. 6.23(A) plots the result of arctangent demodulation. The estimated RMS value is 5.5 degrees, which is close to that calculated in Experiment *B* because both experiments involved walking. Fig. 6.23(B) shows the STFT-based spectrogram that is obtained from Fig. 6.23(A), which is the most perturbed spectrogram of any obtained in these experiments because of the gasping. Abnormally rapid respiration was observed during this experiment. The gasping caused the higher respiration harmonics to interfere significantly with the heartbeat frequency line. Moreover, half of the walking frequency is quite close to the heartbeat frequency, increasing the difficulty of identifying the heartbeat frequency line in Fig. 6.23(B).

Close inspection of Fig. 6.23(B) reveals that the fundamental respiration frequency line starts at 0.43 Hz and the full walking-rate frequency line starts at 1.86 Hz. An examination of Fig. 6.23(B) for the heartbeat frequency within the frequency range 0.75−1.75 Hz yields four candidate starting frequencies: 0.93 Hz, 1.29 Hz, 1.4 Hz, and 1.5 Hz. The frequencies 0.93 Hz and 1.29 Hz are excluded because the former is half of the walking frequency and the latter is the third harmonic frequency of respiration. The frequency 1.5 Hz is also eliminated because it is very close to the heartbeat frequency that was observed in the jogging exercise in Experiment *C*. Hence, from Fig. 6.23(B), the heartbeat frequency is inferred to decline from 1.4 to 1.3 Hz during the period of measurement because the subject is engaged in a cooling-down exercise.

Table 6.1 summarizes the rates of respiration, heartbeat, and walking or jogging that were monitored in the above experiments. The heart rate is the most important information but also the most difficult to obtain. However, since the heart rate depends on the intensity of exercise, as indicated by the RMS value of the arctangent demodulation data, it can be tracked accordingly using machine learning methods, which will be available in the near future.

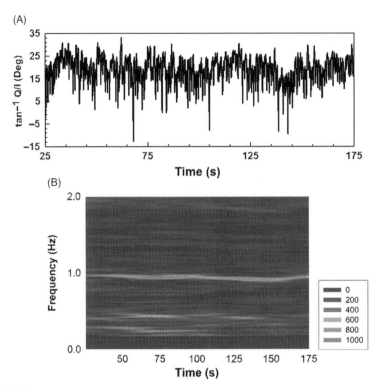

FIGURE 6.23

Monitoring results of Experiment D. (A) Arctangent demodulation data. (B) STFT-based spectrogram.

Copyright © [2015] IEEE. Reprinted, with permission, from Wang F-K, Chou Y-R, Chiu Y-C, Horng T-S. Chest-worn health monitor based on a bistatic self-injection-locked radar. IEEE Trans Biomed Eng 2015; 62(12): 2931–40.

Table 6.1 Comparison of Monitoring Results Among Different Experiments

Experiment	Respiration Rate (breaths/min)	Heart Rate (beats/min)	Step Rate (steps/min)	RMS Phase Variation (degrees)
A	18.6	75	0	4.4
B	12.6	83.4	142.8	8.4
C	21	92.4	277.8	20.1
D	25.8	84–78	111.6	5.5

Copyright © [2015] IEEE. Reprinted, with permission, from Wang F-K, Chou Y-R, Chiu Y-C, Horng T-S. Chest-worn health monitor based on a bistatic self-injection-locked radar. IEEE Trans Biomed Eng 2015; 62(12): 2931–40.

6.3 **ENABLING HEALTH MONITORING IN SMART HOMES**

6.3.1 **IN-HOME HEALTH CARE FOR THE ELDERLY**

The elderly population of individuals 65 years and older has been steadily increasing in the world. Based on the most recent revision of World Population Prospect [36], the number of persons over 65 was more than 1 billion in 2015, reaching up to 2.5 billion in 2050, and 3.6 billion in 2100. This tendency has resulted in a growing need for healthcare approaches that emphasize routine long-term monitoring in the home environment. As nursing homes are expensive and limited in number, there is the incentive and also the wish to stay longer at home. This creates a health risk, especially when the aged persons live alone. Caregivers, furthermore, are often family members or others with minimal specialized training who need professional support to be able to understand and address the needs of the elderly person as well as to better deal with their own problems arising from the stressful task (e.g., depression, sleeplessness, back pain, and other disorders).

In addition to health problems, fall incidents and sustained injuries represent the most dangerous causes of accidents among elderly people [37]. Research pointed out that 30%–45% of the persons older than 65 years fall at least once a year. Approximately one in three to one in two fall several times a year [38]. These fall incidents cause severe injuries in 10%–15% of cases. People who experience a fall event at home and remain on the ground for an hour or more may suffer from many medical complications, such as dehydration, internal bleeding, and cooling, and half of them die within 6 months [39]. The delay in hospitalization increases mortality risk. Studies have shown that the longer the persons lie on the floor, the poorer the outcome of medical intervention is [40,41]. This poses an acute risk to older persons who live alone and may be unable to ask for assistance after a fall occurs. Moreover, a fall not only causes physical injuries such as disabling fractures, but it also has dramatic psychological consequences that reduce elderly people's independence. To address the problem of medical intervention delay, it is imperative to detect the falls as soon as they occur such that immediate assistance may be provided. For those reasons, fall detection is becoming an increasingly important topic attracting the attention of many researchers and companies worldwide.

The first systems and approaches for fall detection were based on devices attached to the patient's body to be used in emergency or dangerous situations. The devices are based on accelerometers and tilt sensor attached to the body, starting in 1991 with Lord and Colvin [42], followed by Williams [43] in 1998, using a belt device which detected the impact of the shock on the ground, and a mercury tilt switch to detect when the person was lying. Depeursinge [44] used three accelerometers to detect the position, speed, and acceleration of the person. Zhang [45] placed a tri-axial accelerometer in a mobile phone to monitor daily activity and fall events. Other systems are based on a combination of a tri-axial accelerometer and gyroscope [46,47]. However, these devices produce discomfort and many false alarms, and older people tend to forget to wear them. Other approaches involve either a wristwatch or a necklace with a button that is activated by the person in case of a fall event. The main problem is that in an emergency situation, the person may no longer have the reflex to press this button or may no longer be able to do so.

In recent years, remote fall detection has become a primary interest in connection to health monitoring both in home and clinical environments. The first systems under investigation were based on video cameras [48,49], floor vibration [50], and acoustic sensors [51]. For the video camera method, people are currently trying to address challenges related to low light, field of view, and image processing, as the number of false positives is still high, but privacy is also a concern. Floor vibration and acoustic sensors have limited success due to environmental interference and background noise. Moreover, they are not effective in case of a "soft" human fall in which the individual collides with an object (table, dresser, chair, carpet, etc.). Due to the disadvantages of existing fall detection technologies, which limit their use under real and practical conditions, there is a need for further solutions. An alternative and promising approach based on radar techniques has been demonstrated in [52−60].

Section 6.3 focuses on a wireless radar-based sensor network (WRSN) architecture, developed at KU Leuven, Belgium, which is able to detect fall events in real-time and to localize a person without the need of any radio-frequency identification tag attached to the person [52−63]. Hardware implementation, data processing techniques, and experimental results are presented and discussed.

6.3.2 WIRELESS RADAR SENSOR NETWORK ARCHITECTURE

An indoor WRSN consists of spatially distributed autonomous radar sensors to monitor physical and physiological conditions which pass cooperatively their data through a network to a physician or caregiver, as shown in Fig. 6.24. The radar sensor is used to transmit an RF signal to a target and to receive the reflected echo, on the basis of which the target's speed, vital signs, and absolute distance can be extracted. The received baseband data is then transmitted wirelessly to the base station for data processing. The latter, in its turn, can be connected to the caregiver by either a wireline (e.g., fixed phone, cable internet connection) or wireless technology. The radar sensor nodes are the central element of the WRSN. The proper number and the optimal physical positioning of the sensors depend on the size and geometry of the room.

WRSNs are ideal for implementing at-home healthcare solutions. Assisted living technologies not only benefit the elderly but also the clinicians and caregivers. WRSNs can in fact detect small changes in vital signs that are not obvious in a one-off visit of a doctor. Measurements can be made unobtrusively over an extended period in a supportive home environment, offering excellent long-term care benefits. WRSNs in fact can detect and respond to emergency situations such as health crises or fall incidents.

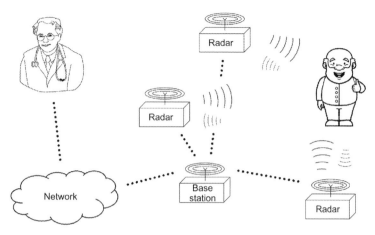

FIGURE 6.24

Smart WRSN for long-term health monitoring.

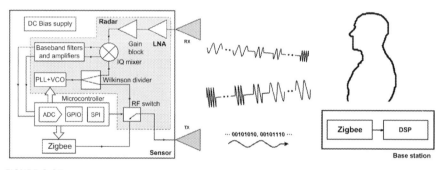

FIGURE 6.25

Block diagram of the WRSN. For better visibility only one sensor is shown.

Copyright © [2015] IEEE. Reprinted, with permission, from Mercuri M, Rajabi M, Karsmakers P, Soh PJ,
Vanrumste B, Leroux P, et al. Dual-mode wireless sensor network for real-time contactless in-door health
monitoring. In: IEEE MTT-S International Microwave Symposium Digest, Phoenix, AZ, USA, May 17–22,
2015, pp. 1–4.

Fig. 6.25 shows the block diagram of a WRSN aiming at real-time fall detection and tagless localization for a room environment [16,58,60]. The approach is scalable to multiple rooms. The WRSN consists of multiple sensors that combine radar, computational, and wireless communication capabilities, and a base station for real-time data processing. Each radar generates and transmits continuously a waveform $T(t)$ toward a target, whose reflected echo $R(t)$, containing speed and absolute distance information, is collected by the receiver. After ADC, the resulting $I(t)$ and $Q(t)$ baseband signals are transmitted wirelessly using Zigbee to the base station, consisting in turn of a Zigbee transceiver and a digital signal processor (DSP), to determine remotely the target's absolute distance and to distinguish a fall event from normal movements (e.g., walking, sitting down, standing up, random movements, no movements, daily activities).

Indoor health monitoring using radar sensors manifests limitations. Depending on the position of a person in a room, his or her reflection as consequence of radar excitation may be obstructed by furniture. Moreover, in the case of fall detection, it is fundamental to assess the changes in speed as experienced by a subject during normal movements or fall incidents. Due to the Doppler effect, it is not possible to evaluate correctly the speed of a target falling/moving perpendicular to the line of sight (LoS) of the antenna. For that reason, the target speed will produce a lower Doppler frequency, and a fall event will be classified as normal movement. These problems can be overcome by means of the WRSN. In fact, by combining data from several sensors, a better estimate of the motion is obtained. Moreover, a single radar can detect absolute distance, so by using multiple sensors it is possible to detect position. These effects combined allow tracking the person as well as detecting fall incidents in all directions.

6.3.3 HARDWARE DESIGN

6.3.3.1 Radar waveform

The radar waveform is based on a hybrid approach [16,61], by which a single tone is alternated with an SFCW waveform (Fig. 6.26). It has been designed to continuously monitor the speed of the target as well as to determine its absolute distance every 2 seconds. More precisely, in order to detect the target's speed, a single frequency signal in the ISM radio band of 5.8 GHz is employed to exploit the Doppler effect. This signal is alternated every two seconds with a stepped frequency waveform working in the UWB band, used to detect the target's absolute distance.

The latter waveform consists of $N = 40$ coherent CW pulses, called *burst* (a radar waveform that contains a sequence of RF pulses), whose frequencies are

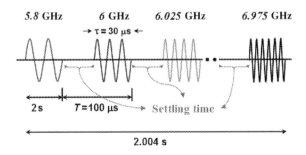

FIGURE 6.26

Radar waveform.

increased from pulse to pulse by a fixed increment Δf of 25 MHz [16]. Each pulse is $\tau = 30\,\mu s$ long, and the time interval between pulses is $T = 100\,\mu s$. This results in a burst duration NT of 2.004 ms, while its total band $N\Delta f$ is 1 GHz positioned between 6 and 7 GHz. This allows an unambiguous range of 6 m with a smallest resolution of 15 cm. The full waveform is 2.004 seconds, and the signal power is 0 dBm. The interval in between CW pulses, defined as *settling time* in Fig. 6.26, is dimensioned to guarantee that a new frequency is correctly generated and ready to be transmitted.

This solution allows satisfying the European and Federal Communications Commission UWB mask requirements while simultaneously having sufficient transmit power to track a person in a typical room setting.

6.3.3.2 Radar sensor node architecture

The radar sensor node consists of three main parts, namely the radar module, the Zigbee transceiver, and the microcontroller (Fig. 6.27) [16,59]. The radar module integrates a wideband VCO with Fractional-N phase-locked loop (PLL), a 5.8−7 GHz wideband Wilkinson power divider, an LNA, three gain blocks, an IQ mixer, an RF switch, and baseband filters and amplifiers. Linear voltage regulators to support adjustable output voltages and to provide excellent power supply isolation are also included.

The microcontroller programs the synthesizer of the PLL to generate the radar waveform, acquires and digitizes the IQ baseband signals, and manages the

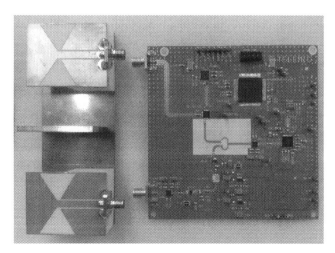

FIGURE 6.27

Radar-based sensor.

Zigbee communication. Moreover, it controls the RF switch to alternately connect the radar transmitter and the Zigbee transceiver to the transmitter antenna (labeled TX in Fig. 6.25). The latter is used by the Zigbee transceiver both to transmit and to receive frames. In fact, to synchronize the system, after power-on the sensor waits until the base station sends a command to initiate operation.

A 5.8 GHz single tone is produced at every 2-second interval, immediately followed by the SFCW waveform, as explained in Fig. 6.26. The sensor requires about 70 μs to generate a new frequency, of which the first 40 μs is to program the PLL, while the other 30 μs is to take into account the VCO maximum settling time. The combined LNA and gain block is chosen to provide a total gain of about 30 dB to avoid saturation of the gain block. This is considering that the two antennas present a cross-coupling of about −35 dB within the operation bandwidth and that the target could move very close to the antennas, producing high reflection and thus high received power. A transmit power of about 0 dBm is sent to the antenna. The bipolar IQ baseband signals are then acquired and digitized by an ADC inside the microcontroller. Since the ADC works with unipolar signals, between 0 and 3.2 V, the outputs of the mixer are bandpass filtered and amplified, and then a 1.6 V DC level is added to position the signals at the center of the ADC's dynamic range. The LPF serves both as an anti-aliasing filter and as a charge reservoir for the ADC's switched capacitor input stage. The 10-bit ADC requires about 14 μs to acquire the samples and to digitize the two IQ components before the next frequency can be generated. These acquired and digitized samples, containing the target's information, are then sent to the base station. This transmission must be properly managed while simultaneously acquiring the baseband monitoring signals.

Since the Zigbee protocol only transmits frames containing bytes, each of the acquired 10-bit samples should be ordered into two bytes. However, since the I and Q samples are acquired at the same time, they can be mapped in three bytes. Sampling the signal containing speed information at 250 Hz for a duration of 2 seconds, the sensor is expected to accumulate 1000 IQ samples. This translates into 1500 bytes. The transmission of these frames must be executed in between speed sampling instants. For that reason, the speed samples are mapped in frames of 76 bytes, 1 byte as sensor identification number (ID), 25 as I samples, and 25 as Q samples. The transmission of such frames requires about 2.6 ms each, much shorter than the duration of the acquisition period, during which the RF switch should connect the TX antenna to the Zigbee module. Each frame is therefore filled after 100 ms and then transmitted before the next sample is acquired. Regarding the SFCW waveform, it results in 80 IQ samples that are transmitted in a single frame of 121 bytes, of which one is the sensor's ID. A new 5.8 GHz tone is generated immediately after the SFCW waveform.

The radar sensor node is combined with two-element dual-band antennas [62], designed to operate at both the radar and Zigbee ISM frequency bands. The main challenge is to reduce the backscattering and crosstalk effects while presenting also a semispherical radiation pattern. A wideband circulator with one antenna

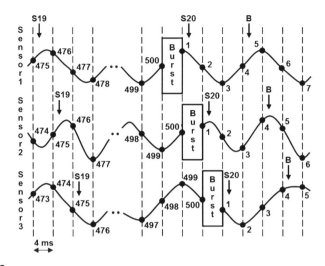

FIGURE 6.28

WRSN synchronization procedure. The solid lines and BURSTs represent the baseband monitoring signals corresponding to the single tone and SFCW signals, which are transmitted to the base station at instants *S* and *B*, respectively. The numbers indicate the *n*th transmit event during a full radar waveform, while the dots indicate the sampling instants.

could represent a more compact solution. However, although it does not experience crosstalk, any mismatch between the antenna and its feed line involves a reflection that easily overwhelms the signal reflected off the body. This effect cannot be reduced below a practical value (i.e., −35 dB) over the whole radar bandwidth by a passive microwave design, unlike what was possible regarding minimizing crosstalk in the case of two antennas [62].

6.3.3.3 WRSN implementation

In the following, it is assumed that the WRSN consists of three nodes, but the approach is scalable to a higher number of nodes. The procedure to synchronize the WRSN is illustrated in Fig. 6.28 [58]. The most important condition to verify is that the radar functionality would not interfere with the operation of the wireless communications module, and vice versa. Moreover, it is fundamental that the sensors do not interfere with each other. Therefore, a time division multiplexing (TDM) technique has been adopted. This means that for each sensor, wireless communications and radar sensing do not execute their functions at the same time. Moreover, all sensors transmit their data to the base station in different time intervals. In addition, to avoiding interference among radar signals, a frequency division multiplexing (FDM) technique has also been exploited. In the considered implementation, Sensors 1, 2, and 3 operate with the single tone at 5.775 GHz, 5.8 GHz, and 5.825 GHz, respectively. These frequency shifts are much higher than the receiver's bandwidth of 100 kHz. Moreover, these shifts produce negligible differences in the resulting Doppler frequencies and consequently in the speed signals, implying that the same processing model can be used for all the sensors. It should be specified that the Zigbee communication does not produce interference with the radar sending since it operates at 2.45 GHz. Moreover, the adopted TDM and FDM procedures ensure also that the bursts corresponding to the SFCW acquired data are transmitted in different time slots and that they do not interfere with the alternating single tone signals. In order to simplify the synchronization, a delay of 4 ms has been inserted among the sensors. However, due to the low human speeds, the target does not significantly change in movement over a time period of 4 ms such that it is possible to assume that the sensors virtually operate at the same time. After power on, the sensors wait until the base station sends a command to initiate operation. Since they are in the same room, one can assume that the command is received by all the sensors at the same instant. After that, Sensor 1 starts to operate, while Sensor 2 and Sensor 3 start to operate after 4 ms and 8 ms, respectively.

The base station consists of a Zigbee transceiver and a DSP [56]. It processes in real-time the digitized IQ radar baseband signals. Each time a radar sensor transmits frames, the Zigbee chip in the base station generates an interrupt that is read by the DSP. The latter collects the frames, sorts them for sensor's ID, split them according to the category to be processed separately, namely whether the samples serve for speed monitoring or for absolute distance detection, and finally applies the data processing techniques described in the next paragraph. The base station receives three Zigbee speed frames from the three sensors every 100 ms, which are separated temporally by 4 ms and are processed independently. To process a continuous flow of data, for each sensor, each incoming speed frame is concatenated with part of the previous signal to create a new complex signal window of 2 seconds with 95% of overlap. Each window is evaluated independently by a machine learning technique. On the other hand, the IQ *burst* baseband signals are transmitted in single frames to the base station, where they are combined with the other sensors' burst data to be processed in real-time using the trilateration technique.

6.3.4 SIGNAL PROCESSING

6.3.4.1 Fall detection

The primary goal of the data processing in the health monitoring system is to classify radar-based speed data into activities. More precisely, the system should discriminate fall events from other normal movements (i.e., walking, sitting down, standing up, random movement, no movements, . . .). This can be done by performing a classification using machine learning techniques.

The machine learning community exhibits different kinds of methods for supervised learning. Among the various methods, the method of least squares support vector machines (LS-SVM) combined with a global alignment kernel (LS-SVM-GA) is highlighted. This framework has been shown to give good results on a wide range of applications and has the advantage that it can be altered easily to be used on different kinds of data [55].

A conceptual overview of the data processing algorithm to distinguish fall events from normal movements is presented in Fig. 6.29. The core part is the detection model, which is a blackbox type of data-driven non-linear model that is estimated using the LS-SVM-GA method. The model aims at discriminating fall incident−related speed signals from all other captured speed signals. Discriminating facts such as, e.g., increasing speed in the case of a fall incident as opposed to a more stationary speed signal for other movements can be automatically learned by a machine learning method and consequently be described in the model. However, since the system should classify signals consisting of multiple movement activities whose start and end time instants are unknown, a sliding window approach should be introduced to process the continuous stream of radar-sensed data.

Any classification method uses a set of features or parameters to characterize each object. For this purpose, methods for supervised classification should be considered, meaning that a human expert both has determined into what classes an object may be categorized and also has provided a set of sample objects with known classes. This set of known objects is called the training set because it is used by the classification program to learn how to classify objects. There are two phases to constructing a classifier, namely the training and testing phases. In the

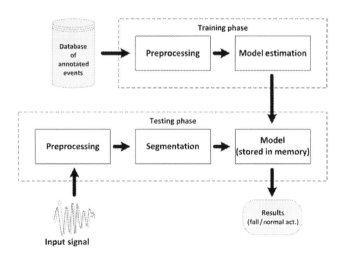

FIGURE 6.29

Fall detection data processing block diagram.

first phase, the training set is used to decide how the parameters ought to be weighted and combined in order to separate the various classes of objects. In the second phase, the weights determined in the training set are applied to a set of objects that do not have known classes in order to determine what their classes are likely to be. This is reflected in Fig. 6.29 by defining two stages, the training phase and the testing phase. Both phases use the digitized speed signals as input.

6.3.4.1.1 Training phase

In the training phase a model is estimated based on a database of annotated and segmented radar signals. This database was compiled using a set of acquired signals, containing a single activity each. All signals were cut in order to retain segments with a fixed length of 2 seconds, which is considered sufficient to cover the details of fall incidents. The positioning of the segment within the raw radar signal is determined by detecting the peak in signal energy and selecting the signal around it. Next, each of these segments are preprocessed by setting the mean to zero and computing the Fourier transform from which only the magnitude spectrum is retained. The resulting set of feature vectors per segment are used to estimate the model. This model is then stored in memory to be used in the testing phase.

6.3.4.1.2 Testing phase

In order to process a continuous stream of radar signals that might describe multiple activities invoked at unknown instants, a sliding window approach has been introduced in the testing phase. Therefore, the incoming baseband radar signal is concatenated with part of the previous signal to create signal windows of 2 seconds with 95% of overlap, which are preprocessed using the same procedure as adopted in the training phase. This corresponds to a sliding window shifted each 100 ms [56]. Setting this overlap involves making a trade-off between the amount of data needed to be processed per time unit and reducing the risk of never having the complete fall event in a single window. After inspecting the data it was seen that atomic events from 100 ms can contain meaningful information, especially in case of a fall incident where the main details are in the last 100 ms of the event. Using this relatively high overlap causes the complete fall event to be in several consecutive windows. This allows a fall detection to depend on more than just a single window, which increases the robustness of the system. For example, a fall incident is detected when the model discriminates three consecutive windows as fall related. Finally, the preprocessed data is fed into the classification model, which judges whether or not the segment is related to a fall incident.

After preprocessing, the CW radar signal windows of 2 seconds are transformed using a Fourier transform from which only the magnitude spectrum is retained. Two alternatives could be used, namely (1) to compute the Fast Fourier Transform (FFT) directly on the complete segment, and (2) to use the STFT, which works well with the LS-SVM-GA method. In the latter case, a window of 2 seconds (i.e., 500 samples) is divided into segments of 32 samples with 50%

overlap. Each segment is windowed with a Hamming window after which a 32-point FFT is computed on each of these frames. The STFT produces 30 vectors of 32 elements, from which only the magnitude spectrum is retained and only the eight most significant coefficients are considered. This means that the results of this operation consists of 30 vectors of 8 elements. On the other hand, FFT is typically used when analyzing data in the frequency domain. One fundamental assumption of FFT is that the data are stationary (or at least wide-sense stationary) with time invariant statistical proprieties. For biomedical applications, most of the signals encountered are nonstationary, and they contain high-frequency components of short bursts and low frequency content of long durations. Therefore, FFT is not sufficient and not suitable to analyze these kinds of signals, and consequently STFT was developed to examine better these kinds of nonstationary signals. A human fall typically starts with a quick movement of falling down, followed by a slow motion of lying on the floor. A CW radar captures the entire fall activity and produces similar dynamic characteristics of short duration of high frequencies and long period of low frequencies in the output. This kind of signal makes STFT a good choice for analysis and feature extractions of CW radar data. These radar signatures are then classified into two classes: fall and normal movements. Fig. 6.30(A) and (B) represents the resulting speed signals of a walking movement and of a fall event, respectively. The frequency of the signals is proportional to the radial velocity of the person during the movement. Fig. 6.30(C) and (D) shows two spectrograms corresponding to the movement activities of Fig. 6.30(A) and (B), respectively. The horizontal axis represents time, the vertical axis frequency, and a third dimension indicating the amplitude of a particular frequency at a particular time is represented by different shades of gray. In this example, a sliding window size of 64 samples with 50% overlap is adopted. In case of a fall, an increase in dominant frequency over time is observed, while for the walking activity the dominant frequency per time window remains within a small specific band.

6.3.4.2 Tagless localization

The SFCW samples are first combined to produce complex SFCW samples by which the target's range profile is determined by applying the Inverse Fast Fourier Transformer to the data [62].

However, this operation has as its main challenge the distinction of the target's reflection from the effects of backscattering and cross-coupling between the two antennas. The latter involve strong reflections that overwhelm the much weaker received signal resulting from reflection on the target, implying inability to acquire any meaningful target information. Besides, the presence of cluttering due to furniture in a practical environment must also be considered. Both factors can be eliminated by a compensation that consists of determining an environmental range profile characterizing the total contribution from both cross-coupling/ backscattering and cluttering [63]. This environmental range profile has been previously determined without any person in the room, and then stored in the base

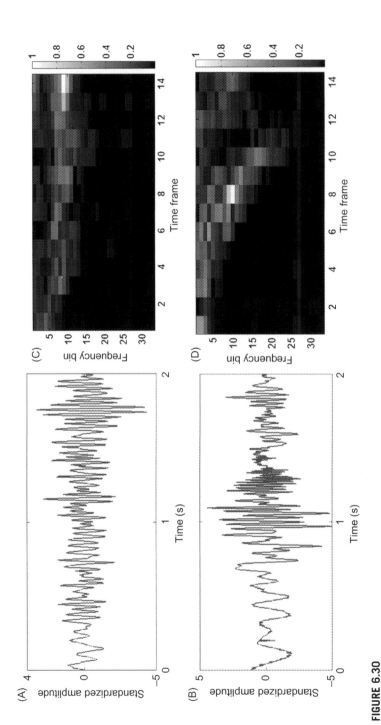

FIGURE 6.30

Speed signal during (A) A walking movement and (B) A fall event. Spectrograms corresponding to (C) The walking movement and (D) Fall event.

Copyright © [2015] IEEE. Reprinted, with permission, from Garripoli C, Mercuri M, Karsmakers P, Soh PJ, Crupi G, Vandenbosch GAE, et al. Embedded DSP based telehealth radar system for remote in-door fall detection. IEEE J Biomed Health Inform 2015; 19(1): 92–101.

station. Its magnitude is subtracted from the range profiles obtained with a target in the room. It is assumed that the environment remains static or that only objects with size smaller than a person are moved. Finally, the new range profile is shifted by a fixed value to compensate for the effect of the fixed angular phase difference between the received signal and the local copy of the transmitted waveform. This value is obtained through calibration using a flat metal plate, placing it at a well-known distance from the antennas, and then evaluating its range profile.

Finally, the target's absolute distances measured by the nodes of the WRSN, are processed by the trilateration technique to determine the target's position.

6.3.5 EXPERIMENTAL RESULTS

Experimental validations have been conducted with human subjects in labs arranged to mimic real living rooms. The scenarios include furniture, a metal shelf, a sofa, tables, PCs, and chairs positioned both around and in the center of the room.

6.3.5.1 Offline fall detection

This validation considers a single sensor fixed to the wall at 1.25 m of height. The classification model has been estimated on the basis of different types of activities, namely falling, sitting down, standing up, walking, random movements, and no movement. These activities are divided into two main groups, namely fall events and normal movements. The model has been generated using 130 signals acquired on two volunteers in different positions in the room. It includes 40 random walking activities, 30 activities of sitting down and standing up, and 40 fall activities, considering both hard falls, by which the person falls directly to the ground, and soft falls, by which the subject tries to avoid the incident by grabbing objects. In case of hard falls a mattress has been used for actor safety. In addition, 20 random movements, such as opening the window and moving a chair, have also been considered.

The estimated model was then validated on a separate validation set, stored on a computer, which contained 30 minutes of radar data, acquired on two persons, which did not appear in the training data. Each of these signals was acquired with a single volunteer in the room at a time, who was free to move with no constraints on the type of movements. The validation data contained 10 fall incidents, invoked at a random instant, and nonfall-related events (i.e., walking, sitting down, standing up, random movement, and no movement).

In order to represent the performance of the classification algorithm a confusion matrix has been set up, defined as follows:

- TP: a fall incident is correctly recognized when three subsequent fall predictions are made on a part of the signal containing the fall event. If two or

Table 6.2 Confusion Matrix of Offline Fall Detection Validation Using the HPT Method

TP	FP	TN	FN
8	415	16891	20

Table 6.3 Confusion Matrix of Offline Fall Detection Validation Using the LS-SVM-GA Method

TP	FP	TN	FN
10	0	17250	0

more detections are made on a single fall then only a single detection is accounted for;
- FP: in case three subsequent fall predictions are made on windows which do not contain fall related data;
- TN: in case three subsequent non fall predictions are made on windows which do not contain fall related data;
- FN: total number of actual falls in data minus TP.

Considering also the power-on situation, this involved 17810 windows that have been compared with the model.

As baseline detection method a High-Pass filter with Threshold (HPT) was included. In this baseline method, first the data is filtered by a high-pass filter with cut-off frequency c, then the energy of the filtered signal was compared to a threshold t. The two parameters c and t were tuned on the training set data using 10-fold cross-validation. Then the algorithm with tuned parameters was used on the validation data. The experimental results of both HPT and LS-SVM-GA are respectively presented in Tables 6.2 and 6.3. The results show that the LS-SVM-GA model was able to detect all of the 10 fall events without generating false positives. As expected, the HPT method performs less well.

Fig. 6.31 shows the classification result on a signal containing multiple activities and a fall invoked at about 42 seconds using a sliding window overlap of 95%.

6.3.5.2 Real-time fall detection

This validation considers one single sensor of the WRSN fixed to the wall at 1.25 m of height. It uses the same model described for the offline implementation. In order to validate it, 40 tests have been performed on four volunteers that have been monitored continuously for 5 minutes each. The volunteers were allowed to move without restrictions in the whole room, meaning that they could mimic all the typical movements that are normally achieved in a domestic environment

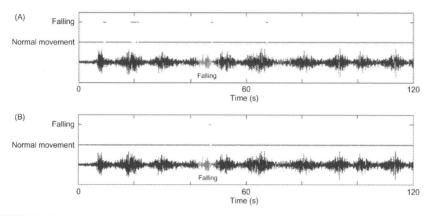

FIGURE 6.31

Example signal processed by both (A) The HPT and (B) LS-SVM-GA methods.

Table 6.4 Confusion Matrix of Real-Time Fall Detection Validation

TP	FP	TN	FN
40	0	117000	0

(i.e., working at the PC, watching a movie, resting on the sofa, walking, eating, etc.). A single volunteer was present in the room at a time and performed only one fall, either hard or soft, during the monitored period. 10 falls have been considered for each absolute distance, from 2 to 5 m, with steps of 1 m. In this validation only situations have been considered where the targets fall yielding angles between 0 and about 45 degrees with the LoS of the antenna. The subjects' weights are between 60 and 80 kg, with heights between 1.6 and 1.85 m.

Since a new window is processed each 100 ms, and considering also the power-on condition, 119240 classifications were performed in this validation. The corresponding experimental results are presented in Table 6.4. The success rate of the system in detecting fall incidents was calculated as the percentage of detected falls. The experimental evaluations have indicated that the fall detector was able to detect all the falls, with a success rate of 100%. No false positives have been reported in this test.

Fig. 6.32 shows the perfect detection of a fall. The event was classified as fall for seven consecutive windows of signal. The alarm was activated after the third window.

6.3.5.3 WRSN fall detection

The final validation on fall detection considers a four-sensor WRSN in a lab of 5×5 m^2, which contains furniture, i.e., a bed, a sofa, tables, and chairs, in order

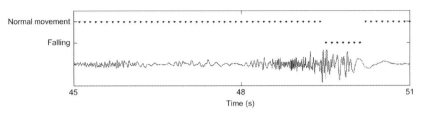

FIGURE 6.32

Real-time fall detection. Each dot indicates the class to which a window of 2 s of signal has been assigned.

to mimic a real room environment (Fig. 6.33(A)). Two nodes have been fixed to the ceiling while the other two are on the wall, whose positions are reported in Fig. 6.33(B).

Fig. 6.34 shows four signals, consisting of random normal movements and a fall event invoked at about 70 seconds, acquired at the same time by the four sensors. They are the results of monitoring a volunteer who was allowed to move without restrictions in the whole room. The signals have been processed in real-time by the base station. Since the frequency of the signal is proportional to the radial velocity of the person during the movement, each sensor node will experience a different signal. Fig. 6.34 shows that Sensors 1, 3, and 4 detect the fall incident, in contrast to Sensor 2. This shows the importance of a WRSN to solve the single-radar limitations. It should be specified that in this work, a fall alarm is triggered when at least a single node detects a fall event.

6.3.5.4 Tagless ranging

This validation considers a single sensor of the WRSN fixed to the wall at 1 m of height. In order to validate the target's absolute distance detection, three different targets, namely two persons and a metal plate, have been located at five different absolute distances from about 1 m to about 5 m, with a step of 1 m. For each absolute distance, 30 range profiles have been measured. These experimental tests have been conducted both in an anechoic chamber and in the real room environment. Since the SFCW waveform has a bandwidth of 1 GHz, the target can be located with a range resolution of 15 cm. This means that the targets' peaks are expected at multiples of 15 cm, respectively at 1.05, 1.95, 3, 4.05, and 4.95 m. This resolution is adequate for the targeted application.

Table 6.5 shows the results of this validation. They show that the percentage in localizing a target correctly decreases with distance. Moreover, they show that the target localization is more challenging in the real room than the anechoic chamber. This owes to the multiple reflections from the environment.

By reducing the crosstalk further, it will be possible to increase the total receiver gain and hence improve the tagless localization, and consequently extend

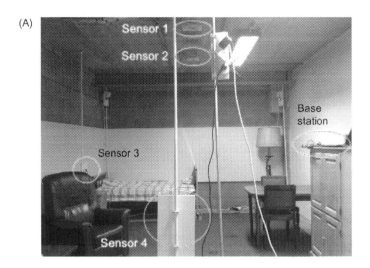

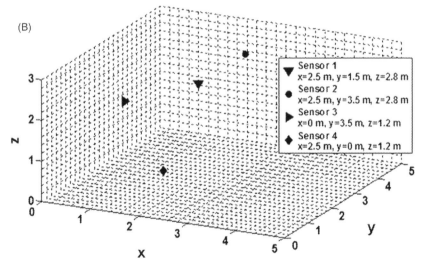

FIGURE 6.33

Four-sensor WRSN. (A) Room setting. (B) Sensors' physical positioning.

Copyright © [2016] IEEE. Reprinted with permission from Mercuri M, Karsmakers P, Vanrumste B, Leroux P,
Schreurs D. Biomedical wireless radar sensor network for indoor emergency situations detection and vital
signs monitoring. In: Proceedings of IEEE Topical Conference Biomedical Wireless Technology (BioWireleSS),
Austin, TX, USA, January 24–27, 2016, pp. 1–3.

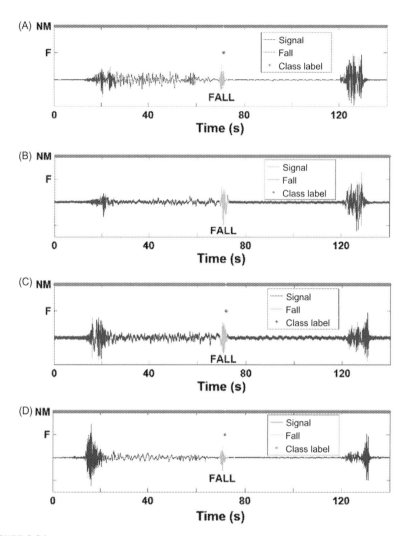

FIGURE 6.34

Real-time classification results of a signal measured, at the same time, with (A) Sensor 1. (B) Sensor 2. (C) Sensor 3; and (D) Sensor 4. Each dot represents the class where a window of 2 s of signal has been assigned. The fall is labeled as "F" while the normal movement as "NM".

Table 6.5 Percentage of Tagless Localization

Distance (m)	Anechoic Chamber			Real Room		
	% Person1	% Person2	% Metal	% Person1	% Person2	% Metal
1.05	100	100	100	100	100	100
1.95	100	100	100	100	100	100
3	93.33	96.67	100	90	86.67	100
4.05	93.33	93.33	100	90	83.33	100
4.95	86.67	90	100	76.67	73.33	100

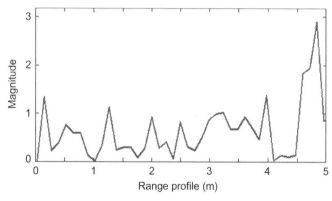

FIGURE 6.35

Range profile of a target at about 4.8 m away from the antennas.

the detection range. Fig. 6.35 shows the range profile of a person next to that of a metal shelf at about 4.8 m away from the antennas.

6.3.5.5 WRSN tagless localization

The final validation on tagless localization considers a three-sensor WRSN. The three sensors are fixed to the wall at 1 m of height, and their physical positions are indicated in Fig. 6.36 in an x versus y coordinate system. Fig. 6.36 shows the results of the 2-D tagless positioning operations with the person in a standing position. The last condition ensures that the target's body part that produces the main reflection lies on the same plane of the radar sensors (i.e., $z = 1$ m), which

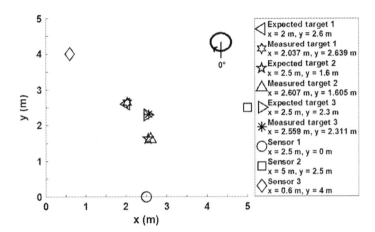

FIGURE 6.36

Radar physical positions and tagless positioning results in a 2-D plane. The figure reports both the real target's position, defined as *expected*, and the target's position determined by the trilateration technique, defined as *measured*.

Copyright © [2015] IEEE. Reprinted with permission from Mercuri M, Rajabi M, Karsmakers P, Soh PJ, Vanrumste B, Leroux P, et al. Dual-mode wireless sensor network for real-time contactless in-door health monitoring. In: IEEE MTT-S International Microwave Symposium Digest, Phoenix, AZ, USA, May 17–22, 2015, pp. 1–4.

is necessary for the 2-D geometry. The expected positions in Fig. 6.36 indicate the target's position measured by a tape ruler. On the contrary, the measured positions are the results of the trilateration with the absolute distances between the target and the respective radar sensors determined by the sensors. It can be noted from the experimental results that the errors, namely the Euclidean distances, between the expected and measured positions are within the radar resolution, which for a bandwidth of 1 GHz is 15 cm.

REFERENCES

[1] Wikipedia, Bioradiolocation. Available from: https://en.wikipedia.org/wiki/Bioradiolocation

[2] Aardal Ø, Hammerstad J. Medical radar literature overview. Norwegian Defence Research Establishment, FFI-rapport 2010/00958, 2010.

[3] Li C, Lubecke VM, Boric-Lubecke O, Lin J. A review on recent advances in Doppler radar sensors for noncontact healthcare monitoring. IEEE Trans Microw Theory Techn 2013;61(5):2046–60.

[4] Lin JC. Noninvasive microwave measurement of respiration. Proc IEEE 1975;63(10). pp. 1530–1530.

[5] Lin JC, Kiernicki J, Kiernicki M, Wollschlaeger PB. Microwave apexcardiography. IEEE Trans Microw Theory Techn 1979;27:618−20.

[6] Chen KM, Misra D, Wang H, Chuang HR, Postow E. An X-band microwave life-detection system. IEEE Trans Biomed Eng 1986;33(7):697−701.

[7] Chen K-M, Huang Y, Zhang J, Norman A. Microwave life-detection systems for searching human subjects under earthquake rubble or behind barrier. IEEE Trans Biomed Eng 2000;47:105−14.

[8] Forster H, Ipsiroglu O, Kerbl R, Paditz E. Sudden infant death and pediatric sleep disorders. Wien Klin Wochenschr 2003;115:847−9.

[9] Hafner N, Mostafanezhad I, Lubecke VM, Boric-Lubecke O, Host-Madsen A. Non-contact cardiopulmonary sensing with a baby monitor. In: 29th Annual IEEE International Engineering in Medicine and Biology Society Conference, Lyon, France, August 2007, pp. 2300−02.

[10] Li C, Lin J, Xiao Y. Robust overnight monitoring of human vital signs by a non-contact respiration and heartbeat detector. In: 28th Annual IEEE International Engineering Medicine Biology Society Conference New York, USA, August 2006, pp. 2235−38.

[11] Alekhin M, Anishchenko L, Tataraidze A, Ivashov S, Parashin V, Korostovtseva L, et al. A novel method for recognition of bioradiolocation signal breathing patterns for noncontact screening of sleep apnea syndrome. Int J Antennas Propagation 2013;2013:1−8.

[12] Massagram W, Lubecke VM, Høst-Madsen A, Boric-Lubecke O. Assessment of heart rate variability and respiratory sinus arrhythmia via Doppler radar. IEEE Trans Microw Theory Techn 2009;57:2542−9.

[13] Kiriazi JE, Boric-Lubecke O, Lubecke VM. Dual-frequency technique for assessment of cardiopulmonary effective RCS and displacement. IEEE Sens J 2012;12(3): 574−82.

[14] Gu C, Li R, Zhang H, Fung A, Torres C, Jiang S, et al. Accurate respiration measurement using DC-coupled continuous-wave radar sensor for motion-adaptive cancer radiotherapy. IEEE Trans Biomed Eng 2012;59(11):3117−23.

[15] Song C, Yazavi E, Lubecke V, Boric-Lubecke O. Smart occupancy sensors. In: Proceedings Asia-Pacific Microwave Conference, Sendai, Japan, November 2014, pp. 950−2.

[16] Mercuri M, Soh PJ, Pandey G, Karsmakers P, Vandenbosch GAE, Leroux P, et al. Analysis of an indoor biomedical radar-based system for health monitoring. IEEE Trans Microw Theory Techn 2013;61(5):2061−8.

[17] Wang F-K, Horng T-S, Peng K-C, Jau J-K, Li J-Y, Chen C-C. Detection of concealed individuals based on their vital signs by using a see-through-wall imaging system with a self-injection-locked radar. IEEE Trans Microw Theory Techn 2013;61 (1):696−704.

[18] Imnioreev I, Sarnkov S, Tao T-H. Short-distance ultra wideband radars. IEEE Aerosp Electron Syst Mag 2005;20(6):9−14.

[19] Immoreev I, Tao TH. UWB radar for patient monitoring. IEEE Aerosp Electron Syst Mag 2008;23(11):11−18.

[20] Zhu Z, Zhang X, Lv H, Lu G, Jing X, Wang J. Human-target detection and surrounding structure estimation under a simulated rubble via UWB radar. IEEE Geosci Remote Sens Lett 2013;10(2):328−31.

[21] Chuang H-R, Chen Y-F, Chen K-M. Automatic clutter-canceler for microwave life-detection systems. IEEE Trans Instrum Meas 1991;40(4):747−50.

[22] Yin Y, Qian J, Lu J, Huang Y. On the operation mechanism of the microwave sensor for measuring human heartbeats and respirations. In: Proceedings IEEE Sensors, Toronto, Canada, October 2003, pp. 22−4.

[23] Chin T-Y, Lin K-Y, Chang S-F, Chang C-C. A fast clutter cancellation method in quadrature Doppler radar for noncontact vital signal detection. In: IEEE MTT-S International Microwave Symposium Digest, Anaheim, USA, May 2010, pp. 764−7.

[24] Horng T-S. Self-injection-locked radar: an advance in continuous-wave technology for emerging radar systems. In: Proceedings Asia-Pacific Microwave Conference, November 2013, Seoul, Korea, pp. 566−9.

[25] Wang F-K, Li C-J, Hsiao C-H, Horng T-S, Lin J, Peng K-C, et al. An injection-locked detector for concurrent spectrum and vital sign sensing. In: IEEE MTT-S International Microwave Symposium Digest, Anaheim, USA, May 2010, pp. 768−71.

[26] Wang F-K, Li C-J, Hsiao C-H, Horng T-S, Lin J, Peng K-C, et al. A novel vital-sign sensor based on a self-injection-locked oscillator. IEEE Trans Microw Theory Techn 2010;58(12):4112−20.

[27] Wang F-K, Horng T-S, Peng K-C, Jau J-K, Li J-Y, Chen C-C. Single-antenna Doppler radars using self and mutual injection locking for vital sign detection with random body movement cancellation. IEEE Trans Microw Theory Techn 2011;59(12):3577−87.

[28] FMCW Radar Sensors Application Notes, Sivers IMA AB, Sweden, 2011.

[29] Downey JM. A stepped frequency continuous wave ranging sensor for aiding pedestrian inertial navigation. PhD dissertation, Department of Electrical Computer Engineering. Pittsburgh, PA: Carnegie Mellon University; 2012.

[30] Adler R. A study of locking phenomena in oscillators. Proc IRE 1946;34(6):351−7.

[31] Ralston TS, Charvat GL, Peabody JE. Real-time through-wall imaging using an ultra-wideband multiple-input multiple-output (MIMO) phased array radar system. In: IEEE International Phased Array Systems and Technology Symposium Digest, Waltham, USA, October 2010, pp. 551−8.

[32] Postolache O, Madeira RN. Microwave FMCW Doppler radar implementation for in-house pervasive health care system. In: IEEE International Workshop Medical Measurements and Application Processing, Ottawa, Canada, May 2010, pp. 47−52.

[33] Stutzman WL, Thiele GA. Aperture antennas Canada Antenna theory and design. 1sd ed. Toronto: John Wiley & Sons; 1981. p. 375−445.

[34] Chung W-Y, Lee Y-D, Jung S-J. A wireless sensor network compatible wearable U-healthcare monitoring system using integrated ECG, accelerometer and SpO2. In: Proceedings 30th Annual International Conference IEEE Engineering Medicine Biology Society, Vancouver, Canada, August. 2008, pp. 1529−32.

[35] Wang F-K, Chou Y-R, Chiu Y-C, Horng T-S. Chest-worn health monitor based on a bistatic self-injection-locked radar. IEEE Trans Biomed Eng 2015;62(12):2931−40.

[36] UN World Population Prospects: The 2015 Revision, http://esa.un.org/unpd/wpp/index.htm, Population Division of the Department of Economic and Social Affairs of the United Nations Secretariat, 2015.

[37] Murray CJL, Lopez AD. Global and regional descriptive epidemiology of disability: incidence, prevalence, health expectancies and years lived with disability. Global Burden Dis 1996;1:201−46.

[38] Dejaeger E, Boonen S, Coussement J, Milisen K. Recurrent falling, osteoporosis and sarcopenia, three major problems, an integrated approach. J Gerontol Geriatr 2009;40(6):262−9.

[39] Lord SR, Sherrington C, Menz HB. Falls in older people: risk factors and strategies for prevention. Cambridge, U.K: Cambridge University; 2007.

[40] Gurley RJ, Lum N, Sande M, Lo B, Katz MH. Persons found in their homes helpless or dead. N Engl J Med 1996(334):1710−16.

[41] Moran CG, Wenn RT, Sikand M, Taylor AM. Early mortality after hip fracture: is delay before surgery important. J Bone Jt Surg 2005;87(3):483−9.

[42] Lord CJ, Colvin DP. Falls in the elderly: detection and Assessment. Annual International Conference of IEEE Engineering Medicine Biology Society, October 31−November 3, 1991, pp. 1938−9.

[43] William G, et al. A smart fall and activity monitor to telecare applications. In: Proceedings of 20th Annual International Conference IEEE Engineering Medicine Biology Society, Hong Kong, China, October 29−November 1, 1998, pp. 1151−4.

[44] Depeursinge Y, Krauss J, El-Khoury M. Device for monitoring the activity of a person and/or detecting a fall. U.S. Patent 6 201 476 B1, March 13, 2001.

[45] Zhang T, et al. Fall detection by embedding an accelerometer in cellphone and using KFD algorithm. Int J Comput Sci Netw Secur 2006;6(10):227−84.

[46] Bourke AK, O'Brien JV, Lyons GM. Evaluation of a threshold-based tri-axial accelerometer fall detection algorithm. Gait Posture 2007;26(2):194−9.

[47] Bourke AK, Lyons GM. A threshold-based detection algorithm using a bi-axial gyroscope sensor. Med Eng Phys 2006;30(1):84−90.

[48] Lee T, Mihailidis A. An intelligent emergency response system: preliminary development and testing of automated fall detection. J Telemed Telecare 2005;11(4):194−8.

[49] Yu M, Naqvi SM, Chambers J. A robust fall detection system for elderly in a smart room. In: Proceedings International Conference Acoustical Speech Signal Process, Dallas, USA, March 2010, pp. 1666−9.

[50] Zigel Y, Litvak D, Gannot I. A method for automatic fall detection of elderly people using floor vibrations and sound − Proof of concept on human mimicking doll falls. IEEE Trans Biomed Eng 2009;56(12):2858−67.

[51] Yun L, Zhiling Z, Popescu M, Ho KC. Acoustic fall detection using a circular microphone array. In: Proceedings Annual International Conference IEEE Engineering Medicine Biology Society, September 4, 2010, pp. 2242−5.

[52] Mercuri M, Schreurs D, Leroux P. SFCW microwave radar for in-door fall detection. In: Proceedings of IEEE Topical Conference Biomedical Wireless Technology, Santa Clara, CA, USA, January 15−18, 2012, pp. 53−6.

[53] Mercuri M, Soh PJ, Boccia L, Schreurs D, Vandenbosch GAE, Leroux P, et al. Optimized SFCW radar sensor aiming at fall detection in a real room environment. In: Proceedings IEEE Topical Conference Biomedical Wireless Technology, Austin, TX, USA, January 20−23, 2013, pp. 4−6.

[54] Mercuri M, Schreurs D, Leroux P. Optimised waveform design for radar sensor aimed at contactless health monitoring. Electron Lett 2012;48(20):1255−7.

[55] Karsmakers P, Croonenborghs T, Mercuri M, Schreurs D, Leroux P. Automatic indoor fall detection based on microwave radar measurements. In: Proceedings of European Radar Conference, Amsterdam, The Netherlands, October 31–November 2, 2012, pp. 202–5.

[56] Garripoli C, Mercuri M, Karsmakers P, Soh PJ, Crupi G, Vandenbosch GAE, et al. Embedded DSP based telehealth radar system for remote in-door fall detection. IEEE J Biomed Health Inform 2015;19(1):92–101.

[57] Mercuri M, Soh PJ, Zheng P, Karsmakers P, Vandenbosch G, Leroux P, et al. Analysis of a fall detection radar place on the ceiling and wall. In: Proceedings Asia-Pacific Microwave Conference, Sendai, Japan, November 4–7, 2014. pp. 102–4.

[58] Mercuri M, Rajabi M, Karsmakers P, Soh PJ, Vanrumste B, Leroux P, et al. Dual-mode wireless sensor network for real-time contactless in-door health monitoring. In: IEEE MTT-S International Microwave Symposium Digest, Phoenix, AZ, USA, May 17–22, 2015, pp. 1–4.

[59] Mercuri M, Karsmakers P, Beyer A, Leroux P, Schreurs D. Real-time fall detection and tagless localization using radar techniques. In: IEEE Wireless and Microwave Technology Conference (WAMICON), Cocoa Beach, FL, USA, April 13–15, 2015, pp. 1–3.

[60] Mercuri M, Karsmakers P, Vanrumste B, Leroux P, Schreurs D. Biomedical wireless radar sensor network for indoor emergency situations detection and vital signs monitoring. In: Proceedings of IEEE Topical Conference Biomedical Wireless Technology (BioWireleSS), Austin, TX, USA, January 24–27, 2016, pp. 1–3.

[61] Mercuri M, Schreurs D, Leroux P. Optimised waveform design for radar sensor aimed at contactless health monitoring. Electron Lett 2012;48(20):1255–7.

[62] Soh PJ, Mercuri M, Vandenbosch GAE, Schreurs D. A dual-band unidirectional planar bowtie monopole for a compact fall-detection radar and medical telemetry system. IEEE Antennas Wireless Propagation Lett 2012;11:1698–701.

[63] Mercuri M, Soh PJ, Schreurs D, Leroux P. A practical distance measurement improvement technique for a SFCW-based health monitoring radar. In: Proceedings of 81st ARFTG Microwave Measurement Conference (ARFTG), Seattle, WA. USA, 7 June 2013, pp. 1–4.

RF/wireless indoor activity classification

7

A. Rahman[1], M. Baboli[2], O. Borić-Lubecke[1] and V. Lubecke[1]

[1]*University of Hawaii, Manoa, HI, United States*
[2]*Columbia University, NY, United States*

CHAPTER OUTLINE

7.1 INTRODUCTION

In the last decade, new front-end architectures, baseband signal processing methods, and system-level integration technologies have been proposed by many researchers in biomedical radar to improve detection accuracy and robustness. The advantages of noncontact detection have drawn interest for various applications, such as energy management in smart homes, baby monitoring, cardiopulmonary activity assessment, and tumor tracking. Advances in biomedical radar technology have also opened the door to many indoor activity monitoring applications, only limited by the imagination [1]. Beyond the measurement of heart and respiratory rates, this technology enables us to measure the effective radar cross section (RCS), yielding information about body orientation, torso displacement, and physical characteristics of human subjects [2]. The data from these

C. Li, M. Tofighi, D. Schreurs and T-Z. J. Horng (Eds): Principles and Applications of RF/Microwave
in Healthcare and Biosensing.

measurements can be analyzed to recognize particular types and qualities of activity. Sleep is widely understood to play a key part in physical and mental health. The quality and quantity of sleep that an individual experiences can have a substantial impact on learning and memory, metabolism and weight, safety, mood, cardiovascular health, disease, and immune system function. Research indicates that in America alone, 40 million people suffer from insomnia and chronic sleep disorders [3,4]. Doppler radar technology shows a promising, noninvasive way of monitoring sleep behavior with automated pattern recognition. This technology can potentially provide a huge cut in the cost of sleep studies. In addition to recognition, artificial neural network (ANN)-based automatic classifiers will be very useful in diagnosis of diseases and physiological conditions. Radar technology may also provide a noncontact, low-cost, and fast unique-identification solution without the privacy concerns associated with camera-based monitoring systems [5]. Doppler radar offers a convenient method for monitoring the activity of animals as well as the recognition of various behavioral events [6–9]. Using multiradar systems provides the ability to classify motion types such as walking, running, or fidgeting, which are important in animal activity monitoring of energetic cost [6]. It is a demographic fact that the elderly population (older than 60 years) has been gradually increasing worldwide. This change calls for greater attention to cost-effective healthcare worldwide. Since falling represents a major risk for elderly people who live independently, innovative techniques are required for long-term monitoring of human subjects [10].

7.2 FUTURE OF RADIO FREQUENCY ACTIVITY CLASSIFICATION

We are at the beginning of an era in networking that has the prospective to define a new phase of human existence. This era will be demarcated by the digitization and connection of data associated with everything and everyone with the goal of automating much of life, effectively creating time, by maximizing the efficiency of everything we do [11].

The emerging future will be characterized by big data. The new vision is that everything will be sensed and data will be turned into knowledge. Computing technology, software, and an immense amount of sensing capabilities will augment human intelligence and the ability to perform more efficiently and achieve more than people could even imagine in the past. Big enterprises have already started implementing some of these concepts.

Samsung Electronics announced Samsung Smart Home, a service enabling Smart TVs, home appliances, and smartphones to be connected and managed through a single integrated platform [12]. Samsung claimed Smart Home's unique functionality enables users to control and manage their home devices through a sole application. The application connects personal and home devices—from refrigerators and washing machines to Smart TVs, digital cameras, smartphones, and even the wearable device GALAXY Gear. Three main service features are

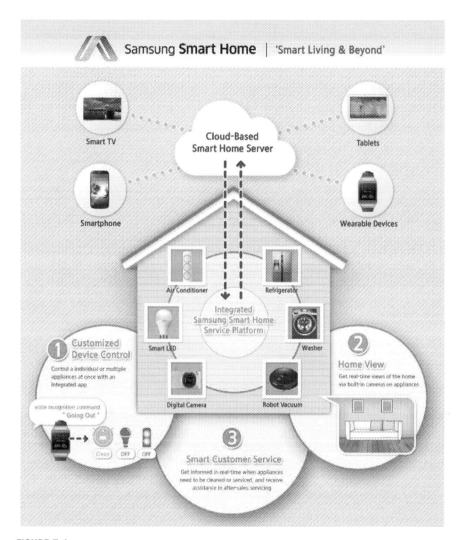

FIGURE 7.1

Samsung's smart home service architecture [12].

enabled that consumers can connect to with their devices from anywhere, anytime: device control, home view, and smart customer service (Fig. 7.1).

This example shows that we are moving towards an internet of things and big data. Cameras are good for security applications and unique identification; however, sometimes they are overkill depending on the application. If we only care about the information, high-resolution data might be redundant and will just add to the cost for storing or transporting data. Some of the applications of cameras can be well replaced by radars. Some features of wearables can also be replaced by radar sensing, providing the benefits of noninvasive sensing.

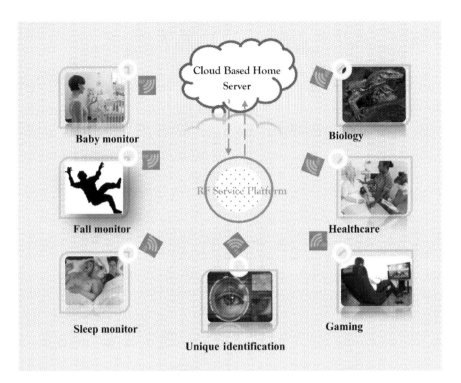

FIGURE 7.2

The authors' vision for indoor RF activity sensing and classification.

It follows that we can envision a future where radio frequency (RF)-based sensing will play a big role in indoor activity monitoring, sharing information with cloud servers. As big data technology is emerging, more and more sensing data will be sent to cloud servers and intelligent decisions will be made. Fig. 7.2 shows some of the possible applications which have already been through several phases of research. A centralized service platform will collect useful data and communicate with a cloud server for decision-making.

7.3 THEORY OF RF ACTIVITY CLASSIFICATION

Using wireless technology in indoor environments for activity classification will require a comprehensive integration of multiple techniques, including RF sensing, signal processing, and artificial intelligence.

7.3.1 SENSING TECHNIQUES

Researchers have reported different types of short-distance radar for different applications. Ultra-wide band radar, continuous-wave (CW) radar, frequency

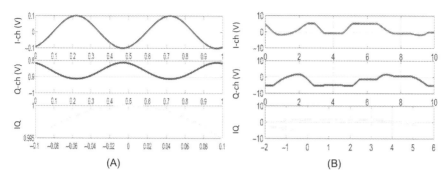

FIGURE 7.3

Time domain data and corresponding IQ plot, for Doppler radar measurement of sinusoidal motion in (A), and (B) shows the same for a series random linear movement.

modulated CW radar, and pulsed radar, to cite a few. Radars also vary based on the operating frequency (X band, K band, etc.). Directivity of sensing also a diversity; this diversity is mainly introduced by antenna design. However, the sensing principle is more or less unique regardless of the type of each radar: they all sense the back-scattered signal reflected from the target.

The single-channel radars are limited by their inability to produce displacement sensing in all positions. However, this limitation is overcome using a quadrature radar system, which in principle provides a stereo vision. The outputs of the quadrature radars are called in-phase (I) and quadrature phase (Q). The combination of I and Q can provide useful insight in characterizing a target's motion. Fig. 7.3 shows the experimental results of two different kinds of motion patterns of a linear moving platform. It is evident from the plots that distinguishing the patterns of I and Q waves holds the information regarding the motion patterns. Additionally, IQ plots are apparently different, indicating that these could be useful parameters in RF activity classification.

7.3.2 EXTRACTION OF USEFUL INFORMATION

Sensing enables us to record data, and these data contain useful information about the target's motion, position, or some other characteristics. However, collected data need to be demodulated with the proper software algorithm. Upon demodulation of data, useful information needs to be extracted from data and those features must be stored in vectors or matrices. For instance, for a long-time series of sinusoidal waves, we may only be interested in the frequency, amplitude, and phase information of the signal, rather than each data point of the time series. The useful features can be statistical or multidomain. Some of the well-known statistical features are the mean and variances. Some other common techniques applied in signal processing are listed in subsequent paragraphs.

7.3.2.1 Mean and variance

The mean measures the central tendency of a given set of numbers. There are a number of different central tendencies including but not limited to mean, median, mode, and range. An average or arithmetic mean is the sum of a list of numbers divided by the number of numbers in the list. Variance is the expectation of the squared deviation of a random variable from its mean. Statistical tools sometimes provide a good understanding of the data.

7.3.2.2 Frequency domain analysis

A frequency domain analysis determines the existence or extent of each given frequency band over a range of frequencies. Fourier analysis and Fourier transforms are the most common techniques of frequency domain analysis. There are many established algorithms for Fourier transforms, each of which is strong in its own set of applications.

7.3.2.3 Fixed and adaptive filters

There are many different bases of categorizing filters, and they may overlap. The common bases are linear or nonlinear, time-invariant or time-variant, causal or noncausal, analog or digital, discrete-time (sampled) or continuous-time, passive or active types of continuous-time filters, and infinite impulse response or finite impulse response types of discrete-time or digital filters. Some filters are inherently good for removing noise, e.g., the Wiener filter, Kalman filter, and Savitzky—Golay smoothing filter. On the other hand, adaptive filters dynamically change their coefficients over time, and therefore, dynamically change their frequency suppression properties.

7.3.2.4 Wavelet transformation

Unlike Fourier transforms, wavelet transform methods provide temporal resolution, i.e., they capture both frequency and location information. The location refers to time. A discrete wavelet transform is any wavelet transform for which the wavelets are discretely sampled.

7.3.2.5 Spectral energy

The distribution of signal energy for each frequency can be mathematically described; also, it can be measured for a given signal. Spectral energy distribution can provide vital information about a signal in the classification and characterization of the signal itself or the device that delivered the output.

7.3.2.6 Spectral entropy

Spectral entropy is a derived parameter. The results of a Fourier transform are usually used to measure this parameter. Spectral distribution of entropy may be a good feature in classifying signals.

7.3.2.7 Fractal analysis

Fractal analysis is a modern method of applying nontraditional mathematics to patterns. In essence, it measures complexity using the fractal dimension. There are many versions of fractal analysis, and each has its own strength and weakness. Fractal analysis assumes beforehand that a signal is fractal in nature. Algorithms for fractal dimension calculations are applied to the signal. This evaluates a form of the signal's complexity.

7.3.2.8 Empirical mode decomposition

The Hilbert–Huang transform (HHT) is a way to decompose a signal into intrinsic mode functions. A trend is also observed. Instantaneous frequency data are obtained alongside it. This method works well with nonstationary and nonlinear data. Unlike a Fourier transform, the HHT is considered an empirical method. Empirical mode decomposition can be applied to a data set, which is a little different than just a theoretical tool.

7.3.3 TURNING DATA INTO KNOWLEDGE

We have discussed the noninvasive sensing techniques and touched on some useful analytical tools that can extract useful parameters from raw data. However, this extracted knowledge can only be useful once we know how it relates to some physical phenomenon that we care about. Often times, these parameters change over time, and we also want to relate these dynamic behaviors to the behavior of certain things. While we care about instantaneous output sometimes, some other behaviors need to be observed over a long time. For instance, we may be interested in knowing someone's heart rate at this moment, which can be measured using the current RF sensing technology. However, if we are interested in someone's long-term heart rate behavior that relates to a certain disease, we need to have more knowledge. This will call for long-term data collection and knowledge of the particular disease. Artificial intelligence may help us make intelligent decisions. Systems can be trained using features extracted from raw data over a long term. Then these trained systems can be used to guide us and warn us using a collective knowledge to classify various phenomena. Essentially, data need to be turned into knowledge. The subsequent paragraphs mention some of the artificial intelligence techniques out of many. These techniques can be used to train and classify data.

7.3.3.1 Neural network

ANNs are a very common family of models in machine learning and cognitive science. ANNs are employed by biological neural networks (the central nervous systems of animals, in particular the brain) which are used to estimate or approximate functions that can depend on a large number of inputs and are generally unknown [13].

7.3.3.2 Probabilistic neural network

A probabilistic neural network (PNN) is a derived algorithm from the Bayesian network. This network is a class of feedforward neural networks. It was introduced by D.F. Specht in the early 1990s. In a PNN, the processes are organized into a feedforward network with four layers. These layers are the input layer, hidden layer, summation layer, and output layer [14].

7.3.3.3 Support vector machine

Support vector machines (SVMs) are a kind of supervised learning model. Each SVM has its learning algorithms for analyzing data. SVMs are used for classification and regression analysis. Given a set of training instances, each assigned to one of two categories, an SVM training algorithm constructs a model that assigns new examples into one category or the other. SVMs essentially follow a non-probabilistic binary linear classifier algorithm [15].

7.3.3.4 k-Nearest neighbor

The k-nearest neighbors (or k-NN for short) algorithm is a nonparametric method. This is one of the most common algorithms used for classification and regression. An object is classified by a popular vote of its neighbors, with the object being allocated to the class most common among its k-NN. Usually, k is a small positive integer.

7.3.3.5 Bayesian classifier

In machine learning, naive Bayes classifiers are a family of simple probabilistic classifiers based on applying Bayes' theorem with strong (naive) independence assumptions between the features. Some of the concepts of naïve Bayes classifier are explained in Ref. [16].

7.4 CASE STUDY/APPLICATION

7.4.1 SLEEP STUDY

Doppler radar has a high potential for the remote monitoring of human subjects [17]. Robust sensing of heartbeat and respiration rates is attractive for many applications such as healthcare, military, and security. Several works have focused on accurate measurement of the heartbeat and respiration rates [18,19]. However, recognizing other characteristics such as body orientation is still challenging. Whereas the phase of the signal recorded from a Doppler radar is related to chest movement, the power of the received signal is dependent on the interaction of the wave with the subject's body. The amount of the back-scattered signal from the moving part can be characterized with the Doppler RCS. The RCS is proportional to the physical area of the chest as well as the surface's reflectivity and directivity of radar at the operating frequency. Due to the asymmetry of the body around the vertical axes and the different types of motion on each side, the chest orientation with respect to the line of sight between the subject and the transceiver also affects the RCS.

7.4.1.1 Sleep position variation with RCS

A simple and effective method is presented in Ref. [2] to distinguish between different types of sleep positions using the RCS, since the RCS of the back of the subject was about nine times bigger than that of the front of the body. Plotting the Q component versus the I component of the signal back-scattered from the back leads to an arc with a large radius of curvature. Moreover, the experiment shows that the weight of the body has an effect on reducing the amplitude of the respiration motion [2].

To calculate the RCS, after canceling the DC component [2], the complex data vector is broken into several segments in such a way that each segment corresponds to about a full respiration cycle. The radius of each arc is then calculated by applying the center estimation [20,21] and circle fitting algorithm [2].

The center-tracked arc from a subject in supine position is depicted in Fig. 7.4. Only the arc from one segment of data is displayed.

As shown in Fig. 7.4, when the subject lies in a prone position, the motion of the chest and stomach is the same uniform motion as that of the back. However, the motion depth is reduced because of the weight of the subject [2]. Such information is essential for accurate calibration when assessing how deeply a subject is breathing.

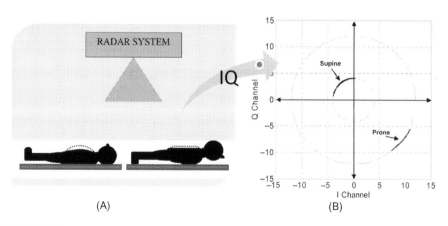

(A) (B)

FIGURE 7.4

Center-tracked arcs for a subject breathing whie lying in two positions (A). The estimated circle radius is 4.04 and 12.13 V for body front and back surfaces, respectively (B), providing a method for tracking a subject's position changes [2].

7.4.1.2 Sleep disorders

Sleep disorders are a type of medical illness in which typical sleep is disrupted or abnormal sleep behavior occurs. Sleep disorders can lead to physical, mental, and emotional dysfunction. There are diverse types of sleep disorders. Statistics indicate 15 million Americans suffer from obstructive sleep apnea, one of the most common types of sleep disorders. Polysomnography (PSG) is the current gold standard to study sleep apnea. In PSG, several body functions such as brain activity (EEG), eye movements (EOG), muscle activity (EMG), heart rhythm (ECG), respiratory airflow, and respiratory effort are monitored during sleep. During this test, the patient spends a night in a sleep lab and his physiological parameters are recorded using sensors attached to the body. All of this makes PSG time consuming, complex, inconvenient, and expensive. A noncontact physiological radar monitoring system (PRMS) for sleep disorder monitoring is introduced in Refs. [22,23] as a simple and more affordable device to improve the efficacy and accessibility of sleep tests. It uses CW Doppler radars along with a real-time algorithm to detect sleep apnea.

7.4.1.3 PRMS architecture

A PRMS detects the movement of the chest and abdomen during sleep by transmitting a RF signal and detecting the phase modulated signal reflected by them. To acquire the information on both amplitude and direction of motion, a PRMS utilizes a quadrature Doppler radar with two orthonormal outputs, I and Q. This also eliminates any RF "blind spots" that could possibly exist in a single-channel system. The respiration trace is built by combining the two channels using center estimation and arc-tangent demodulation. Assuming the respiratory movement is given by $x(t)$ and the DC component is compensated, the two quadrature outputs can be defined as:

$$B_I(t) = sin\left(\theta + \frac{4\pi x(t)}{\lambda} + \Delta\varphi(t)\right) \tag{7.1}$$

$$B_Q(t) = cos\left(\theta + \frac{4\pi x(t)}{\lambda} + \Delta\varphi(t)\right) \tag{7.2}$$

where $\Delta\phi(t)$ is the residual phase noise, and θ is the constant phase shift related to the nominal distance to the subject, including the phase change at the surface of a target and the phase delay between the mixer and antenna [10,12]. The respiratory movement is obtained by applying arc-tangent operation to the outputs as shown in Eq. (7.3).

$$\varphi(t) = \arctan\left(\frac{B_Q(t)}{B_I(t)}\right) = \arctan\left(\frac{\cos(\theta + p(t))}{\sin(\theta + p(t))}\right) = \theta + p(t) \tag{7.3}$$

where

$$p(t) = \frac{4\pi x(t)}{\lambda} + \Delta\varphi(t)$$

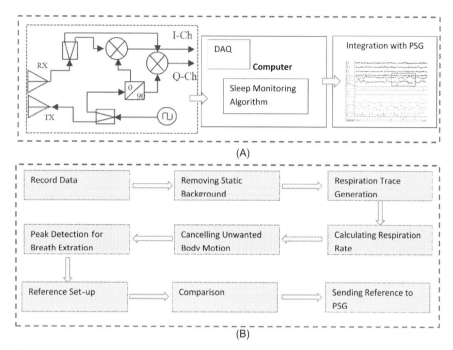

FIGURE 7.5

(A) The block diagram of radar's integration in sleep study; (B) the sleep study algorithm.

The schematic of a PRMS is shown in Fig. 7.5. The respiration trace recorded from the radar is converted from analog to digital and then analyzed using the sleep monitoring algorithm. The results of the analysis can be integrated with a standard PSG device.

7.4.1.4 Sleep disorder monitoring algorithm

Sleep monitoring software which includes a real-time algorithm is implemented to accompany the hardware. The steps of the algorithm are shown in Fig. 7.5. In the first step, the data were recorded from a 2.4 GHz Doppler radar that was suspended in about 1 m above the patient, to cover his chest movement. Then the static background is removed by subtracting the average of the data from each channel:

$$x(t) = r(t) - \frac{1}{n}\sum_{k=0}^{n} r(t) \tag{7.4}$$

where $r(t)$ is the received signal in each channel, and n is number of samples in each data set. In the third step of the algorithm, the respiration trace is built by combining the I and Q components of the baseband data from the quadrature Doppler radar. The baseband signals and respiration trace are shown in Fig. 7.6.

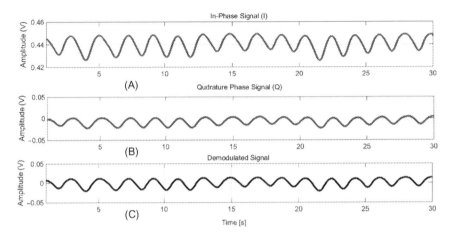

FIGURE 7.6

Generating respiration trace (C) from baseband signal (A) and (B) using linear demodulation method.

The respiration rate is the average of the rates calculated in the frequency and time domains. In the frequency domain, the Fourier transform is applied to the signal, and the frequency of the peak in the frequency spectrum is selected as the respiration rate. In the time domain, the local maxima is detected using peak detection algorithm, and the time between two consecutive maxima is defined as one breath. The respiration rate then was the total number of breaths per minute. Any motion other than respiratory related movement must be detected and eliminated before detecting the sleep apnea. The algorithm considers a motion to be unwanted if there is a sudden increase in amplitude and/or a sudden change in the respiration rate. Fig. 7.7A shows the radar data when the patient moves his hand. A peak detection algorithm is implanted in the sixth step of the algorithm to identify a single breath in the respiration trace. The data between two local maxima are considered as a complete breath, as shown in Fig. 7.7B.

In the seventh step of the algorithm, the references are defined from normal breathing for comparison. Two types of references were considered. In static threshold mode, it is considered that the patient breaths normally during the first 30s of data. The following parameter is calculated for comparison:

1. The average of energy of each breath. The energy of a signal is calculated by

$$E = \sum_{k=0}^{n} x_k^2 \tag{7.5}$$

2. The average of duration of each breath.
3. The average area enclosed in the IQ plot by the arcs of each breath.

The arc is determined by plotting the Q component of the radar data for each breath versus its I component. In dynamic threshold mode, the references

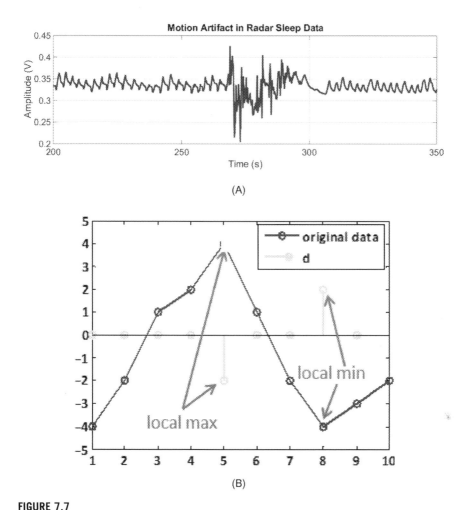

FIGURE 7.7

(A) Unwanted hand movement causes a sudden increase in amplitude; (B) peak detection for breath rate measurement.

are updated continuously using the last four normal breaths. The signal is then compared to the references to detect the abnormal breathing. Comparing the radar data of nap studies with the sleep technician-scored data from Sandman shows the signal amplitude falls under 80% and 60% of the reference for apnea and hypopnea, respectively, because of the decreasing respiration depth. Fig. 7.8A and B shows a sample of the signal changes when an apnea or hypopnea occurs. Moreover, the duration of each breath during apnea and hypopnea is 40% and 20% longer or shorter than normal breathing, respectively.

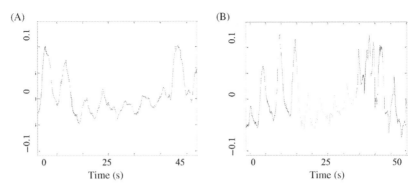

FIGURE 7.8

Occurance of sleep apnea: (A) apnea and (B) hypopnea.

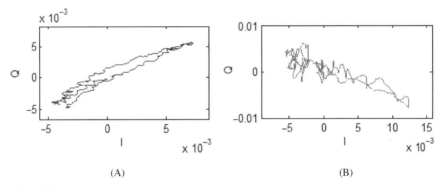

FIGURE 7.9

Normal (right column) versus apnea (left column) determination. (A) IQ area for normal breathing condition, and (B) the IQ plot for one breathe where apnea is observed. The enclosed area by the curves is calculated by integration method. The area of the arc increases from 1.0227e-05 to 2.1737e-05 in apnea situation.

The third comparison criterion is the area of the arc. As mentioned before, plotting the Q component versus I component forms a portion of a circle. Based on the information gathered from nap studies, the area of this arc will also change during an apnea or hypopnea (Fig. 7.9).

The above criterion is calculated and compared with the references for each breath. Two outputs are generated in this step. The first output is the result of the comparison for each breath. Outputs are labeled as "1" or "2" if hypopnea or apnea has happened. Hence, the delay for this output is equal to the duration of one breath. The second output shows the occurrence of apnea or hypopnea that lasts more than 10 s. So it has a 10 s delay relative to the respiration trace. In the final step of the algorithm, the results of the respiration trace, respiration rate, comparison result for

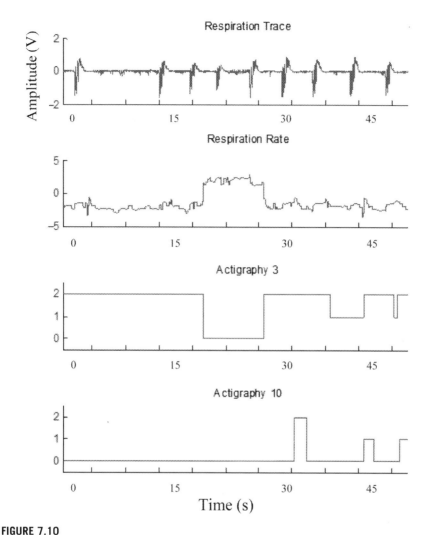

FIGURE 7.10

A snapshot of the outputs that are sent to the standard PSG [22,23].

each breath (actigraphy 3), and 10 s comparison (actigraphy 10) are sent to be integrated with a standard PSG as the results of the sleep disorder monitoring algorithm (Fig. 7.10).

7.4.2 PERSONAL IDENTIFICATION

Adult human breathing patterns at rest are slightly different from one another in terms of tidal volume, inspiratory and expiratory duration, and airflow profile.

Besides this diversity, in every recording of ventilation at rest in a steady-state condition, breath to breath variabilities are observed in ventilatory variables [5]. This unevenness is nonrandom and may be explained either by a central neural mechanism or by instability in the chemical feedback loops. It is not unusual that each individual appears to select one particular pattern among the infinite number of possible combination of ventilatory variables and airflow profiles [5]. If enough data are recorded and people are observed over time, these parameters may lead to unique-identification of people solely based on vital signs. This personal identification may add another dimension to the traditional fingerprint systems. The advantages of RF-based unique identifications using vital signs are manifold. In one way, this system can be used as a noninvasive, low-complexity system with less burden in terms of data transmission requirements; on the other side, these data contain useful health diagnostics (Fig. 7.11).

A recently proposed Doppler radar-based, noncontact, unique-identification method is capable of uniquely identifying a small set of people based on their vital sign parameters [5]. A noninvasive, CW radar-based system can monitor respiration parameters in a similar way to that discussed in previous sections. Useful features can be extracted and a neural network can be trained to classify the signals. This classification can represent a unique identification of a person or a group of persons.

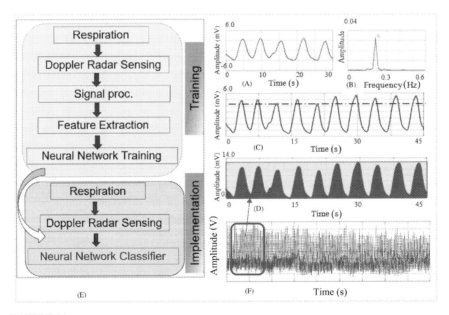

FIGURE 7.11

Epoch features for neural network training are illustrated. (A) Time series of respiration record; (B) PSD; (C) peak line fit and error; (D) packing density; (E) flow chart of unique identification; and (F) a sample of data collected over a long time that was segmented for NN training [5].

A reasonably large data set is recommended, as proof of concept data from three individuals were collected for 15 min. There were no notable strong physiological differences among the subjects. Each person's collected data were segmented into 45 s epochs for feature extraction. For each person, 19 windows, each containing 45 s of data, were used for training the neural network. Three features were extracted from each window/epoch as described in following text:

Peak power spectral density (PSD): PSD describes how the power of a signal or time series is distributed over the different frequencies. The frequency of the peak was selected as one of the features. A parameter named the respiratory signal's packing density is calculated as the normalized area covered by the respiration signal in a bounding box. The third parameter was named the linear envelope error, which is essentially the sum of the errors of the peaks from peak-fit line (Table 7.1).

The trained network was then evaluated for its correctness. Data were collected from the same three persons at another time. A confusion matrix shows the classification performance. Person 1 and Person 3 were classified accurately. However out of 17 samples of data from Person 2, four were misclassified as Person 3. Doppler radar is examined as a unique human identifier with a fair rate of success. One of the training-based classifiers, such as a neural network classifier, is used to identify individuals based on their breathing patterns (Table 7.2).

Table 7.1 Part of Respiration Segment Feature Data [5]

Packing Density	Frequency of Maximum Spectral Power (Hz)	Linear Regression Error in Peak Fit (mV)	Person
0.521341	0.213298	0.170702	P1
0.517554	0.191968	0.038110	P1
0.406263	0.213298	0.010226	P1
...	...	...	
0.312932	0.277287	0.019954	P2
0.369265	0.277287	0.067173	P2
0.354366	0.234627	0.090660	P2
...	...	...	
0.499702	0.234628	0.027535	P3
0.473084	0.234628	0.019782	P3
0.4183060	0.2346208	0.0136630	P3

Table 7.2 Testing Results Confusion Matrix [5]

	Person 1	Person 2	Person 3	Success (%)
Person 1	7	0	0	100
Person 2	0	13	4	76.4
Person 3	0	0	17	100

7.4.3 OCCUPANCY SENSING

It is hard to distinguish multiple persons in a single space using Doppler radar because the signal is ambiguous. However, using improved architecture such as a multiple antenna system, the sensor may count individuals and isolate them based on different cardiovascular signatures beyond detecting life presence only. This may be possible due to the diversity of people's heart rates. In cases of very similar vital signs, multiple antenna systems—single input, multiple output or multiple input, multiple output—may distinguish different subjects based on angle of arrival. Approximating the direction of arrival can help separate subjects spatially in a multiantenna environment.

Yavari et al. experimented with using a 2.4-GHz homodyne Doppler quadrature receiver with off-the-shelf coaxial components to detecting presence. The motivation is to devise a better way for turning appliances off when not in use, thereby saving a huge amount of power and contributing to power sustainability. For quantifying presence over an empty room, a short time window is used to slide through the time domain data. The window length N is found by calibration of the occupancy detector. For a windowed signal with the length N, the average of the squared signal is considered [24]. The window length in this experiment is 5 s, with 0.5 s steps. The sliding window and average calculations are reiterated over the entire data set. The experiment was conducted by recording data for 12 min. The order of the different levels of activity that was measured as follows: periodic movement of a mechanical object, an empty room, and then a human subject in a stationary position (respiration and heart only), walking in diverse directions, and engaged in high-intensity activity like running. Fig. 7.12 shows the results of the experiment. A mechanical target was used in this test to symbolize periodic motion in a space, such as motion by fans, that can interfere with sensors and cause false alarms.

At the start of the test, the only movement was the periodic motion of the mechanical target. The target frequency was 1 Hz, with 1-mm amplitude. The amplitude of the periodic motion was set to be small so that it would be closer to the heart and respiration range, making it harder to discriminate. The mechanical target and human subject for the stationary position were each 1 m away from the antennas.

7.4.4 ANIMAL ACTIVITY MONITORING

Observing the activity of animals in their natural environments can yield important information about energy expenditure, thermoregulation, behavioral patterns, and even population health. Researchers working with animals often observe their subjects indoors by creating an artificial environment simulating the subject's natural habitat.

7.4.4.1 Motion classification of animals (lizard)

Some investigation has been performed to monitor the activity and associated energy conservation strategies for a lizard and reported in Ref. [6]. Radar methods

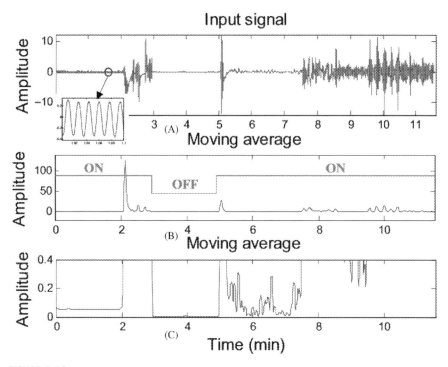

FIGURE 7.12

(A) Twelve minutes of activity measurements: mechanical object with periodic movement, empty room, human subject in stationary position (respiration and heart only), walking in different directions, and engaged high-intensity activity; (B) moving average output— rectangular line shows detected activity over no motion; (C) zoom in y-axis [24].

promise to reduce the visual and data collection burden associated with real-time observation and video recording. The researchers used two quadrature 10.525 GHz radar modules that were placed perpendicular to each other to record motion in two axes. The schematic of the experiment set-up is shown in Fig. 7.13 [6].

A plot of the quadrature components (in-phase component I versus quadrature phase component Q) shows a curve in the IQ plane when a motion in front of the radars is sensed.

The direction of the movement of the lizard was resolved by scheming the phase angle of this curve. Rotation of phase angles in clockwise indicates an exit away from the radars, while the counter clockwise is suggestive of entrance towards radars. Furthermore, the extent of the motion perpendicular to the plane of the radar antenna is directly related to the length of the arc and can be calculated from [19]:

$$\Delta x = \frac{\lambda}{4\pi} \Delta\Omega \qquad (7.6)$$

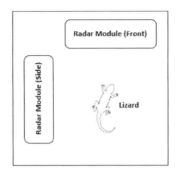

FIGURE 7.13

Experiment set up.

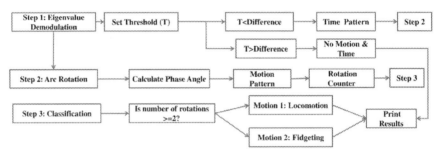

FIGURE 7.14

Classification algorithm used for each radar to characterize motion [6].

where λ is the wavelength of the radar's operating frequency, and $\Delta\Omega$ is difference between the two angles scanned by the arc due to the displacement of the target [6]. A numerical example is as follows: the radar's transmitted signal is 10.525 GHz ($\lambda = 2.85$ cm, derived from frequency, velocity, and wavelength relation), and a movement of approximately 1.85 cm results in a complete circle ($\Delta\Omega = 2\pi$). The motion is classified as fidgeting or locomotion by counting the number of circles or closed loops in the IQ plane [6]. Two-dimensional movement is determined by comparing the data between the two sensors. The algorithm of detecting and classifying motion is shown in Figs. 7.14 and 7.15.

7.4.5 FALL DETECTION

Signal processing algorithms and techniques involved in elderly fall detection using radar have been studied by different research groups. The back-scattered reflections from humans differ in their Doppler characteristics depending on the nature of the human activities. The nonstationary nature of these signals can be further analyzed using time-frequency methods [25]. Some published papers also claimed the usefulness of wavelet transform methods [26].

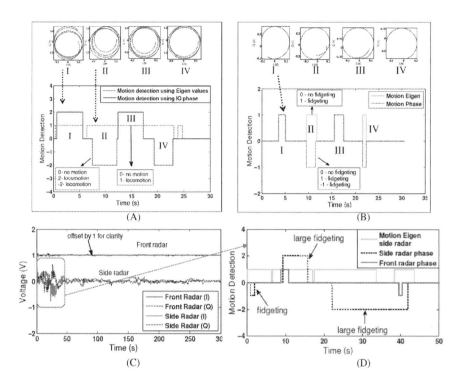

FIGURE 7.15

(A) Plot showing motion detection using an Eigenvector and classification using angle estimation and rotation count for locomotion in front of the front radar module. The IQ plot "I" corresponds to the motion period "I." For motion detection using the phase, a value of 2 indicates motion and −2 indicates motion in the opposite direction, as described in the text; (B) plot showing motion detection using an Eigenvector and classification using angle estimation and rotation count for fidgeting in front of side radar module. The IQ plot "I" corresponds to the motion period "I." For motion detection using the phase, a value of 1 indicates motion and −1 indicates fidgeting in the opposite direction. (C) Raw data from front and side radar showing changes in amplitude due to motion, and (D) result of the detection algorithm for front and side radar. The swaying of the body is detected as locomotion by the side radar and as fidgeting by the front radar as expected. A few spurious alerts were generated by the Eigenvector algorithm but were revealed as nonmotion by phase analysis and video reference [6]. Such automated classification techniques could be used for stand-alone analysis, or to indicate time periods for which to save or view video recorded data.

Often radar-based fall detection follows two stages. At the prescreen stage, the coefficients of wavelet decomposition at a given scale are analyzed to identify the time locations of a fall. The next stage is the classification stage, performing rigorous analysis to identify and compare diverse scenarios. Fig. 7.16 shows a time-frequency analysis of the radar's record data to identify a fall.

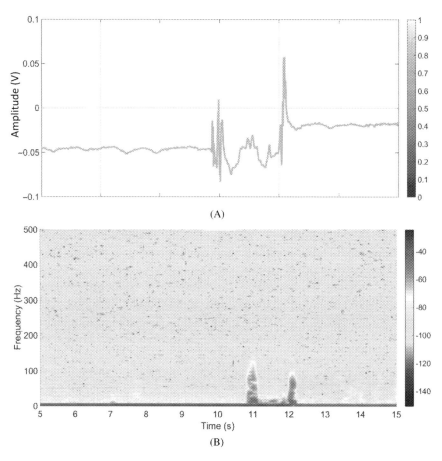

FIGURE 7.16

(A) Time series data from a radar receiver and (B) time-frequency analysis. A fall was observed at the 11th second, which is evident in the plot [26].

7.5 CONCLUSION

Radars can work as low-complexity imaging systems for indoor activity monitoring. The advantages of using radars are manifold, including automation, small database requirements, low latency, simple processing, and reduction of privacy concerns compared with cameras. Radars can noninvasively measure very small displacements without using imaging markers and provide clear diagnostic information without requiring a heavy load of personnel for analysis and interpretation. With such radars, intelligent systems can be developed. Practical applications extend from dedicated monitoring applications in medical sleep centers to home use and wearable applications given recent research showing the possibility of vital signal detection from mobile platforms [27,28].

REFERENCES

[1] Li C, Lubecke VM, Boric-Lubecke O, Lin J. A review on recent advances in Doppler radar sensors for noncontact healthcare monitoring. IEEE Trans Microw Theory Tech 2013;61(5):2046−60.

[2] Kiriazi J, Borić-Lubecke O, Lubecke V. Radar cross section of human cardiopulmonary activity for recumbent subject, In: Engineering in Medicine and Biology Society, 2009. EMBC 2009. Annual international conference of the IEEE, pp. 4808−4811, 3−6 September 2009.

[3] Lee J, Lin J. A microprocessor-based arterial pulse wave analyzer. IEEE Trans Biomed Eng 1985;BME-32(6):451−5.

[4] Lin JC. Microwave sensing of physiological movement and volume change: a review. Bioelectromagnetics December, 1992;13:557−65.

[5] Rahman A, Yavari E, Lubecke VM, Lubecke O-B. Noncontact Doppler radar unique identification system using neural network classifier on life signs. In: IEEE topical conference on biomedical wireless technologies, networks, and sensing systems (BioWireleSS), pp. 46−8, Austin, TX, January 2016.

[6] Singh A, Lee SSK, Butler M, Lubecke V. Activity monitoring and motion classification of the lizard Chamaeleo jacksonii using multiple Doppler radars. In: Engineering in Medicine and Biology Society (EMBC), 2012 annual international conference of the IEEE, pp. 4525−8, August 28, 2012−September 1, 2012.

[7] Lin JC. Microwave sensing of physiological movement and volume change: a review. Bioelectromagnetics 1992;13:557−65.

[8] Martin PH, Unwin DM. A microwave Doppler radar activity monitor. Behav Res Methods Instrument 1980;12(5):517−20.

[9] Heal JW. An animal acitivty monitor using a microwave Doppler system. Med Biol Eng 1975;13(2):317.

[10] Mercuri M, Karsmakers P, Leroux P, Schreurs D, Beyer A. Real-time fall detection and tag-less localization using radar techniques. 2015 IEEE 16th annual wireless and microwave technology conference, April 2015, pp. 1−3.

[11] Weldon MK. The future X network, a bell labs perspective. Boca Raton, FL: CRC Press; October, 2015.

[12] Samsung News Room, "https://news.samsung.com/global/samsung-unveils-new-era-of-smart-home-at-ces-2014", accessed June 26, 2016.

[13] MacKay DJC. Information theory, inference, and learning algorithms (PDF). Cambridge: Cambridge University Press; 2003, ISBN 9780521642989.

[14] Specht DF. Probabilistic neural networks. Neural Netw 1990;3:109−18.

[15] Shawe-Taylor J, Cristianini N. Support vector machines and other kernel-based learning methods. Cambridge: Cambridge University Press; 2000.

[16] Rish I. (2001). An empirical study of the naive Bayes classifier (PDF). IJCAI workshop on empirical methods in AI.

[17] Lohman B, Boric-Lubecke O, Lubecke VM, Ong PW, Sondhi MM. A digital signal processor for Doppler radar sensing of vitalsigns. IEEE Eng Med Biol Magaz 2002;21:161−4.

[18] Massagram W, Hafner NM, Park B-K, Lubecke VM, HostMadsen A. and Boric-Lubecke O. Feasibility of heart rate variabilitymeasurement from quadrature Doppler radar using arctangentdemodulation with dc offset compensation. In: Proc. 29th annu. conf. IEEE EMBS, Lyon, France, 2007, pp. 1643−6.

[19] Droitcour AD, Seto TB, Park B-K, Yamada S, Vergara A, El-Hourani C, et al. Non-contact respiratory rate measurement validation for hospitalized patients. 31st Annu. conf. IEEE EMBS, Minneapolis, 2009, to be published.

[20] Park B-K, Vergara A, Boric-Lubecke O, Lubecke VM, and Host-Madsen A. Quadrature demodulation with dc cancellation for a Doppler radar motiondetector, submitted to IEEE MTT Transactions.

[21] Host-Madsen A, Petrochilos N, Park B-K, Boric-Lubecke O, and Lubecke VM. Optimum demodulation for a Doppler radar system for vital signextraction, submitted to IEEE Signal Processing Letters.

[22] Baboli M, Singh A, Soll B, Boric-Lubecke O, Lubecke VM. Good night: sleep monitoring using a physiological radar monitoring system integrated with a polysomnography system. Microw Mag IEEE 2015;16(6):34−41.

[23] Singh A, Baboli M, Gao X, Yavari E, Padasdao B, Soll B, et al. Considerations for integration of a physiological radar monitoring system with gold standard clinical sleep monitoring systems. In: Engineering in Medicine and Biology Society (EMBC), 2013, 35th annual international conference of the IEEE, pp. 2120−3, 3−7 July 2013.

[24] Yavari E, Jou H, Lubecke V, Boric-Lubecke O. Doppler radar sensor for occupancy monitoring. In: Biomedical wireless technologies, networks, and sensing systems (BioWireleSS), pp. 139−41, Austin, TX, January 2013.

[25] Moeness G. Amin, Yimin D. Zhang, Fauzia Ahmad KC, Dominic Ho. Radar Signal Processing for Elderly Fall Detection: The future for in-home monitoring in IEEE Signal Processing Magazine 2016;33(2):1053−5888.

[26] Su BY, Ho KC, Rantz MJ, Skubic M. Doppler radar fall activity detection using the wavelet transform. IEEE Trans Biomed Eng 2014;62(3):865−75.

[27] Rahman A, Yavari E, Gao X, Lubecke V, and Boric-Lubecke O. Signal processing techniques for vital sign monitoring using mobile short range Doppler radar. In: IEEE radio wireless symp., San Diego, CA, 2015, pp. 1−3.

[28] Rahman A, Yavari E, Lubecke V, Lubecke O-B. A low-IF tag-based motion compensation technique for mobile doppler radar life sign monitoring. IEEE Trans Microw Theory Tech 2015;63(10):3034−41.

Index

Note: Page numbers followed by "*f*" refer to figures.

Printed in the United States
By Bookmasters